No. 1265
$18.95

Giant Handbook of 222 Weekend Electronics Projects

TAB BOOKS Inc.
BLUE RIDGE SUMMIT, PA. 17214

FIRST EDITION

FIRST PRINTING

January 1981

Copyright © 1981 by TAB BOOKS Inc.

Printed in the United States of America

Library of Congress Cataloging in Publication Data

Main entry under title:

Giant handbook of 222 weekend electronics projects.

    Includes articles by various authors originally appearing in School shop magazine.
    Includes index.
    1. Electronics—Amateurs' manuals. I. School shop.
TK9965.G5   621.38   80-28688
ISBN 0-8306-9641-5
ISBN 0-8306-1265-3 (pbk.)

Contents

Preface

Every electronics hobbyist looks for more projects to build and use. Here are over 200 of them covering 13 different subject areas.

Step-by-step instructions explain how to make computers, construction aids, solar projects, power supplies, transmitter and receivers, test equipment and workshop devices. Plenty of diagrams, schematics, photographs and parts lists assist you.

All of the outstanding projects in this book have been made available by the editors of *School Shop* magazine, the how-to-do-it publication that has printed articles relating to education since 1941. Without their efforts and cooperation, this book would have been impossible.

Chapter 1
AM/FM Radio

The projects in this chapter cover both transmitter and receivers. All make excellent beginner's projects.

LOW-POWER BROADCAST BAND TRANSMITTER

This portable, low-power transmitter can be used to broadcast over short distances to any broadcast band receiver (Fig. 1-1). Since it uses only one transistor and some easily located parts, it is an attractive, worthwhile project. It can be used legally as long as the antenna length is kept below 30 inches (Fig. 1-2).

The Circuit

The transmitter's tapped capacitance identifies it as nothing more than a Colpitts oscillator (Fig. 1-3). The frequency of the tank circuit is changed by adjusting L1 with a screwdriver. Modulation is supplied by the output of the dynamic microphone. The dynamic microphone is a magnetic type earphone converted to a microphone simply by speaking into it at close range. The 3-volt power supply comes from two penlite cells.

The transmitting antenna is made of a length of #12 copper wire, which can be obtained from house-wiring cable. Power output from the antenna is very small and provides no danger of radiation beyond the room unless a long antenna is connected. *Keep the length of the antenna below 30 inches to insure legal operation.*

Fig. 1-1. Transmitter, microphone and antenna plug.

Construction Hints

The case can vary with the materials available (Fig. 1-2). The broadcast coil, L1, is available in a number of crystal set circuits and from local radio parts supply houses.

The microphone can be mounted directly onto the cover, thus eliminating the connector, the jack assembly and a length of cable. A slide switch can be used for SW1 and mounted on the cover under the microphone. Wire can be soldered to the penlite cells and thus eliminate the battery holder.

A five-lug terminal strip, TS1 (Table 1-1), is a handy device to keep the transistor leads and resistor and capacitor leads from shorting. A simpler method of mounting the antenna can be devised to eliminate the banana plug and binding post. However, the illustrations are designed to show a preferred method of making a finished piece of electronic gear with a professional appearance.

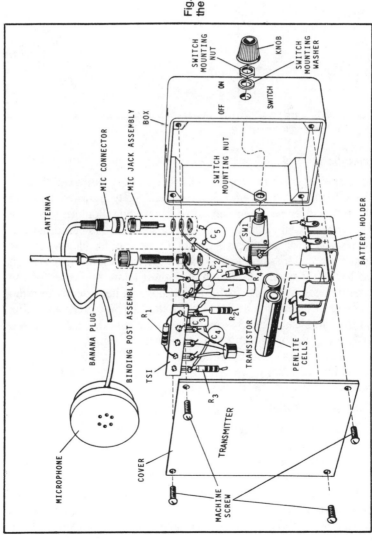

Fig. 1-2. Parts placement for the transmitter.

9

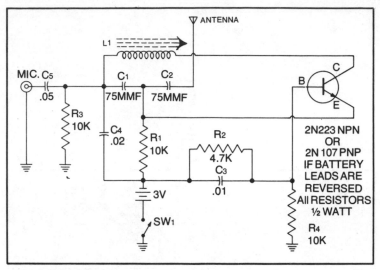

Fig. 1-3. Transmitter schematic.

Operation

The set is ready to operate as soon as the penlite cells are in, the antenna is connected and the switch, SW1, is turned on. Hold the transmitter near a broadcast receiver and talk slowly into the microphone while adjusting the tuning screw in the broadcast coil to a *dead spot* on the receiver. Once the frequency of the transmitter is the same as the radio receiver, a whistling noise will be heard. Sharper adjustment will then allow the voice of the operator to be heard. (If the microphone is held near the radio speaker, a squeal will be heard when the two pieces are operating on the same frequency.) Now move away from the radio and see just how far the transmitter will actually send a signal.

Many hours of fun can be had by communicating from one room to another or from downstairs to upstairs. A portable transistor receiver can be used as the other half of the set.

TRANSISTOR RADIO

Here is a project that is easy to build and costs less than $5 including headset.

Notice the similarity between the crystal radio (Fig. 1-4) and transistor radio (Fig. 1-5). The important feature of these two radios is the visual circuitry. By adding a 9-volt battery and a PNP power transistor to the crystal radio, you can easily see how a transistor amplifies. The germanium diode acts as a valve at the base of the transistor, turning the 9-volt circuit off and on. This action, in turn, is heard in the headset by making the sound louder.

To operate the radio, first hook up the ground, the antenna and the headset. Move the dial across the coil to bring in stations. To shut off the radio, simply remove one of the headset leads. See Table 1-2.

To build the transistor radio, follow this procedure:

☐ Line off the plywood board into 1″ squares, as indicated in the drawing. A #2 pencil works well. It is recommended that you first sand the board, line it off, and varnish it to protect the pencil lines.

☐ Set the ¾″ wire brads ⅜″ into the plywood board in the seven terminal locations.

☐ To make the coil, wrap magnet wire approximately ½″ from each end of the paper roll. To fasten the coil to the radio, simply take two upholstery tacks and fasten each end of the coil down to the 1″ line on the back of the radio. The magnet wire on the right end of the coil should be cleaned by scraping the varnish off with a knife. Attach this end to the terminal as shown. The end of the magnet wire on the left side remains unattached. Sand the coil of magnet wire across the top to remove varnish to insure conduction of the dial to the coil.

☐ In making the dial, first remove the paint from a straight piece of hanger wire. Using pliers, bend a loop in the end of the wire. Make two more bends—one to align the loop to the screw hole to fasten the dial to the board, and the other should be 1¼″ up the dial, permitting the dial to overhang and touch the coil. To

Table 1-1. Parts List for the Low-Power BB Transmitter.

1 Cover, 1/16″ × 2-9/16″ × 3-¾″	2 Switch Mounting Nut
1 Box, 1-9/16″ × 2-⅞″ × 4″	1 Switch Mounting Washer
1 Microphone (magnetic headset, 1000 ohm)	4 Machine Screw, 4-40 NP, ¼″
	1 TS1, Terminal Strip, 5 soldering lugs
1 Microphone Connector, cord mounted, female	1 L1, Loopstick, broadcast band
	2 C1, C2, Capacitor, 75 MMF any WVDC
1 Microphone Jack, panel mounted connector	1 C3, Capacitor, .01 MF, any WVDC
1 Binding Post	1 C4, Capacitor, .02 MF, any WVDC
1 Banana Plug	1 C5, Capacitor, .05 MF, any WVDC
1 Solid Copper Wire, #12, 27″	1 Transistor, 2N333 (or 2N107 if battery
1 SW1, Switch	leads are reversed)
2 Battery Holder, AA cell	3 R1, R3, R4, Resistor, 10K, ½ W
2 Penlite Cell, 1.5 volts, AA size	∓10%
1 Knob (to fit switch shaft)	1 R2, Resistor, 417K, ½ W, ∓10%

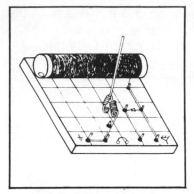
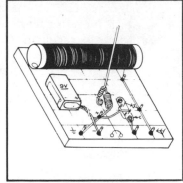

Fig. 1-4. Crystal radio.   Fig. 1-5. Transistor radio.

finish the dial, use #18 bell wire. First make a T splice onto the dial and wrap the bell wire eight times around the pencil. The loops will make a spring that will eliminate the breaking of the bell wire when the dial is moved. Attach the dial to the plywood board, using one tapping screw. The last step is to hook up the bell wire to the positive terminal and to the ground terminal.

☐ Attach the transistor to the plywood board, using two tapping screws as the attaching devices. You will notice that a short piece of #18 bell wire is bent into a terminal loop and is placed

Table 1-2. Parts List for the Transistor Radio.

quantity	description	size
1	plywood board	½" × 6" × 6"
7	wire brads	17 ga, ¾" long
3	tapping screws	#8, ½" long
1	cardboard tube	6" × 1¼"
1	tin plate	¾" × 2 ⅝"
1 set	2000 ohm headphones	
100'	magnet wire	#24
1	PMP "space-saver" power transistor	
1	germanium diode	
1	9-v rectangular battery	
3'	#18 bell wire	
1	battery connector	
2	upholstery tacks	
2	carpet tacks	
1	hanger wire	7" long

12

under one tapping screw. The other end of the wire is attached to one of the headphone terminals. This will hook up the collector part of the transistor.

☐ Attach battery clip, but remember where the positive and negative terminals are. To make a battery holder, use the tin plate. On each end of the tin plate put a ¼" hem, then bend it around the battery with the hems in against the side of the battery. Use two carpet tacks to attach battery holder to the plywood board.

☐ Use #18 bell wire to hook up the positive terminal to the emitter of the transistor. Now hook up the terminals from the antennas to the coil.

☐ Attach the germanium diode to the base of the transistor, keeping the polarity of the germanium diode in mind. Solder all terminals and the T splice on the dial. Draw the symbols with a felt tip marker to identify terminals.

☐ To attach the headset to the radio, you may do one of several things: Solder alligator clips onto the leads of the headset. For headsets with pins on the leads, solder one end of a piece of #18 bell wire 1" long, to each headphone terminal on the radio. Bend the other end around each pin on the headphones three times. You now have two sockets that will take pins. Use Fahnestock clips.

The radio is now completed and ready to be hooked up. Any number of things can be used for an antenna: bedsprings, television antenna, homemade antenna and so on. This radio will work without a ground; however, it will work much better with one.

A "STRAP IRON" RADIO

Here is a radio that can be built—complete with earphone—for only a few dollars. This low price is made possible by substituting a piece of black strap iron (hot rolled steel) in place of a crystal diode usually used in radios of this type. This radio is simple to construct, making it an excellent beginner's project. It requires no batteries—just an aerial and a ground. At least three or four local stations will come in loud and clear if the set is properly constructed, with clean, tight connections.

A crystal diode conducts electricity in only one direction and thus serves as a detector (Fig. 1-6). In this strap-iron radio, the radio signal is forced to pass through a piece of pencil lead held *lightly* against a piece of strap iron. This combination—pencil lead, strap iron, and light pressure—produces a circuit which will conduct electricity in only one direction (Fig. 1-7). Thus, we have a detector.

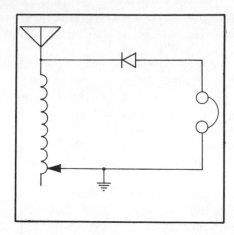

Fig. 1-6. Strap
iron radio schematic.

Start by cutting all the pieces to the sizes shown on the parts list (Table 1-3). Cut 1/16" from the corners of the tinplate with tin snips (Fig. 1-8). This avoids getting cut on the sharp corners. Use a metal punch to punch the three 3/16" diameter holes in each piece

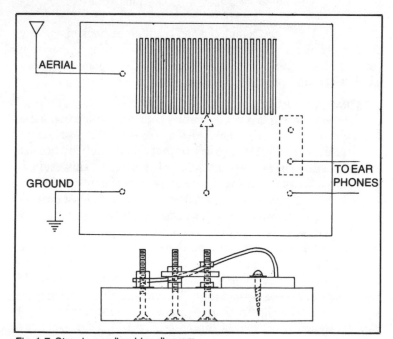

Fig. 1-7. Strap iron radio wiring diagram.

No. of Pieces	Item	Size
2	36-Gage Tinplate	½″ × 5″
1	Crs	1/16″ × ½″ × 4″
1	Black strap iron	1/16″ × ½″ × 1¼″
1	Coil form—maple	⅜″ × 1¾″ × 5″
1	Base—popular	¾″ × 5″ × 6″
2	Rh blue woodscrews	#4—¾″ long
6	Fh stove bolts	⅛″ × 1¼″
14	Nuts	5-40
4	Steel washers	5/32″ i.d.
	Formex magnet wire	36′

Earphones (Graymark #303 or the equivalent; available from Graymark Electronics, 798 Kelton Ave., Los Angeles 24, Calif.)

of tinplate located as shown in the drawing. Using a file, round both ends of the crs and drill a ⅛″ hole ½″ from one end. Place the crs in a vise and bend it slightly as shown in the drawing. Make certain that the end of the crs is not sharp because it will cut the wire on the coil form. Drill two 3/16″ diameter holes, ¼″ from each end, in the piece of 1/16″ × ½″ × 1¼″ black strap iron.

Be careful not to rub, file, polish, or otherwise remove the black surface on this piece. *The radio will not work if this surface oxide is removed.*

The corners of the coil form are rounded, and a ⅛″ hole is placed ½″ from each end as shown in the drawing. The coil form is then sanded smooth, and a suitable finish applied. Leave a 1″ lead. Wind a single layer of #22 magnet wire, starting and ending 1″ from the end of the wood. The 1″ lead is needed at only one end of the coil because the other end of the coil is not connected to anything. The coil should be tightly wound, with no spaces between the wires.

Notice that all holes in the base are located 1″ from an edge. Drill these ⅛″ diameter holes as shown on the drawing; countersink from the bottom and finish the base.

Position the coil form, with the wire wound as explained above, on the base ⅜″ from the edge. Center and drill two 1/16″

lead holes and use two ¾" round head wood screws to fasten the coil form in place. Washers can be used here if desired. Put the tinplate and strap iron in place and fasten, using five ⅛" × 1¼" flat head stove bolts and nuts. Thoroughly clean the coil lead with knife or sandpaper, and fasten to the nearest bending post. This cleaning must be thorough so that the end of the wire shines like a new copper penny and so that there is no varnish at all left on the end of the wire.

Place the 1/16" × ½" × 1¼" piece of black strap iron in place as shown in the drawing. Notice that the strap iron is held secure with two nuts below and one nut above on one end. On the other end, the strap iron does not touch any nut or screw (the screw passes through the 3/16" hole in the strap iron, but does not touch the strap iron). The strap iron, instead, touches a piece of ordinary

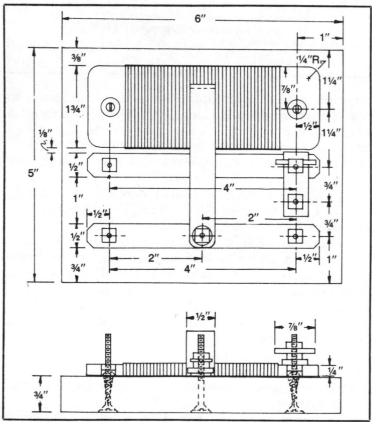

Fig. 1-8. Dimensions of the strap iron radio.

pencil lead and the pencil lead touches the nut. This arrangement forces the radio signal to pass through the pencil lead and through the surface of the strap iron. For this arrangement to "work," the pressure at this point must be light and the black surface oxide must be left on the strap iron. Details of how the signal is produced follow.

The radio signal goes through the lead without being shorted by the strap iron and the bolt that holds it in place, because this bolt at no point touches the strap iron. The bolt goes up through an oversized hole in the strap iron but does not touch the edges of this hole nor the surface of the strap iron. Therefore, the signal must pass through the lead and then through the surface of strap iron to reach the earphones. The lead provides the only electrical connection from the bolt to the strap iron.

It can be seen from this that if we attempted to take the signal off the bolt holding the lead that the signal would not pass through the lead, nor the strap iron, and would have no detection.

Attach to your radio an aerial, a ground and a set of earphones as shown on the wiring diagram. Any type of earphones will work. A length of wire, 50' or longer, can be used as an aerial. A ground usually consists of a short length of wire connected from the receiver to a cold-water pipe. Move your tuning arm back and forth several times on the coil form to wear off the enamel covering from the wire and permit the tuning arm to make contact. Move the tuning arm and you should be able to receive three or four stations provided you are within 15 miles of a metropolitan area.

You can increase the volume on your radio by making the aerial longer or by replacing the strap iron and pencil lead with a germanium diode. If the aerial is made too long, your stations will be difficult to separate. If it is too short, you will not be able to receive any signal. Start with 75' and experiment to find the length that suits you best.

If your radio does not work, check to see that the wires have all been scraped *perfectly clean* and that all connections are tight. Check also to see that the electricity flows from the upper piece of tinplate, through the stove bolt, through the pencil lead, and then through the surface of the black strap iron and into the earphone.

Remember! The pressure on the pencil lead must be light.

In all cases, the radio works and receives one station loud and clear. Satisfactory reception of up to four stations is determined to a large extent by such factors as length and location of aerial, the distance to the broadcasting stations, and whether all the electrical connections are clean and tight. The radio will work immediately if it is made according to directions.

TRANSISTOR TRANSMITTER

Here's a simple, low-cost transmitter suitable for the beginning or advanced hobbyist. The components are available at supply houses. Close scrutiny of the junk box may reduce the cost, since many of the parts are common components usually encountered in this type of work (Fig. 1-9).

The transmitter is built around a 2N170 transistor, a center-tapped loopstick, and a carbon microphone. The carbon microphone is a Western Electric F-1 unit. A high-quality unit improves the quality of transmission greatly.

A look at the schematic diagram (Fig. 1-10) shows the transmitter is an oscillator (Hartly fashion) being modulated by the carbon microphone in series with the power supply. Frequency is controlled by turning the slug in and out of the coil. This unit will tune across the entire broadcast band.

Although this method is seldom seen in transistor circuitry, the base is biased with a 22KΩ resistor across the 100pF capacitor (Table 1-4). This grid-leak-type arrangement seems to be very stable and requires a minimum of parts.

Outside of the usual precautions taken with transistors, such as controlling heat, the battery polarity should be observed. This is a NPN unit, so the positive side of the battery should be connected to the center top of the coil. This makes the collector positive. Do not exceed 6 volts as this is the maximum rating for this unit.

Figure 1-9 does not show the antenna connected, because this unit worked well without one. An antenna will increase the range, but one should not exceed FCC regulations. One method that works well is to solder a short antenna wire with an alligator clip to the proper terminal on the loopstick. Clipping the wire to a lamp

Fig. 1-9. The transistor transmitt
in its plastic case.

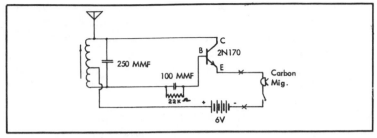

Fig. 1-10. Schematic of the transistor transmitter.

cord or similar wire will couple the signal to the power lines. Ample signal is transferred to pick it up for a considerable distance.

If a meter is available, check initial connection of the battery to see if current does not exceed 10 mA. Tune the receiver to a dead spot on the dial and then, holding the transmitter close to the receiver, tune the slug until it is heard on the receiver. Once the proper settings have been found, the transmitter can be taken some distance from the receiver. Several places may be found with the transmitter close to the receiver. In this case, move the transmitter away from the receiver a short distance. These harmonics will be less apparent at this distance. Only one setting should work best.

The dry cells should last for many happy hours of transmission, since the transmitter requires only 10 mA.

HIGH-PERFORMANCE RADIO RECEIVER

This circuit is a high-performance, low-cost radio receiver complex enough to offer a real challenge to the beginner. Certain criterions were set up in advance during design: no AC-DC circuits would be used; tubes using standard octal sockets which would provide uncrowded, easily soldered supports for resistors and condensers, would be used throughout (Figs. 1-11 and 1-12); all parts were to be standard items easily available from any wholesale

Table 1-4. Parts List for the Transistor Transmitter.

1	Ferri-tenna (Center tapped)
1	Carbon microphone (W.E.F-1)
1	NPN transistor (GE 2N170)
1	100-micromicrofarad capacitor (mica)
1	250-micromicrofarad capacitor (mica or ceramic)
1	6-volt battery (4 Penlite dry cells)

parts house at nominal cost (Table 1-5); the circuitry selected should permit ready modification into additional training projects; and provision was to be made for operation from either an external or built-in power supply.

Several pilot models were built from a preliminary design based on a 6SK7 tube as a tuned-radio-frequency amplifier, a 6F5 tube as a bias detector and a 6F6-G as a power output tube. Although these preliminary models showed good sensitivity and excellent audio quality, the selectivity left a great deal to be desired. Needed was another tuned circuit to enhance the sharpness of tuning.

Since the addition of another tuned-radio-frequency amplifier stage seemed undesirable because of the probability that extreme care would be necessary to prevent self oscillation, it was decided to obtain this increased selectivity through the addition of a band-pass tuner connected between the antenna and the grid of the 6SK7 radio-frequency stage. Study of various parts catalogs revealed a number of commercial coils that would be ideal for such a purpose. Unfortunately, they were expensive.

Consequently, a somewhat unorthodox band-pass system was designed from two Miller type 20-A antenna coils. Manufactured for replacement service, these coils are sturdy, low-priced, and easily repaired if damaged. The first coil, L1L2, is used without change; the second coil, L3, is easily modified by clipping the two primary leads and slipping off and discarding the primary winding. L1L2 is mounted solidly to the chassis while L3 is temporarily allowed to pivot on its mounting bracket to permit varying the degree of coupling between the two coils.

A matching Miller type 20-rf interstage coil couples the output of the 6SK7 radio-frequency stage to the grid of the 6F5 detector. It is located under the chassis, immediately below the tuning condenser. A volume control of the "antenna-shunt, C bias" type controls the gain of the 6SK7 rf amplifier.

Condenser C9, connected between plate and cathode of the 6F6-G output stage, performs two functions. As a fixed tone control, it "rolls off" the higher audio frequencies, producing a more agreeable audio response; and by introducing a small amount of degenerative feedback in the output stage, it minimizes harmonic distortion.

A full-wave power supply is provided, using a 6X5 tube as a rectifier. A "brute-force" filter system comprised of a small filter choke and two high-capacity electrolytic condensers reduces hum to below audibility.

Fig. 1-11. Top of the high-performance receiver. Parts are as follows, reading from left to right and from top to bottom: SW, C11, C10, C1C2C3, 6X5 tube, 6FG tube, L3, L1L2, T2, T1 and the 6F5 tube.

Construction is very simple and can be completed by the average hobbyist in six hours or less. After the larger parts are

Fig. 1-12. Bottom view of the high-performance receiver. Parts are as follows, reading from left to right and top to bottom: R1, 6SK7 tube, L6, C11, C12, SW, L4L5, 6F6G tube, 6X5 tube and the 6F5 tube.

Table 1-5. Parts List for the High-Performance Receiver.

C1 C2 C3—3-gang tuning condenser, 365 mmfd
per section.
C4, C5—.1 mfd paper condensers, 400 v
C6—.1 mfd 25-v, electrolytic
C7—270 mmfd, mica or ceramic
C8, C9, C10—.01 mfd paper condensers, 400 v
C11, C12—20. mfd, 375-v, electrolytic
L6—15 henry filter choke, 75 ma
L1, L2, L3—Miller Type 20-A antenna coils
L4 L5—Miller Type 20-RF interstage coils
R1—50 k ohms volume control, C bias taper
R2—470 ohms, ½-watt resistor
R3—100 k ohms, ½-watt resistor
R4—27 k ohms, ½-watt resistor
R5, R6—1.0 megohm, ½-watt resistor
R7—470 ohms, 1-watt resistor
T1—Output transformer, 6F6 plate to voice coil
T2—Power transformer, 460 v ac, center-tapped
  50 ma 6.3 v ac   2.5 amps.
Sw.—Toggle switch, S.P.S.T.
  Miscellaneous: Chassis, 2″ × 7″ × 13″;
Grommets, ⅜″; Small hardware.

mounted, it is suggested that wiring start with the filament circuit. Pin #2 of each tube is grounded to the chassis through soldering to a lug secured under the socket mounting screws. All #7 pins are joined together and connected to the A+ side of the filament supply.

Then starting with the 6F6-G output tube, the various condensers and resistors are wired in place, using tube socket pins and insulated tie points for proper support. The connections between the various coils and the stator sections of the three-gang tuning condenser are then made. Care should be taken in the dressing of these leads and those connecting to the control grids of the 6SK7 and 6F5 tubes to minimize stray coupling which might result in instability. All leads through the chassis are run through ⅜″ rubber grommets.

Since the power supply requires only three connections to the receiver, a B+ connection to #4 pin of the 6F6-G, an A+ connection to #7 pin of the same tube, and a chassis ground, an external power supply can easily be used instead, connections being made by a three-wire cable to the three points mentioned. Power requirements are 250 volts DC at 50 mA, and 6.3 volts AC at 1.3A.

Initial tune-up and adjustment are easily performed, and require no test equipment. Temporarily adjust the coupling between L2 and L3 to provide a spacing between the two windings of ¼″ or less. Attach a good outside antenna to the antenna terminal of L1 and a permanent-magnet type of dynamic speaker to any two of the secondary terminals of the output transformer, T1. After turning on the power supply, turn the tuning condenser to approximately half capacity, and tune in a comparatively weak out-of-town station. Using an insulated alignment tool, carefully peak each of the three trimmer condensers on the tuning gang for maximum response.

Now check the preliminary alignment by slowly tuning from the high- to the low-frequency end of the broadcast band. Proper band-pass coupling and proper alignment are indicated if strong local stations tune sharply and out-of-town stations are received with good volume and clarity. Remember that increased coupling between L2 and L3 will increase the sensitivity but decrease the selectivity, while decreased coupling will result in decreased sensitivity but increased selectivity. Each time the coupling between L2 and L3 is varied, the trimmer condensers across C1 and C2 should be again peaked on a weak station of about 1000 kHz. With certain tuning condensers it may be necessary to shunt the existing trimmer connected across C1 with a small additional

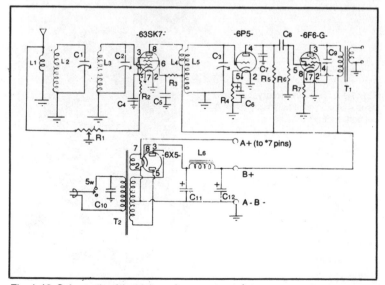

Fig. 1-13. Schematic of the high-performance receiver.

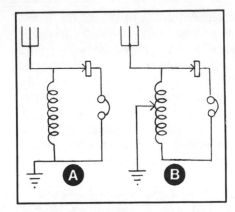

Fig. 1-14. Either wiring diagram (A or B) will do for the simple crystal radio.

trimming condenser to compensate for the lack of tube input capacity normally connected across such a tuned ciruit.

When optimum coupling is obtained, tighten the machine screw used to mount L3 to the chassis. Also vary the taps on the secondary of T1 for best volume and most agreeable tonal quality.

By using miniature tube sockets mounted on small aluminum plates, the octal tubes can be replaced, one at a time, with their

Fig. 1-15. This simple crystal radio uses a plastic tube as its coil form. A pill bottle will do just fine.

minature equivalents, and comparisons in their relative performance drawn. It has been found experimentally that the 6SK7, 6F5, and 6F6-G, can be readily replaced by a 6BA6, 6AT6 and 6AQ5, respectively, with no changes in resistor or capacitor values.

Another interesting experiment is to try various detector circuits in place of the bias detector used originally with the 6F5 tube. The 6F5 will make a good diode detector if the control grid is tied to the plate, and the two joined elements considered as a diode plate; and with the conventional triode connection, the 6F5 performs well as a grid-leak and infinite-impedance detector.

SIMPLE CRYSTAL RADIO

This simple crystal radio will bring in local stations with ample volume. A coil of wire, germanium crystal, phones, antenna, and ground are all that are needed to bring in a good signal from nearby transmitters. The wiring diagrams are shown in Fig. 1-14.

The coil is easily made by winding approximately 60 turns of #22 magnet wire on any nonmagnetic tube such as plastic (shown above), cardboard, wood, etc., approximately 1½" diameter by 3". Use banana or pin jacks and plugs for antenna, ground, and phone connections. Connect the crystal inside the tube (Fig. 1-14). To make the set more selective, tap the coil every 10 turns as shown in B.

Chapter 2
Audio

The following projects involve sound in electronics—or audio, They cover everything from tape recorders to small amplifiers.

TAPE RECORDER SOUND SWITCH

The operation of turning on and off a tape recorder is to some people just plain work! To solve this problem there are two ways out. One is to hire someone else to do the "dirty" work; the other is to build an automatic switch like the one described here.

The first thing to do is to check and see if there is a jack for an "Aux. Spkr." If not, find a spot on the tape recorder to install two ¼" phone jacks. One of these is to be used for the "Input" into the switch (or, when not in use, it may be used for external speakers), and the other is to be used for the motor control. Figure 2-1 shows how to wire these two jacks.

The next step is to assemble the connecting cables that will be used to connect the switch to the recorder. It doesn't matter how long these cables are, but I suggest they be about 4' so that the switch can be located away from the tape recorder if desired. Use two ¼" phone plugs for the connection to your tape recorder, one RCA phone plug, and one Cinch-Jones No. 300 series plug for the connections to the sound switch (Table 2-1). Be sure to use shielded cable for the "Ext. Spkr." to "Input" cable. The type of wire used in the other cable is optional.

Now that you have both jacks installed and the connecting cables finished, you are ready to start on the main part of the

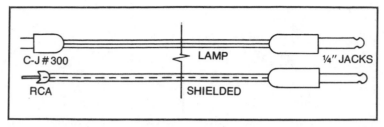

Fig. 2-1. How to wire the two jacks required for the tape recorder sound switch.

project (Figs. 2-2 and 2-3). Begin by deciding where you wish to place your main parts and mark all the holes.

Then drill all holes and mount the major parts, such as VI, RLT, TI, P1, P2, J1, J2 and the fuse holder. Be careful not to mount the relay too close to the tube or the heat generated might affect the operation of the relay.

Now you are ready to start wiring the unit (Fig. 2-4). Take your time and do a good job. Watch out for cold solder joints and be sure to use only rosin-core solder.

To operate the unit connect it to the tape recorder using the cables you have already made. Turn the tape recorder on. The tubes should light, but the motor should not operate. If the motors do operate, check the wiring of the jack. Now turn on the sound switch and wait for it to warm up. Turn the volume on the tape recorder all the way up and do the same with the sensitivity control on the unit. Speak into the microphone and the motors should start to turn.

Adjust the volume on the tape recorder to normal operation. This should be done by setting the control on its numbers, 1 to 10.

Table 2-1. Parts List for the Tape Recorder Sound Switch.

1 Octal socket	1 100 ohm lw resistor
1 6SN7 GT Tube	1 1000 ohm 1w resistor
1 50 mA rectifier	2 dial knobs
1 Transformer—117v pri., 125v 15 ma, 6.3v 6amp Stancor #PS-8415	14' length of shielded cable
1 1N34 crystal diode	1 4' length of lamp wire
1 3 ag lamp fuse	2 ¼" phone jacks
1 Fuse post	2 ¼" phone plugs
2 20/20 mfd 150 v electrolytics	1 RCA phono plug
3 .01 mfd paper capacitors	1 RCA phono jack
1 5000 ohm ½w resistor	1 Cinch-Jones two connector plug (300 series)
1 4700 ohm ½w resistor	1 Cinch-Jones two connector jack (300 series)
1 270k ohm ½w resistor	2 1.5 meg pots
	Miscellaneous hardware, grommets, etc.

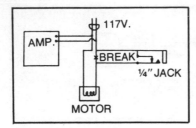

Fig. 2-2. The motor control setup for the tape recorder sound switch.

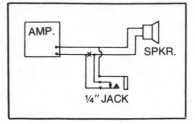

Fig. 2-3. The auxiliary speaker setup for the tape recorder sound switch.

As you turn down the volume, speak the numbers into the microphone.

Adjust the sensitivity on the sound switch so that it will pick up what you want it to record.

Your sound switch is now complete and should be of great value to you in your recording. It also may be used for numerous other gadgets around the house by just hooking it up to the ouput of an amplifier and using the relay to open doors or just about anything you want it to do.

If you run into any trouble, go back over the wiring job (Fig. 2-4) with a fine-tooth comb. Look especially for wiring errors, cold-solder joints and excess solder shorting out two or more components.

A NEXT-TO-NOTHING AMPLIFIER

This amplifier project consists of next to nothing—two transistors and a 6V battery. It too becomes a nearly indestructible amplifier, capable of driving almost any size speaker at satisfactory listening level.

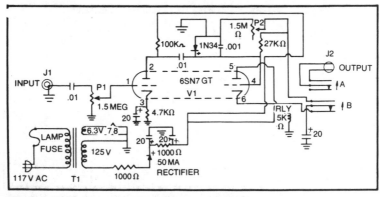

Fig. 2-4. Schematic of the tape recorder sound switch.

The project involves two transistors, one must be an NPN and the other a PNP. One of these must be a power transistor. There is nothing critical about the transistors. Almost any of the now inexpensive ones available should work. They must be capable of amplification of audio signals, one capable of high voltage gain and the other of high power gain. Refer to Figs. 2-5 and 2-6 to see the two possible arrangements of the NPN and PNP transistors.

The battery voltage can be anything from 1.5 to 9 volts. The higher the voltage is, the greater the volume is. Six volts works quite well.

The basic amplifier configurations used are a common emitter voltage amplifier coupled direct to a common collector power amplifier. The common emitter has low-input impedance and high-output impedance. The common collector has high-input impedance and low-output impedance, which allows us to directly couple the two stages together and also directly couple to a speaker without the use of an output transformer.

The power transistor should be mounted on a heat sink which can be made from a small scrap piece of aluminum. The higher the battery voltage is, the more heat that will be developed. It may be necessary to stack two pieces of aluminum and bend them to produce a finned heat sink to provide more heat dissipation. The power transistor may operate warm. It should not operate at temperatures too hot to touch or it will destroy itself.

Once the basic amplifier is operational, several modifications can be made. In Fig. 2-7, a capacitor and a resistor are connected across the output and tied to the input. This will cause the circuit to oscillate in the audio range, thus providing a code oscillator, assuming a key is inserted to break the current flow of the battery.

The basic amplifier now becomes a very simple RC oscillator and the tone is created by the time constant set up by the value of R × C. By varying the value of R or C you can vary the tone produced

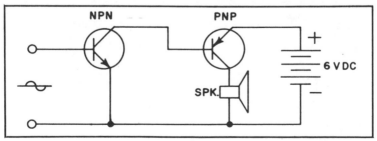

Fig. 2-5. A next-to-nothing amplifier schematic using NPN and PNP transistors.

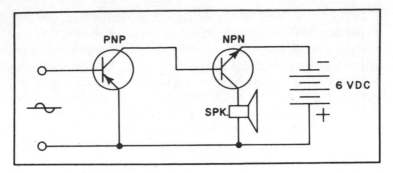

Fig. 2-6. Another next-to-nothing amplifier schematic using NPN and PNP transistors.

at the speaker. Start out with a 100K Ω resistor and a 0.5μF capacitor. After the student sees that it works, he can vary these values to any time constant he wants from one beep per minute to 1000 Hz higher.

Figure 2-8 shows a slightly further variation of using sensitive pushbutton switches to create a transistorized musical device. Experimentation as to the tones produced by the values of R and C would be left up to the individual.

Figure 2-9 shows still another variation to the basic amplifier of coupling a simple diode detector to product a *trf* (tuned radio-frequency) receiver. A long-wire antenna is needed for reception. This receiver is capable of tuning the strongest and nearest station.

VERTICAL COLUMNS OF SOUND

A sound system that is economical, portable, and rugged has many uses. Integrated amplifiers and loudspeakers may be availa-

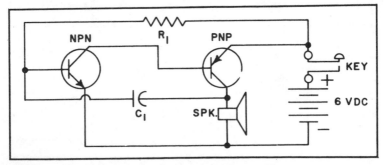

Fig. 2-7. A modification of the basic next-to-nothing amplifier.

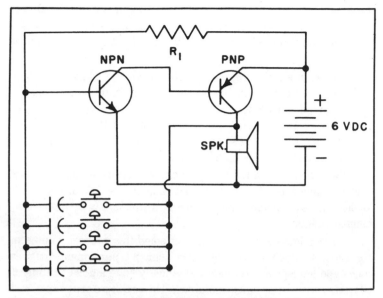

Fig. 2-8. A slightly different modification from that in Fig. 2-7. This one uses sensitive pushbutton switches to create a musical device.

ble, but the ability of a conventional loudspeaker to project to the limits of a group may be lacking.

Most loudspeakers in an efficient enclosure have uniform vertical and horizontal dispersion of sound, resulting in waves being reflected from ceilings and floors causing reverberation and feedback. The sound column, however, is capable of greater

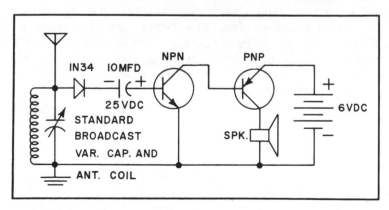

Fig. 2-9. This modification of the next-to-nothing amplifier uses a simple diode detector to create a tuned-radio-frequency receiver.

31

horizontal dispersion because of the stacking of the speakers, which results in a deeper penetration into the hearing area and a minimum of sound radiated vertically to reflect from ceilings and floors.

The enclosures are easily fabricated and make excellent projects for hobbyists who have an interest in sound reproductions. A specialized knowledge of electronics is not necessary as the plans (Fig. 2-10 through 2-12) and text have sufficient detail, so that any individual possessing a fair degree of woodworking skill may feel safe in undertaking their construction.

Two cabinets, which may be constructed from slightly more than one piece of 4' × 8' material, and the wide variety of 8" loudspeakers available, offer a selection which may be tailored to fit most budgets.

The enclosure is an infinite baffle and the internal volume is not critical. Each enclosure uses four 8" permanent magnet cone-type loudspeakers which individually have a power rating of 10 watts and an impedance of 8 ohms. The speakers are phased and wired in a series parallel arrangement to give a total impedance of 8 ohms. Construction should be as airtight as possible and of ¾" plywood of reasonably sound core to eliminate any possibility of rattle or buzz.

Construction

Begin by ripping and jointing the stock to dimension. Rabbets ¾" × ¼" are cut in three sides of the top, bottom and one edge of each side panel. Dadoes ¾" × ¼" are then cut on the inside of the top, bottom and side panels to receive the speaker mounting board.

Locate the centers of the speaker cutouts and use a compass to draw the required 7" diameter circles. These are then cut out with a sabre saw or compass saw. Sooth all edges and give the speaker mounting panel a coat of flat black paint to decrease the visibility of the speakers through the grill.

Assemble the cabinet top, bottom speaker board and side panels with 1¼"-#8 flathead wood screws and glue using sufficient screws to insure an airtight structure.

Drill the speaker mounting board for four 1½"-#8 roundhead machine screws at each speaker location. Make a provision in the back panel to run the connecting cable. On the original enclosures, pin plugs and jacks were utilized to enable the separate handling of the long cables needed in portable installations. The cables could

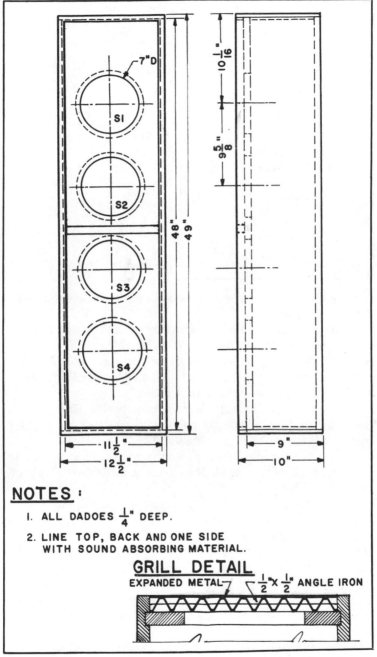

NOTES :

1. ALL DADOES $\frac{1}{4}$" DEEP.

2. LINE TOP, BACK AND ONE SIDE WITH SOUND ABSORBING MATERIAL.

GRILL DETAIL

EXPANDED METAL $\frac{1}{2}$" X $\frac{1}{2}$" ANGLE IRON

Fig. 2-10. Construction dimensions of the vertical columns of sound.

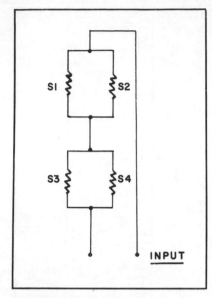

Fig. 2-11. Wiring diagram for the vertical columns of sound.

be brought out directly from the enclosure if an internal strain relief provision is provided.

The cabinet should be primed inside and out and, since equipment of the type is subject to considerable handling, a finish which is durable and easily retouched is desirable. For the enclosure shown in Figs. 2-13 and 2-14, a multicolor spray finish, such as is used in marine interiors, was used.

To finish the enclosure, mount each speaker with four #8 machine screws, washers and nuts. Take care that the speaker frame is not warped by over-tightening (Fig. 2-15).

Wire the speakers as shown on the schematic. It is wise to leave about 18″ of wire inside the cabinet to facilitate the removal of the cabinet back.

Since the speakers operate in a common cabinet, they should be phased for optimum performance. Some units are polarized so all that is necessary is to wire the positive pole to the positive pole and so on. If the units are not so marked, a small dry cell may be momentarily connected across each set of terminals and the direction of cone movement observed. The cell polarity may then be marked on the frame to indicate which way the connection should be made to insure that the cones all move in the same direction.

Line the back, top and one side with a sound absorbing material such as fiberglass to reduce the possibility of standing waves developing during operation. The back may then be attached with 1¼"-#8 flathead wood screws on 6" centers.

To protect the fiber speaker cones, a grill of expanded metal was fabricated and reinforced with ½" angle iron. This was primed and given a coat of enamel. This type of grill is strong and decorative. Cover the face of the speaker board with black crinoline and mount the grills with pan head sheetmetal screws through the mesh.

Singularly, or in pairs, these enclosures have proven useful for a variety of gatherings. Two units will fill an average gymnasium with sound for a film showing or lecture with the output from a modest amplifier.

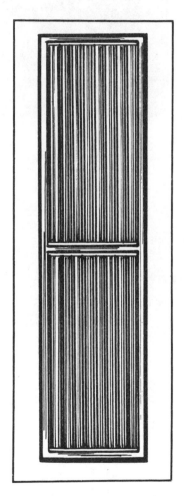

Fig. 2-12. The front of one of the speakers.

Fig. 2-13. Enclosure is airtight and must be constructed to eliminate any possibility of rattle or buzz. The grill is made of expanded metal reinforced with ½" angle iron. Grill mounts to speaker board, which is covered with black crinoline.

Fig. 2-14. Openings in the back panel accommodate the connecting cable.

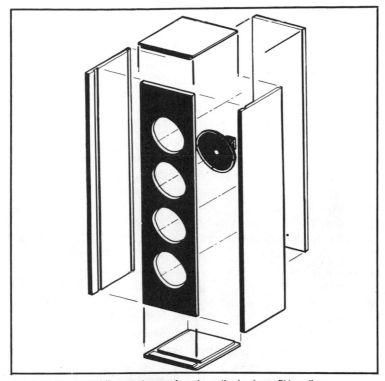

Fig. 2-15. How the different pieces of each vertical column fit together.

Low cost, portability, straightforward fabrication and unique acoustical properties make these enclosures a welcome addition to anyone's sound equipment.

A 2W AMP FOR HIGH-IMPEDANCE INPUTS

With just a few changes in the connections to a 2W amplifier (see Fig. 2-16), the circuit shown can be used for high-impedance inputs. The input can be either + (non-inverting) or − (inverting). However, the best practice is to use the − input and ground the + input. If the + input is not grounded, it may have just enough output signal coupled to it to cause feedback, with resultant oscillation. In Fig. 2-16 the potentiometer R1 serves as the volume control. Only one connection to the source (input) is shown in the diagram. The other connection to the signal source is the ground terminal on the module.

High-impedance leads are prone to hum pickup. For this reason, if the source is any distance from the amplifier module, use shielded wire. A typical high-impendance signal is a simple crystal phonograph pickup. The connections from the pickup to the amplifier should be shielded wire with the center wire connected to the input terminal and the braided shielding connected to the ground.

When used as a phonograph amplifier, the output is rather harsh due to the presence of the high-frequency components of the sound. This can be improved through the addition of a tone control as shown in Fig. 2-17. As the potentiometer knob (*R2*) is turned in a clockwise direction, the tap on the potentiometer is moved toward *C3*, causing the capacitor to have a greater effect on the frequency response. At the maximum setting of the tone control potentiometer, the capacitor serves as a path of low-impedance to the high-frequency signals and therefore tends to allow only the lower frequency parts of the signal to be amplified. This is the same type of tone control used in most AM radio receivers in automobiles and vans.

The tone control circuit (Fig. 2-17) is not a true tone control as used in high fidelity sound devices. For this application, two controls are required, one for the bass frequencies, and one for the treble. This type of tone control tends to attenuate (make smaller) the total signal, so a preamplifier is necessary. If the device is to have two channels as in a stereo system, then two sets of preamplifier-tone control-amplifier are required.

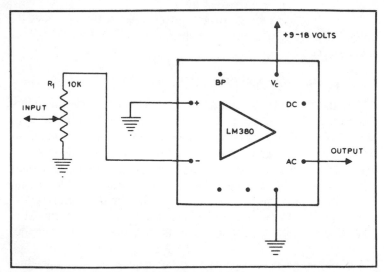

Fig. 2-16. This basic 2W amplifier can be used for high-impedance inputs.

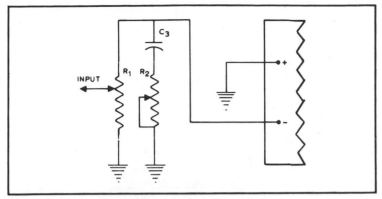

Fig. 2-17. A tone control circuit using the basic 2W amplifier.

MORE GAIN AND POWER FROM AN AMPLIFIER

The fixed gain of a basic 2W amplifier can be boosted by the use of a positive feedback circuit. This is shown if Fig. 2-18. Note that the input for the circuit is still through the negative input with a feedback circuit consisting of *R4, R2* and *C3* connected from the output to the positive input and ground. These values have been carefully selected to provide an increased gain in the amplifier without causing a loss of stability which could result in oscillation. To further guard against oscillation, *R3* and *C4* have been added.

Of course, there has to be some disadvantage in changing the circuit to obtain increased gain. That disadvantage is a reduction in the frequency response of the amplifier. Thus, when this method of increasing the gain is used, the amplifier is no longer capable of high fidelity operation. However, the overall gain of the amplifier has been increased by a factor of 4 from the fixed gain or 50 to its new gain of 200.

When more power output is required, two amplifier modules can be used in a bridge connection as shown in Fig. 2-19. Through the use of this circuit the power output becomes the sum of the power outputs of the two amplifiers, or 4 watts. The input to the two amplifier modules is cross-connected, that is the positive input of one amplifier is connected to the negative input of the other. One of the two input lines is grounded and the other is used for the signal through the volume control *R1*.

In the bridge amplifier, the loudspeaker is connected to the two DC outputs of the amplifiers. Resistance *R2* is the balance control connected to the bypass connection of each amplifier as shown in the diagram. To balance the amplifiers, place a sensitive

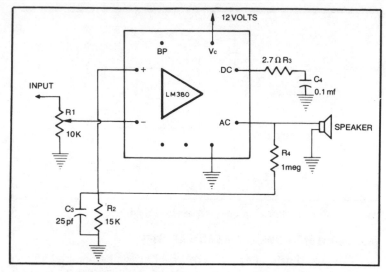

Fig. 2-18. The 2W amplifier with positive feedback. C3 and C4 are rated at 50 volts, and all resistors are ½W resistors.

DC voltmeter (VTVM) across the loudspeaker connections (without the loudspeaker being connected in the circuit) and adjust the control for a zero reading between the outputs of the two amplifiers. When this adjustment has been completed, connect the loudspeaker and do not move the potentiometer knob.

If the application of the basic amplifier module requires more gain, it can be accomplished with positive feedback if the resultant loss in frequency response does not interfere with the intended use of the amplifier. If more power is required, two amplifiers can be connected in a bridge circuit. In the next application—the sound relay—if more sensitivity is needed, the positive feedback technique can be used.

A SOUND RELAY

This application of the basic amplifier—a sound relay—uses many of the same components used in other circuits. The additions (see Fig. 2-20) are the diode *D1*, the crystal microphone and the sensitive relay. The relay is Calectro #D1-962 with a 10 mW, 1000 Ω coil. The contacts for the relay are s.p.d.t. Any similar relay can be used with this circuit.

Note that the crystal microphone is used as the signal source. The AC output is connected to the 8 Ω winding of the transformer, with the 1000 Ω winding serving as the secondary. The transformer functions as an impedance matching devise to match the low

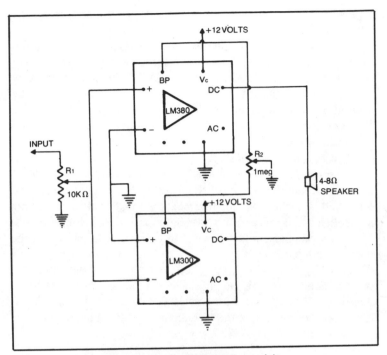

Fig. 2-19. A bridge amplifier using the 2W amplifier module.

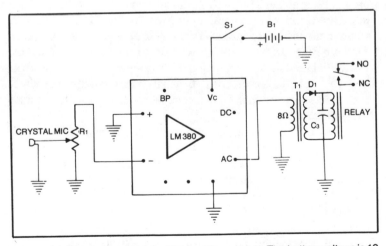

Fig. 2-20. A sound relay using the 2W amplifier module. The battery voltage is 12 volts. C3 is a 0.1μF, 50V Capacitor, R1 is a 10K potentiometer, and T1 is a 1000Ω to 8Ω output transformer.

impedance output of the amplifier to the 1000 Ω impedance of the relay coil.

The relay functions on DC and the sound output of the amplifier is AC. Therefore the diode serves as a simple half-wave rectifier to convert the AC to pulsating DC. This pulsating DC is filtered with capacitor *C3*. The value of *C3* can be changed as desired. If made larger, it will delay the release time of the relay. This in effect causes the relay to remain energized for a certain time after the sound has ceased. Too small a value of *C3* will cause the relay to chatter needlessly.

The relay has a limited load capacity of 0.25A at 12V for its contacts. If using another relay, observe the specification sheet furnished with the relay for the maximum load that the relay can safely handle. If more power is needed in the switching contact, this relay can cause a larger relay to operate, one which has the contact capacity desired.

In using this sound relay, when the microphone detects sound it is amplified and this amplified sound causes the relay to be energized. It may be used for a variety of purposes. For the deaf it can detect the ringing of the telephone and light a lamp. It can be used to detect sound level in a group.

A PREAMPLIFIER WITH VERSATILITY

This circuit board uses the LM382, a low-noise dual preamplifier manufactured by National Semiconductor Corp. This integrated circuit has several unique features that cause it to be of value in the modular approach to the design of electronic projects. This is a rugged device with short-circuit protection. Also like the LM380, it is designed to function from a single voltage supply from +9V to +18V. This causes it to be easily adapted to devices which can function in the van or automobile, or from a self-contained battery source.

A method of biasing the amplifier with pin connections to the internal bias system of resistors makes it possible to choose several available levels of maximum amplification. To set the amplification level, capacitors are connected between the pins and ground as indicated in the chart of Fig. 2-21. The biasing capacitors used are 20 μF at 16V and provide a choice of 40 dB, 55 dB, and 80 dB maximum gain.

The use of additional resistors and capacitors connected to the pins of the integrated circuit will adapt the frequency response of the preamplifier to the several types of equalization necessary

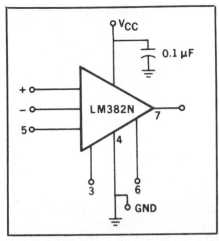

Fig. 2-21. Preamplifier schematic.

when using it as the input from a tape head or a phonograph pickup. Without an equalization network, the frequency response of the preamplifier is essentially flat over the entire audio spectrum.

The preamplifier on the circuit board whose foil patter is shown in Fig. 2-22, when biased for a maximum gain of 40 dB, has an output of actually 42 dB or a voltage gain of 133 (Table 2-2). The maximum input voltage without distortion (sine wave at 1000 Hz) was measured as 0.0023V and the output voltage with this input

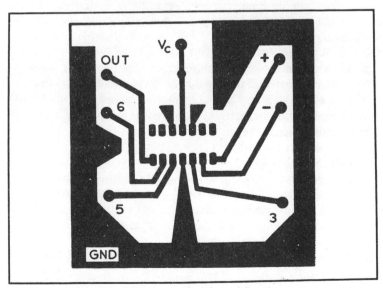

Fig. 2-22. Circuit board of the preamplifier shown full scale.

Table 2-2. Maximum Gain Chart for the Preamplifier.

Gain	Required Capacitors
40 dB	20 μF, 16 V from pin 6 to ground only.
55 dB	20 μF, 16 V from pin 3 to ground only.
80 dB	Both capacitors as specified for 40 and 50 db.

was 3.0V. This output is more than adequate to provide the input voltage for a power amplifier such as the LM380.

Because the LM382 is a DC amplifier, both the input and output signals need to be capacitive coupled to the integrated circuit. In the test circuit, the input signal was connected through a 0.1 μF ceramic disc capacitor to the positive (+) input and an identical capacitor was connected to the output terminal for output voltage. With the exception of the bias capacitor connected to pin 6, not one of the terminals had connections.

An examination of the foil pattern for the circuit board will show the unusual design of having the grounded portion of the foil almost totally circling the integrated circuit. This is for shielding and is necessary because of the sensitivity of the amplifier. In addition to this foil pattern shield, it is essential that the input and output leads be physically separated as much as possible. At maximum amplification, any small portion of the output voltage which is coupled to the input will result in feedback and resultant oscillation.

Like any integrated circuit, the LM382 must be properly positioned in the socket to make certain that the power leads are correctly connected. If this particular integrated circuit is inserted in its socket with pins 4 and 11 interchanged, a large current will surge through the chip and instantly cause it to burn out. The small dot or indentation to the left of the notch on the top of the integrated circuit indicates the position of the #1 pin.

As examples of the uses to which this module is suited, it can be used as a microphone preamplifier in the construction of a CB power microphone, it can be used to amplify the signal from a photocell for light-beam communications, and, when both sides of the preamplifier are used, it can be the input amplifiers for a stereo system for the automobile or van.

USING THE PREAMPLIFIER WITH THE 2W AMPLIFIER MODULE

The preamplifier (LM382) module is easily connected to the power amplifier (LM380) module if several precautions are observed. First, the two modules can be connected together only be capacitive coupling. This capacitor is shown in Fig. 2-23 as C_c.

Capacitive coupling must be used since the two ICs are direct current amplifiers ahd have different voltages to ground at the points that must be interconnected. In this application, the output connection of the preamplifier has a voltage of 6.38V while the input connection of the power amplifier has a voltage of 0.06V. Obviously, if these two points were connected the difference in the voltages would cause the malfunction of both integrated circuits. An ideal value for the coupling capacitor is 0.1 μF. (Table 2-3). This value is sufficient for an adequate frequency response at the audio frequencies being used. The ceramic disk capacitor is our best choice to minimize the hum pickup which would result from the use of another type of capacitor.

Combined with the coupling capacitor is the variable resistance element (the 1.0Meg Ω potentiometer) used in the input circuit of the LM380 as a volume control. This permits a variable amount of the output voltage of the preamplifier to be used for the input to the power amplifier.

Adequate shielding is a second precaution in the use of these two modules in amplifier applications. Therefore, the case of the potentiometer *must* be connected to the common ground connection. Without this grounding, a serious hum will result that prevents the proper operation of the circuit.

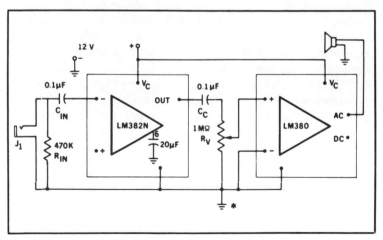

Fig. 2-23. Asterisk indicates a common ground point between J1, each circuit board and the metal cover of R_V. A metal chassis will provide the necessary "common ground" for adequate shielding of the system.

Table 2-3. Parts List for Using the Preamplifier with the 2W Amplifier.

		Radio Shack Part No.*
C_{in}	$= 0.1\ \mu F$, ceramic disk capacitor	272-135
C_c	$= 0.1\ \mu F$, ceramic disk capacitor	272-135
J_1	$=$ miniature phone jack 2 conductor	274-251
R_{in}	$= 470\ K$, ¼ W	271-1300
R_v	$= 1.0\ M\Omega$ potentiometer	271-211

*Part source supplied by authors implies no endorsement by the publisher.

Shielding and the separation of the input and output connections of the amplifier are critical for the operation of an amplifier with this amount of gain. Since the gain of the 2W amplifier module is 50 and that of the preamplifier is 40, the total gain is 90 for the circuit shown in Fig. 2-23. This exceeds the total gain required for many applications, but will be used in several of the future articles in this series.

Only a metal cabinet will be satisfactory for housing this amplifier, and all ground connections must be made to this cabinet. Input and output connections must be as far removed from each other as possible with the shortest possible leads to prevent feedback which will result in unwanted oscillations. If the volume control is mounted to this cabinet with a good electrical contact, then an additional connection to the case of the volume control will be made through its mounting.

The circuit of Fig. 2-23 has been designed to function on 12V but will work satisfactorily with a supply voltage as low as 9V or as high as 18V. As shown, the input can be any type of magnetic or piezoelectric transducer such as a crystal microphone, a telephone pickup coil, or the magnetic head of a tape recorder. However, no equalization network has been incorporated in this circuit to match the response of the amplifier to that of the tape recorder or to ceramic phonograph pickups. This addition to the circuit will be incorporated in a future article in which both sections of the LM382 and two LM380 ICs will be used to construct a stereo amplifier with equalization and dual tone controls.

Chapter 3
Automotive

Cars have become more electronic in the past few years. These projects involve automotive maintenance and modification.

CLOTHESPIN TIMING LIGHT

This inexpensive timing light for Volkswagens can be built for pennies. Study the diagram carefully in Fig. 3-1. The bulb is held by the clothespin jaws. A panel or ceiling bulb will do.

Gap the distributor points and rotate the engine by hand until the distributor rotor points to the No. 1 mark on the distributor housing. Clamp one clothespin to the No. 1 (primary) terminal of the coil. Clamp the other to a ground.

Turn on the ignition. Rotate the crankshaft pulley clockwise until the timing notch is opposite the crankcase jointing faces. The light should go on as the notch passes the jointing faces. If it doesn't, the timing is off.

SOLID-STATE ELECTRONIC IGNITION

Solid-state electronic automotive ignition systems have a lot to offer—better gas mileage, easier cold weather starting, fewer tuneups and virtual freedom from maintenance since there are few mechanical parts to wear out. And converting to electronic ignition doesn't have to be expensive. The method described here, for V-8 or six-cylinder distributors, uses parts available at auto salvage yards. For the combination of new and used components to convert a six-cylinder Chevrolet, the cost was $11.

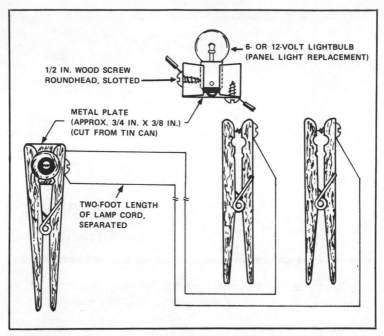

Fig. 3-1. Wiring diagram for the clothespin timing light.

The conversion consists of modifying the distributor, mounting the dual ballast resistor block and control unit, and connecting wires to the ignition switch starting bypass switch, coil positive terminal, and coil negative terminal. Required equipment includes a lathe, timing light, soldering gun, ignition wrenches, large hammer, wire cutters, screw-driver, micrometer, machine taps (10-24, 8-32) drill bits (#25, #29, 1/8 in., 1/16 in., 5/32 in.), 1/4" capacity hand drill, .008" nonmetallic feeler gauge, hacksaw, chisel and telescoping (snap) gauge.

In addition to the distributor to be modified, you will need a Chrysler electronic control unit or an Echlin unit (NAPA #TP50), wire harness, dual ballast resistor, reluctor, pickup coil unit, and 1/4"length of 1/8" brass rod. Although any 12V ignition coil will work with this system, a Chrysler coil is recommended. The reluctor [NAPA part #MP802 (six-cylinder) or #MP800 (eight-cylinder)] and pickup unit 2MP801 are available from Echlin through local NAPA parts jobbers for about $7.

To convert the six-cylinder Chevrolet, first we remove the distributor from the engine, disassemble it, and thoroughly clean it. Then arrange the parts (Fig. 3-2) in preparation for the following sequence of steps.

☐ Mount the distributor cam in a lathe (Fig. 3-3) and machine the rubbing surfaces to match the inside diameter of the reluctor; a very slight interference fit is desirable. The cam may not have to be

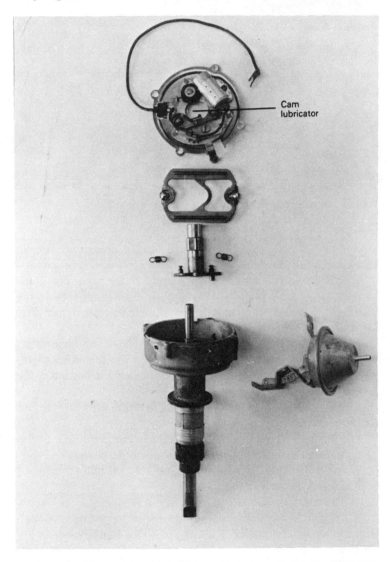

Fig. 3-2. Distributor parts must be disassembled, cleaned and arranged before the conversion to electronic ignition begins.

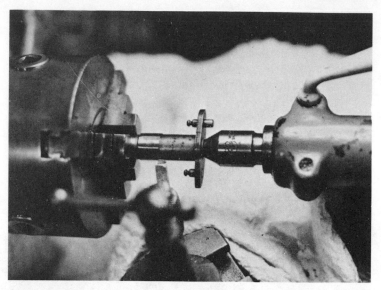

Fig. 3-3. Distributor cam mounted in lathe for turning operation.

machined completely round; some of the flat areas may remain when the final diameter is reached.

☐ Remove the cam lubricator and cut the pivot pin 1/16" above the mounting plate while the base plate is attached to the distributor housing. Temporarily reassemble the distributor (Fig. 3-4) to facilitate this operation.

☐ Separate the mounting plate and the distributor base plate.

☐ Use a hammer and anvil to flatten the mounting plate, taking care not to damage the nylon bearings on the bottom of the plate.

☐ Place the distributor housing securely in a vise for reassembly.

☐ Scribe a line through the center of the pivot hole to the vacuum advance connecting point on the mounting plate (Fig. 3-5).

☐ Put the flattened mounting plate onto the shortened pivot shaft and set it in the housing.

☐ Press the reluctor into position.

☐ Position the pickup unit so the tip of the pickup is aligned with the scribed line on the mounting plate and touching one of the reluctor teeth.

☐ Mark the position of the alignment pin hole for the pickup unit.

□ Drill a ⅛" hole for an alignment pin. Make a pin by cutting a ¼" length from a ⅛" brazing rod. Remove the mounting plate and place the pin in the drilled hole. Rivet the pin to the mounting plate with a sharp center punch or chisel.

□ Replace the mounting plate and reluctor in the distributor and place the pickup unit on the alignment pin.

□ Put a .008" feeler gauge between a reluctor tooth and the tip of the pickup unit; then drill and tap a hole in the mounting plate at the center of the adjustment slot of the pickup unit for the 8-32 hold-down screw.

□ Upset the end of the shortened pivot pin to prevent the mounting plate from slipping off the pivot pin. Check to be sure there is a .008" gap between the reluctor and the pickup unit.

□ Place the distributor cap on the housing; then mark on the housing the location of the #1 spark plug terminal. The screw identified in Fig. 3-6 is in alignment with # 1 spark plug terminal.

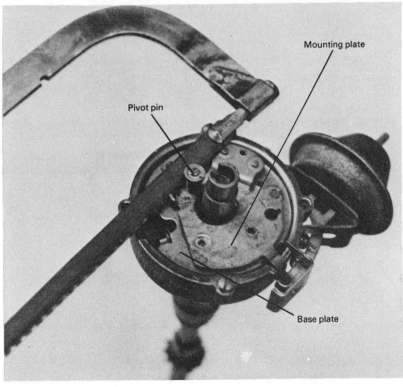

Fig. 3-4. To cut the pivot pin, distributor is temporarily reassembled.

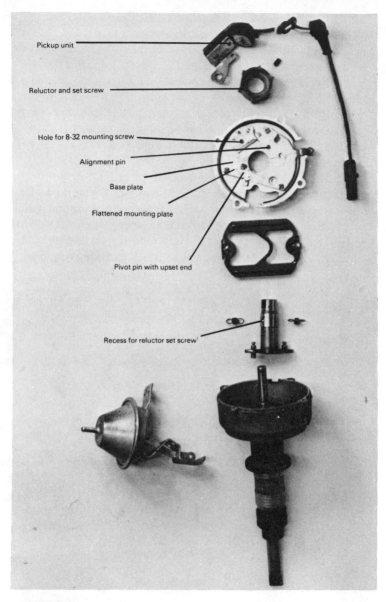

Pickup unit

Reluctor and set screw

Hole for 8-32 mounting screw

Alignment pin

Base plate

Flattened mounting plate

Pivot pin with upset end

Recess for reluctor set screw

Fig. 3-5. A line must be scribed through the center of the pivot hole to the vacuum advance connecting point on the mounting plate.

☐ Replace the rotor on the distributor shaft, and rotate the shaft until the rotor aligns with this terminal mark. While holding the shaft and rotor to keep them from turning, align one reluctor tooth with the tip of the pickup unit. (The spark occurs just as the reluctor and pickup unit go out of alignment.)

☐ Install a set screw in the reluctor. Drill a 1/16" pilot hole through the reluctor and into the distributor cam shaft 1/16" to establish relative position. Remove the reluctor, and drill and tap it for a 10-24 by ¼". Allen set screw using the pilot hole. With a 5/32" drill, bore an alignment recess in the cam shaft at the pilot hole. Both a stock distributor cam shaft and a turned shaft are shown in Fig. 3-7. Assemble the unit and check that the set screw does not protrude from the reluctor; shorten the set screw as necessary.

☐ Install the distributor in the engine.

☐ Mount the dual ballast resistor on the firewall or fender well near the coil, making sure the resistor is not near combustible

Fig. 3-6. The screw indicated by the arrow marks the No. 1 cylinder.

Fig. 3-7. Stock distributor camshaft (left) and turned camshaft with drilled recess hole (right).

objects, since it can become very hot if the engine stops with the ignition on.

☐ Mount the control unit on the metal panel near the ballast resistor. The control unit must have a good ground to the engine. (Specific locations in Fig. 3-8 are for a 1968 Chevrolet van.)

Refer to Fig. 3-9 in the following steps:

☐ Connect the wiring harness to the control unit, ballast resistor and pickup unit. The green/red wire from the control unit must go to the 5 Ω (filled in) side of the dual ballast resistor. (Chrysler control units use polarized plugs which will fit only in the correct way.)

☐ Connect a wire from the ignition switch to the blue wire near the ballast resistor.

Fig. 3-8. The control unit, which must have a good ground to the engine, is mounted on the metal panel near the ballast resistor.

☐ Connect the black wire with a yellow marker stripe from the control unit to the coil negative (dist.) terminal. A tachometer lead may be connected here as well.

☐ Connect the brown wire from the ballast resistor and the resistor bypass (starting bypass) wire to the coil positive (bat.) terminal (see Fig. 3-10, wiring diagram).

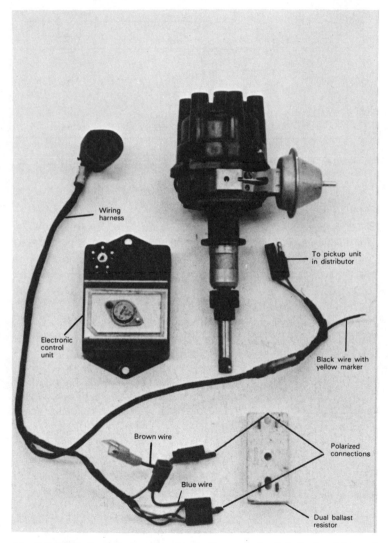

Fig. 3-9. Wiring harness is connected to the control unit, ballast resistor and pickup unit.

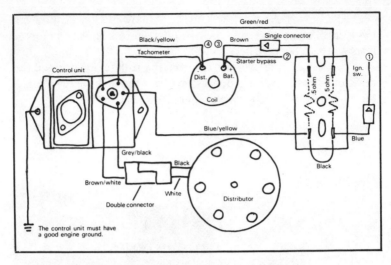

Fig. 3-10. Wiring diagram for the solid-state electronic ignition.

☐ Time the engine in the usual manner. The installation is now complete.

To test the system quickly, switch the ignition on, disconnect the dual connector at the distributor pickup unit, and then, while holding the coil wire about ½" from ground, strike the connector on the grey-black wire to ground. A fat, blue spark should result. An auto engine oscilloscope is not required, but it does demonstrate and clarify the operation of the system.

Once the system is installed and operating, further service should be unnecessary. Timing will remain as set, plug life will be extended and performance should be greatly improved.

Chapter 4
Club Teaching
Demonstrators and Group Activities

Radio and computer clubs are a big part of hobby electronics. If you stay in electronics long enough, someday you'll be asked to teach a course on amateur radio or computers or whatever. The demonstrator projects at the beginning of this chapter are exactly what every club instructor needs. Each demonstrator decreases the necessary learning time for its particular fundamental. And the group activity projects toward the end of the chapter are excellent for adding a little flavor to a club's program.

DEMONSTRATION TWO-SPEED INDUCTION MOTOR

This demonstration electrical motor shows clearly how changes in speed may be achieved by changing the number of poles. An ingenious arrangement of leads to a plate on the outside of the motor helps you demonstrate the theory behind this operation.

The motor described here is a three-phase, two-speed, squirrel-cage induction motor adapted from a war-surplus ½-hp motor. With the squirrel-cage induction motor any change in speed must be made by changing the number of poles or the frequency. The former is the practical method, of course, because changing the frequency would involve expensive frequency-changing equipment and some method of changing the conductors per path, per phase simultaneously.

Two speeds were obtained in early induction motors by employing two separate windings in the stator. The modern two-speed induction motor, however, does not employ two separate windings. The two speeds are obtained from one winding which is arranged to give two different numbers of poles. By external manipulation of the leads, the polarity is changed from *salient* to *consequent*.

The difference between salient and consequent poles is that the salient pole connection gives one magnetizing pole per phase for each group of coils while the consequent pole connection gives two. By arranging a winding so that the polarity can be changed from salient to consequent, or vice versa, a change in poles is affected and two speeds result. All standard induction motor connections are of the salient type and, of course, give one speed.

Plan of Procedure

The old winding is stripped from the stator. The core dimensions are recorded for use in designing the new winding.

The slots are insulated. New coils are placed in the core. Coils are connected in pole groups.

The start and finish ends of the pole groups are connected to #16 wire extra flexible leads. Retaining wedges are placed over slots. The winding is tested for electrical defects (short circuits, open circuits, grounds and reversed coils). Then the winding is sprayed with high-grade insulating varnish.

Twenty-four 3/16" holes are drilled equidistantly in the end bell opposite the pulley end of the motor. The start and finish ends of the pole groups are brought out through the holes referred to in the previous step when the motor is assembled.

A clear ½" plastic disc, 1' in diameter, with the diagram of the pole groups drawn on a piece of paper and cemented to its back, is mounted on the motor where the leads came out. The start and finish ends of the 12 pole groups are fastened to the jacks. Then a set of flexible jumpers with pin jacks fastened to both ends are made up for connecting purposes. The two speeds, 900 and 1800 rpm, are demonstrated by making the prescribed external connection to the line (Fig. 4-1).

Computation of Winding Data

The stator has 35 slots (Wss. .375; Dss. .625), a bore diameter of 4.625, length of 2.75, tooth width of .1875, and .75 of iron below the teeth. The squirrel-cage rotor winding was intact. The winding selected for the stator was lap for four-pole salient and eight-pole consequent, the salient 1800-rpm winding to be two

Fig. 4-1. Leads from the poles inside the meter make it possible to demonstrate changes in meter speed by changing the number of poles on the external plate with the wiring diagram showing.

parallel-star for four poles and the consequent 900-rpm winding series-star for eight poles.

The diagram used at the bench for interconnecting the pole groups of the three phases is constructed by following a specially arranged schematic sketch. This sketch has to be so arranged that the voltage and current per turn, per coil, per group, per path, and per phase remains the same when the motor is operating at either speed. Unless this prearrangement is considered, all the basic design factors such as tooth density, flux density in the air-gap, circular mils per ampere, etc., will change. This, of course, will affect the power output, torque, and efficiency. This specially arranged schematic sketch is shown in Fig. 4-2.

Figure 4-3 is a shop diagram constructed according to the schematic sketch of Fig. 4-2. Notice that the current, flowing from the line leads into the motor leads 1, 2 and 3, splits and forms two paths through the phases. This is salient current and is connected to the pole groups so that they form salient poles. Salient poles are alternate. That is, the current reverses instantaneously on adjacent pole groups. Each pole group establishes one magnetic pole. By tracing this salient current through the three phases, using Fig. 4-2 as a guide, the salient polarity of the pole-groups is readily observed.

Figure 4-4 is a reconstructed schematic sketch of Fig. 4-2. Compare Figs. 4-2 and 4-4. If in Fig. 4-2 we moved motor leads 4 up to the left, 5 up to the right, and 6 straight down, we would have

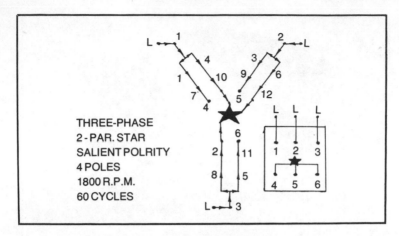

Fig. 4-2. Two-speed induction motor wiring schematic.

constructed Fig. 4-4. without breaking a connection. The winding then becomes series star, the polarity consequent; thereby doubling the number of poles with the speed dropping to 900 rpm. This is accomplished, in a practical way, by simply making external changes in the motor leads 1-2-3-4-5-6. For the four pole, two parallel star, salient connection, the line leads are brought into motor leads 1-2-3 and 4-5-6 are connected together. For the

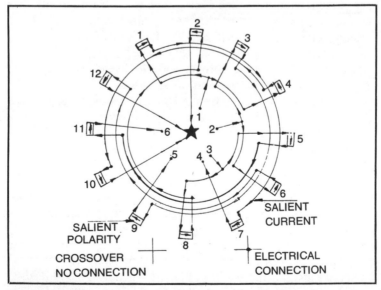

Fig. 4-3. Diagram constructed according to the schematic of Fig. 4-2.

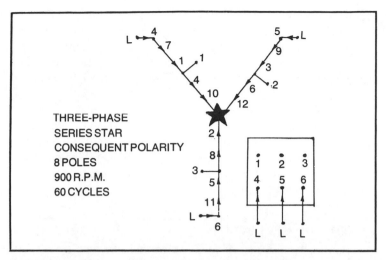

THREE-PHASE
SERIES STAR
CONSEQUENT POLARITY
8 POLES
900 R.P.M.
60 CYCLES

Fig. 4-4. Another two-speed induction motor wiring schematic.

eight-pole, series star, consequent connection, the line leads are changed to 4-5-6 and 1-2-3 are left unconnected.

Figure 4-5 is a shop diagram that is a duplicate of Fig 4-3. The only difference is that the consequent current is traced through the phases by following Fig. 4-4 and placing the consequent polarity on the pole-groups. Notice now that polarity does not alternate as it

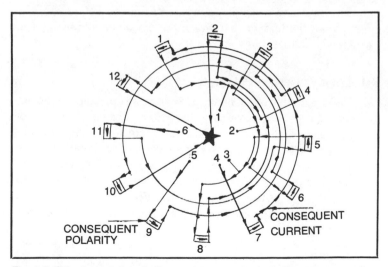

Fig. 4-5. This diagram is a duplicate of the one shown in Fig. 4-3, except the consequent current is traced through the phases by following Fig. 4-4 and placing the consequent polarity on the pole groups.

does for salient polarity, but is the same for all groups. So, although Figs. 4-2 and 4-4 are exactly the same diagrams, they can be changed from salient to consequent, from two-parallel star to series star, or vice versa, by simply making external changes in the motor leads coming from the winding. Then, as already explained, the poles double or halve giving the two speeds.

Technical information regarding the two-speed induction motor, using the principle of salient and consequent polarity, is rather meager. It is necessary, therefore, to compile a list of the requirements that must be followed rigidly for its development. We will then take each important requirements and illustrate the principle and its application.

Two-Speed Motor Requirements

Both salient and consequent poles must be employed. The coil-span must be chorded 30 percent to 50 percent. This is done by using the larger number of poles in computing the coil-span.

The number of pole-groups is computed by using the smaller number of poles. Six leads are brought out of the stator. This provides for a change in polarity. They are labelled 1-2-3-4-5-6.

Proper selection of connections must be made (series to two-parallel, delta to two-parallel star, etc.) This keeps the generated back emf constant for both speeds.

The first and second requirements on the list need to be illustrated if the learner is to grasp the fundamental principles of two speed windings.

Salient and Consequent Polarity

To explain this phenomena, we can resort to a direct-current analogy. Suppose we have three direct-current motor frames of exact mechanical dimensions which have four field poles. We will call them Figs. 4-6 through 4-8. Figure 4-6 is the conventional type of four-pole salient connection and produces four magnetic poles. Figure 4-7 is the same frame with only two field coils connected salient and produces two magnetic poles. Figure 4-8 is the same as Fig. 4-7 except that the connection is consequent and produces four magnetic poles.

Here we see the law of the magnet operating. Like poles repel. So to produce two and four poles from this frame, all that is necessary is to change the jumper between the two field coils. By changing this jumper, the polarity changes from salient to con-

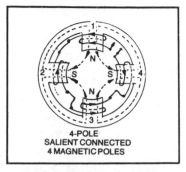

4-POLE
SALIENT CONNECTED
4 MAGNETIC POLES

Fig. 4-6. The conventional type of four-pole salient connection which produces four magnetic poles.

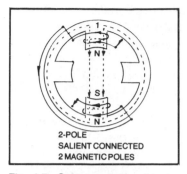

2-POLE
SALIENT CONNECTED
2 MAGNETIC POLES

Fig. 4-7. Only two field coils are connected salient here. This produces two magnetic poles.

sequent and doubles the number of magnetic poles. An inspection of the figures will bear this out.

Coil-Span Must be Chorded

Another requirement for two-speed induction motor connections is that the coils be chorded 30 percent to 50 percent. This means that, when calculating the coil-span, the number of poles used in the coil-span equation is the larger of the combination. If full-pitch coils are used, that is, the smaller number of poles of the combination, zero magnetization will result for the consequent connection. By chording the coils, iron space is provided for the establishment of the additional poles.

Figures 4-9 and 4-10 illustrate the use of full pitch for the four poles. Each coil represents a pole-group in a phase. Figure 4-9 is

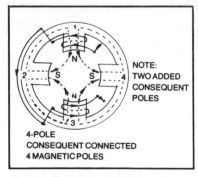

4-POLE
CONSEQUENT CONNECTED
4 MAGNETIC POLES

NOTE:
TWO ADDED
CONSEQUENT
POLES

Fig. 4-8. This frame differs from Fig. 4-7 in that the connection is consequent and four poles are produced.

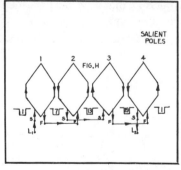

Fig. 4-9. The use of full pitch for the four poles. The connection is the conventional salient manner.

connected in the conventional salient manner, using the finish-to-finish and start-to-start relationship between pole groups. This connection results in an instantaneous reversal of the excitation current in adjacent successive groups. Notice the current in the two coil-sides in slots 1-7-13-19 are in the same direction so that their magnetic effects are additive and produce four salient poles.

Figure 4-10 is the same arrangement as Fig. 4-9 except that the connection is consequent. The relationship between pole groups is finish-to-start and finish-to-start. This connection causes the excitation current to flow instantaneously in the same direction in adjacent successive pole groups. Notice that the current in the two coil-sides in slots 1-7-13-19 oppose each other. This results in the complete neutralization of the magnetic effect of each pole group, producing no poles. If the coil-span had been computed on a basis of the larger number of poles (Slots ÷ P) + 1 = (24 ÷ 8) + 1 = 4 or 1 and 4, consequent poles result. This provides the necessary iron space for the establishment of the consequent poles. A visual inspection of Fig. 4-11 will support this theory.

CATHODE-RAY DEMONSTRATOR

Here is a simple project which makes an excellent teaching aid to show the operation of the cathode-ray tube. The components are all standard and are readily obtainable. The cathode-ray tube may be purchased from many surplus houses for about $5.

The cathode-ray tube is a type 902 which uses a relatively low potential on the deflecting plates, therefore making it a safe teaching aid. Another advantage is that the tube (902) has a standard octal base so an ordinary octal socket may be utilized. A shielded transformer is necessary to prevent sray radiations from affecting the tube performance. The transformer is a standard receiver type.

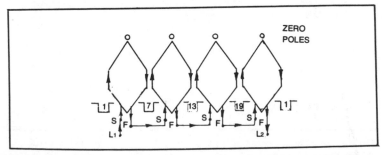

Fig. 4-10. Another use of full pitch for the four poles. The connection here is consequent.

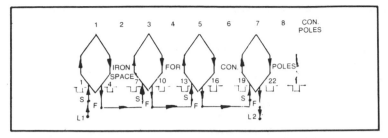

Fig. 4-11. Diagram showing how necessary iron space for consequent poles is provided.

The one used here had a primary of 110V and secondaries of 250-0-250-V, 6.3-V, and 5-V. We used half of the high voltage secondary and taped the extra hot leads. The 6.3-V winding is used for the filament supply of the CRT, and the 5-V winding is used for the filament supply of the type 80 rectifier tube. The type 80, although a dual diode, is utilized as a half-wave rectifier in this circuit (Fig. 4-12).

Construction

Various materials were considered for the mounting board and it was decided that pegboard offered the most advantages. It facilitated easy mounting of the components with a minimum amount of drilling and cutting. The CRT was placed as far as possible from the power transformer to insure against any possible interference. The transformer was mounted at the bottom because of its weight and because it also acts to stabilize the demonstration board.

Wiring is simplified by carefully planning the component layout. All components, except the variable resistors, are mounted on the surface end to eliminate long leads. Thus, all parts are clearly visible and their function can be easily explained (Fig. 4-13 and 4-14).

The voltage supply to the horizontal and vertical plates may be interrupted to substitute an external oscillator to show the action of sweep circuits. In normal use, a jumper is employed to connect the vertical and horizontal plates to their respective controls. These jumpers on the face of the board are connected by means of snap connectors which are easily removed. One of the horizontal and one of the vertical plates are internally connected to pin 1 of the CRT. This point serves as a common connection for the external oscillator.

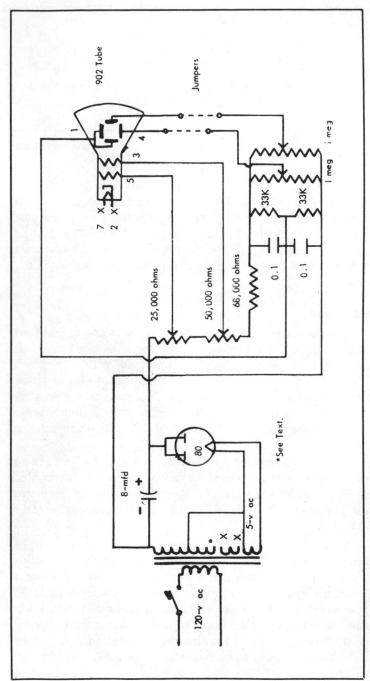

Fig. 4-12. Cathode-ray demonstrator schematic.

Tube socket holes were punched through the pegboard with standard chassis punches. The CRT tube socket should not be secured until the apparatus is completed, so that the tube may be revolved and oriented with respect to horizontal and vertical. Enough wire should be allowed to the tube socket for this orientation.

The sides of the panel may be made from scrap lumber and are screwed on to the pegboard. A switch should be incorporated to facilitate turning the apparatus on and off. This may be an integral part of any of the four potentiometers or may be placed in the line cord.

Applications

This demonstrator may be used without any further equipment to show electrostatic deflection of the electron beam. With the aid of two permanent bar magnets, magnetic deflection may be shown. With an external oscillator, the beam can be swept either horizontally or vertically and various wave forms can be demonstrated.

Fig. 4-13. Front view of the cathode-ray demonstrator.

Fig. 4-14. Rear view of the cathode-ray demonstrator.

TELEVISION SIGNAL DEMONSTRATOR

In teaching the fundamentals of television, it is important to show the relationship between the televison signal and the picture that it produces on the screen. You can do this effectively with the television signal demonstrator (Fig. 4-15).

The signal transmitted by a television station has a waveform which provides us with the picture information line by line, intermixed with an assortment of necessary blanking and synchronizing pulses. The picture information, or video, and these pulses vary the amplitude of this signal. Within the video portion, the lowest parts of the waveform produce the white areas of the picture; the highest, the black areas; and in between the various shades of gray.

In the actual transmission of television, an uninterrupted series of such lines and pulses are broadcast, and it is the television receiver that divides this continuous signal into individual lines, then electrically places them one beneath the other to form the picture. We do just that with our model.

For the desert scene pictured in the model (Fig. 4-16), jig-saw each horizontal line section out of ¼″ × 1″ moulding and stack them one upon the other between ¼″ dowel rod guides. No templates are used. The waveforms representing each line are drawn directly on the blanks. In appearance, the completed assembly resembles a picture carved in three dimensional bas-relief.

Although an actual television picture is made up of 525 lines, you can get your points across with this 24-line model. This demonstrator can be constructed in any desired size and divided into any number of horizontal lines to increase the amount of detail in the picture.

DEMONSTRATION MOTOR

In our complex civilization, with the highest standard of living in the world, our lives are influenced to a marked degree by a small

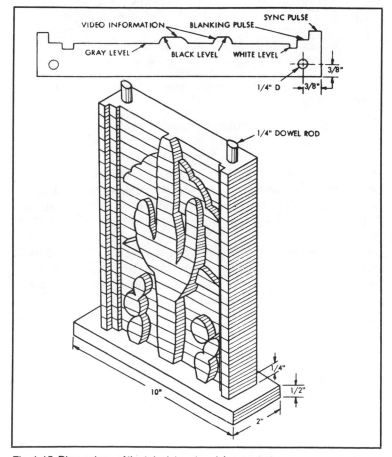

Fig. 4-15. Dimensions of the television signal demonstrator.

Fig. 4-16. A technique for drawing the desert scene shown on the television signal demonstrator.

mechanical contrivance which we commonly refer to as a *fractional horsepower motor*. Washing machines, clothes dryers, refrigerators, furnace blowers, fans, mixers, pumps, and numerous other household equipment appliances are powered by single-phase motors of less than one horsepower. Pause for one moment to name three common products found in everday life that have not been manufactured with motor power somewhere along the course of production.

The usual fractional horsepower motor for residential use is of the split-phase or capacitor type. Most washing machine or appliance repair stores will be able to supply a used motor—preferably, a capacitor motor with a thermal overload device.

This first step in the building of this demonstration motor is to carefully mark the end bells and stator with a center punch. This is done so that the motor parts may be assembled in their original position.

Disassemble by removing the four bolts throught the motor. Tap off the front end bell, and, at the same time, remove the rotor. The back end bell may now be removed and the connection block and starting switch disconnected from the end bell.

The stator contains two sets of windings. These are the running winding and the starting winding. In a 1725-rpm motor, the

70

four starting coils are connected as one circuit and the four running coils are connected as another.

Each coil is disconnected from the next and leads of 18-gage stranded wire (plastic insulation), approximately 24″ long, are connected to both ends of each coil. Short lengths of No. 12 "spaghetti" make excellent insulation over the splices instead of tape. Each lead is tagged with adhesive tape and marked with a ballpoint pen. Tags are marked as follows: starting coil 1, IN; starting coil 1, OUT; starting coil 2, IN; starting coil 2, OUT; etc. Two leads are taken off the starting switch, capacitor, and thermal overload and tagged as such.

The 22 leads are brought out the end bell and the motor assembled in its original position according to the punch marks. Care must be taken to arrange all leads inside the motor in such a position that they will not be damaged by the rotor.

A circle of tempered Masonite, 12″ to 16″ in diameter, with 22 equally spaced segments, is used as a connection panel. One-quarter inch holes are drilled ¾″ from the edge of the panel on each segment line. Roundhead stove bolts, ¼″×1″, are used as binding posts or terminals as illustrated in Fig. 4-17. The panel should be mounted on brackets above the motor. Each of the leads are now connected to the under side of the panel in position as shown in Fig. 4-18. See also Fig. 4-19.

A few suggested ways for using the demonstration motor are:

Demonstrate method of connecting leads to form alternate poles—North, South, North, South; show how rotor may be started by hand when only the running coils are connected; demonstrate bucking action of coils when one coil is improperly connected; connect starting coils and switch in one circuit,

Fig. 4-17. The connection panel and arrangement of terminals of the demonstration motor.

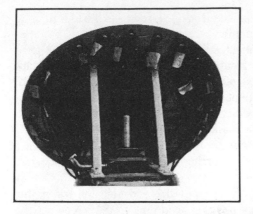

Fig. 4-18. The back of the connection panel of the demonstration motor shows the connection of the marked leads.

illustrating action of starting switch (hand starting of rotor is necessary); connect running and starting coils together in regular split-phase motor; reverse rotation of motor by reversing leads; addition of capacitor in starting-winding circuit to produce a capacitor-type motor with higher starting torque; and addition of thermoguard unit to motor circuit for motor protection.

Excellent results have been obtained in using the motor unit as a practice medium for coil connection and motor wiring before starting a motor rewinding job.

Fig. 4-19. A side view of the demonstration motor.

HOUSE WIRING PANEL

When someone is ready to start work on the panel (Fig. 4-20 and Table 4-1), he should be shown how it is possible to select any one of 32 different circuit combinations, depending on the arrangement of single-pole switches, three-way switches, duplex outlets, and a four-way switch in the three 2″ × 2″ × 4″ handy boxes. Keep in mind that each combination can be fed at each of the five boses.

He should next be required to draw his wiring diagram with colored chalk on the space provided opposite the boxes. He can use either schematic or pictorial symbols, whichever you prefer. When you have approved the drawing, the student should pull the necessary wires into the boxes, splice where necessary, and then ask you to inspect his work before soldering or using solderless connectors. Following this, he should tape all joints and mount the various parts. The last step is to close up the boxes.

When the student has finished a combination, he should be required to completely disassemble his work with the exception of the boxes and the conduit, which should be left on the board. The exception to this would be in cases where you require the student to use flexible, metal-sheathed cable or nonmetallic, sheated cable between the boxes instead of thin-wall conduit.

TELEVISION DEMONSTRATOR

Because of the economics of time and money, this TV demonstrator is intercarrier series-string circuit. The circuit is illustrated in block diagram form (Fig. 2-21).

Because of the realization that one of the major obstacles in the teaching and understanding of electronics is the lack of mechanical movement, the circuit was laid out in a manner which would be expedient to both you and club student. The TV demonstrator is constructed on a plywood panel approximately 4′ × 6′.

For purposes of clarification, the panel is divided into sections. Each section represents a section of the television receiver, and each section is divided from an adjoining section by aluminum stripping. These sections are: the front end; the low-voltage power supply and heater connections; video I-F, video detector, and video amp. and sync separator; sound I-F, ratio detector, and A-F amplifier; vertical deflection; horizontal deflection and high voltage; and pix tube (CRT) and speaker.

This division of the circuit makes possible a closer study of the theory of operation as well as of the function of each of the

Table 4-1. Parts List for the House Wiring Panel.

2 pcs.	¾" × 5" × 43", clear white pine
2 pcs.	¾" × 5" × 18", clear white pine
1 pc.	⅜" × 17¼" × 42×" plywood, GIS
4	⅜" × 1¾" × 1¾", inside corner irons
16	¾" No. 6 fh wood screws
2	2" right-angle screw hooks
1 pc.	3/32" × 6" × 36" green chalk board, or equivalent.
1	4 oz. bottle of contact cement
1	Chalkboard eraser
	Yellow, white, red, and black chalk
2	4" octagon boxes
2	4" octagon-box, round coverplates with lamp sockets
2	120-v, low-wattage lamps
4	2" × 2" × 4" handy boxes
12	3/16" × ¾" rh stove bolts, nuts, and flat washers
3	Handy-box switch cover plates (metal)
3	Handy-box duplex outlet cover plates (metal)
1	Handy-box cover plate with lamp socket
11	½" thin-wall conduit connectors
2	½" 90° thin-wall conduit connectors
9 pcs.	½" × 1½" thin-wall conduit
1	4-way switch
2	3-way switches
2	Single-pole switches
2	Duplex outlets
1	5-amp Minnibreaker
15'	16-2 rubber-covered cord
1	Clamp-type heavy-duty male plug
4'	2-wire parallel, rubber-covered wire
	Scrap pieces of No. 14 red, black, and white plastic-covered wire *

components for each section, and of the relationship of the section to the complete receiver. The rectifier, a semiconductor, each resistor, each capacitor, with the exception of electrolytics, and each RF choke plugs into the circuit. Because of the great losses which would result in the spreading out of the circuit, the front end was left intact as a unit. Type FT-241-A crystal holders were purchased and disassembled and the components were soldered to the "inside" pin connections as illustrated (Fig. 4-22).

This type of demonstrator has almost unlimited possibilities for the dissemination of information regarding the various circuits

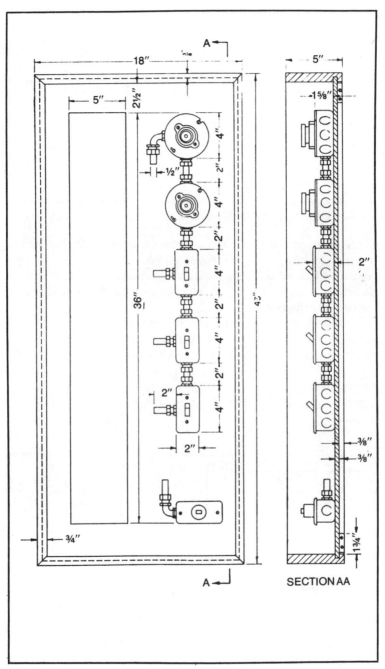

Fig. 4-20. Dimensions of the house wiring panel.

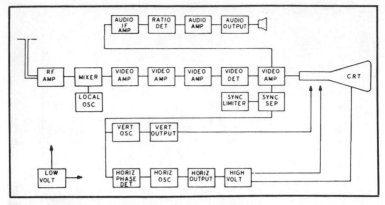

Fig. 4-21. Block diagram of the television demonstrator.

found in a television receiver. For example, a circuit may be caused to malfunction by using improper values or faulty resistors and capacitors (these have been previously prepared by being mounted on extra crystal holder bases), plugged into the circuit in place of the correct component.

Other features showing the versatility of this type of demonstrator are the many strategically located banana jacks in each of the various circuits. This makes possible both voltage and oscilloscope checks throughout the entire circuit. In an I-F stage, for example, it is possible to take measurements of signal input, signal output, plate voltage, screen voltage, bias voltage, AGC voltage, etc.

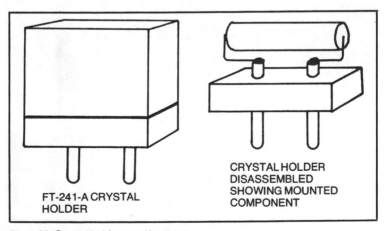

Fig. 4-22. Crystal holder specifications.

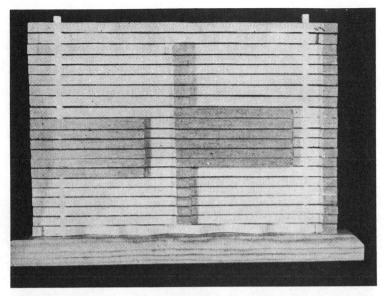

Fig. 4-23. The composite synd signal demonstrator assembled to show the hammerheads of two interfaced fields as they are seen on a TV screen.

COMPOSITE SYNC SIGNAL DEMONSTRATOR

The synchronizing section of a television receiver is one of the most difficult to explain without some form of demonstrator. The wooden model shown (Fig. 4-25) effectively demonstrates how the

Fig. 4-24. Demonstrator assembled to show hammerheads of alternate fields separately. Line-strips set end-to-end show an oscilloscope waveform.

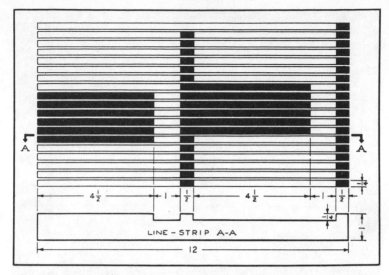

Fig. 4-25. Dimensions of the composite sync signal demonstrator.

synchronizing pulses are transmitted, received, separated and put to use.

The composite sync signal demonstrator is constructed of 12″ lengths of ¼″ x 1″ lattice wood, each strip of which represents a single line on the screen of a television receiver. By setting these individual line-strips on edge and placing them end-to-end, you can show how the horizontal blanking and synchronizing signals appear at the end of every line of video information; how the video stops at the end of the last line, or in the middle of the last line on alternate fields; and how the equalizing pulses appear in either case to provide identical conditions in the vertical integrator network before the arrival of the serrated vertical pulses.

Viewed in this manner the line strips show the transmitted waveform as it would appear if seen on an oscilloscope. The vertical amplitude is to scale, but the horizontal dimensions are purposely exaggerated to make such important features as the front porch and leading edge of the horizontal sync pulse more readily noticeable.

Stacked one above the other, the line-strips easily de-monstrate how one horizontal sync pulse, equalizing pulse, or serrated vertical pulse will occur at exactly the same point near the end of each line in time to maintain horizontal synchronization with the transmitting station during the vertical retrace of the picture tube beam and the blanking period that occurs before each new field of video. The model also shows how interlacing is achieved by delaying the start of the serrated vertical pulses a half-line in alternate fields.

The line-strips in the model may also be arranged to show the appearance of the various pulses in combinations such as might appear at half the regular vertical sweep rate or, as they are more commonly seen, interlaced to appear as the familiar "hammerhead" pattern, an arrangement which may be viewed on the screen of any television set by simply adjusting the vertical hold control until the picture begins to roll down slowly. Between successive pictures a black horizontal bar about an inch high will be seen, and by turning up the brightness and turning down the contrast; the observed pattern, representing the various timing pulses, becomes plainly evident.

On a television screen, unforunately, the jitter or line movement, makes it extremely difficult to examine the detail in the individual lines. The model literally spares the headaches that inevitably come from observing a TV screen too closely.

The individual line-strips are jigsawed, drilled and assembled on a base into which ¼″ aluminum guide rods has been set. The model is sanded and painted with model airplane enamel so that all parts of each waveform or pulse up to the blanking level are painted gray, and those areas above that level are painted black.

BATTERY CHEMISTRY DEMONSTRATOR

Using a magnet board, however, as illustrated here, allows you much freedom in the teaching battery chemistry to a club. The battery board is made of metal and the various components of the discharge-charge cycle are mounted on magnets so that you can move through the process logically, one step at a time. Thus, the chemistry of battery cycling can be understood easily by radio club students.

The presentation should be prefaced by a discussion of battery nomenclature, materials and constructions. Charts, filmstrips, models or other aids are very effective for this portion of the lesson. After the club students understand the basic fundamentals of the lead acid battery, you can prepare them for a lesson in battery chemistry by explaining the chemical symbols and the formula involved in the cycle. By writing the words and symbols together with the formula on the blackboard, your students will be able to keep each term in proper perspective (Table 4-2).

After the basic steps have been understood, you can use the magnet board to show visually how the chemical interaction functions. To keep the presentation as simple as possible, the minimum number of molecules are used. They are as follows: 1 each, lead—Pb; 1 each, lead dioxide—PbO_2; 2 each, water—H_2O; 2 each, sulphuric acid—H_2SO_4. A charged cell is illustrated (Fig. 4-26) with this minimum number of molecules.

The Discharge Cycle

The first step is to move the sulphate ion in proximity with the lead plate. The positive lead ion attracts the negative sulphate ion to cause a bonding, freeing two electrons as a result of the interaction. The two electrons thus freed can be moved to the

Table 4-2. Battery Reactions.

Lead=Pb   Hydrogen=H_2
Sulphuric Acid=H_2SO_4 Lead Dioxide=PbO_2
Oxygen=0   Lead Oxide=PbO
Water=H_2O Lead Sulphate=$PbSO_4$
Supphate=SO_4
Fully Charged Battery: Fully Discharged Battery:
Negative Plate=Pb   Negative Plate=$PbSO_4$
Positive Plate=PbO_2 Positive Plate=$PbSO_4$
Electrolyte=H_2SO_4 plus H_2O Electrolyte=H_2O

CHEMICAL FORMULAS
Discharge

1. $PbO_2+2H++2e$ ——————————— $PbO+H_2O$
2. $PbO+H_2SO_4$ ——————————— $PbSO_4+H_2O$

Negative Plate:
1. $PB+SO_4=$ ——————————— $PbSO_4+2e^-$

Electrolyte:
1. $2H_2SO_4+2H_2O+2O=$ ———————— $4H_2O+2SO_4=$

Charge

Positive Plate:

1. $PbSO_4+SO_4=$ ——————————— $Pb(SO_4)_2+2e^-$
2. $Pb(SO_4)_2+2H_2O$ ——————————— $PbO_2+2H_2SO_4$

Negative Plate:

1. $PbSO_4+2H++2e^-$ ——————————— $Pb+H_2SO_4$

Electrolyte:

1. $4H_2O+2SO_4=$ ——————————— $2H_2O+2H_2SO_4+2O=$

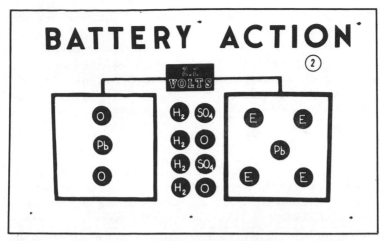

Fig. 4-26. A charged cell with a minimum number of molecules.

terminal of the negative plate. This process, multiplied billions of times, creates a surplus of electrons at the negative terminal causing an electrical potential (Fig. 4-27).

By placing a load in the circuit, the two electrons are able to move through the circuit (turning the starter motor as an example of a load) and deposit themselves on the positive plate. As two hydrogen ions are available from the splitting of the sulphuric acid molecule, a bond takes place between the oxygen ion and the hydrogen ions to form water. The two electrons from the negative

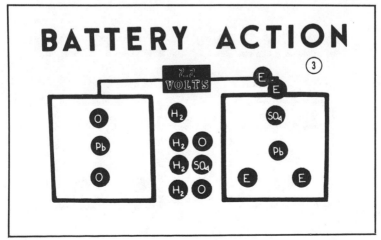

Fig. 4-27. A charged cell with the electrons gathered at the negative terminal.

plate replace the oxygen ion to form a molecule of lead oxide on the positive plate (Fig. 4-28).

As lead oxide reacts readily with sulphuric acid, the sulphate ion easily displaces the oxygen and forms lead sulphate on the positive plate. The oxygen ion bonds to the hydrogen ions to form more water. Figure 4-29 illustrates this. The battery is completely discharged now with lead sulphate on both plates and water in the solution.

The Charge Cycle

By placing a generator in the circuit, the process can be reversed. The two electrons on the positive plate can be removed from the lead sulphate and returned to the negative plate as the lead sulphate ionizes on both plates as a result of the generator current flow. The sulphate ion on the negative plate combines with the hydrogen ion in solution to form sulphuric acid. The lead ion plus two electrons forms lead on the negative plate. The hydrogen ion in solution combines with the sulphate ion from the positive to form sulphuric acid in the electrolyte. The lead ion on the positive plate combines with the oxygen ions to form lead dioxide on the positive plate. As the charging current causes electrolysis to take place, the extra molecule of water can be shown to be released as hydrogen and oxygen gas—explaining the reason for the need to add water to the battery. The recharging step is shown in Fig. 4-30. The completely recharged battery is illustrated, too (Fig 4-26). A parts list is shown in Table 4-3.

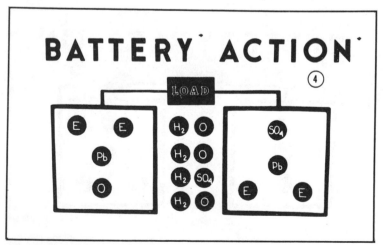

Fig. 4-28. A partially discharged cell.

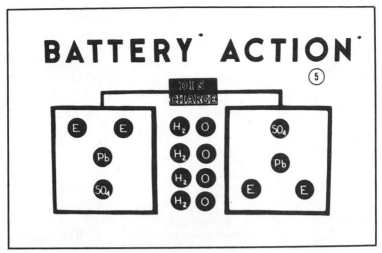

Fig. 4-29. A completely discharged cell.

RC CONSTANT PANEL

The demonstrator described here has proven a valuable help in illustrating the effect of both resistance and capacity on the charge and discharge rate of a capacitor. India ink is used to draw the schematic diagram on a plywood panel ¼″ x 24″ x 32″. Latex wall paint provides an excellent surface for inking in the diagram. A pastel shade of paint is used to add interest. Afterwards the panel is

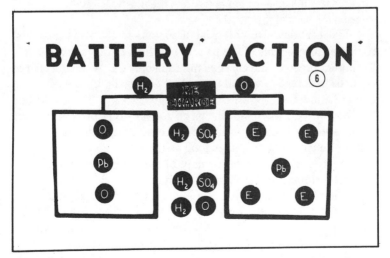

Fig. 4-30. A recharged cell with hydrogen and oxygen escaping due to electrolysis.

Table 4-3. Parts List for Battery Chemistry Demonstrator.

No. of Items	Item	Size
1	Metal-covered ¼" plywood	14" × 18"
2	Dowel with symbol Pb	1"
4	Dowel with symbol E	1"
4	Dowel with symbol O	1"
4	Dowel with symbol H_2	1"
1	word "Charged" Plywood with	2" × 4" × ¼"
1	word "Discharged" Plywood worded	2" × 4" × ¼"
1	"2.2 Volt Potential" Plywood worded	2" × 4" × ¼"
1	"Load"	2" × 4" × ¼"

sprayed with a flat lacquer to provide a more durable finish and to resist soiling.

Banana jacks, on 1" centers, are installed adjacent to the schematic symbol to which they applied. The required components are mounted on bases of ¼" plexiglass. Electrical contact is effected through banana plugs fixed to the plastic base. All wiring is done on the back of the panel.

In operation, a DC potential of 50 volts to 150 volts is applied to the input terminals of the panel. The current flows through R1, the potentiometer, and charges the capacitor C1. When the charge exceeds the breakdown voltage of Ne1 (a neon bulb), the lamp will fire and allow the capacitor to discharge across it (Figs. 4-31 and 4-32). However, when the discharge voltage drops below the firing point of the neon bulb the lamp will cease to conduct and go out. This cycle will be repeated as long as the circuit is energized, and the effect is that of a continuously blinking light—the blink rate being determined by the RC time constant.

The time required to fully charge the capacitor at a given voltage will depend on the setting of R1 and the value of C1. By altering the value of either C1 the RC time constant will be changed, and the blink rate of the lamp will change proportionally.

A variety of capacitors, from .01 μF to 4 μF will provide adequate range for this panel. The potentiometer easily permits

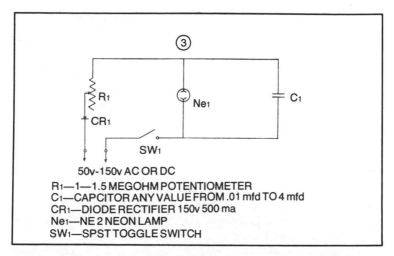

R1—1—1.5 MEGOHM POTENTIOMETER
C1—CAPCITOR ANY VALUE FROM .01 mfd TO 4 mfd
CR1—DIODE RECTIFIER 150v 500 ma
Ne1—NE 2 NEON LAMP
SW1—SPST TOGGLE SWITCH

Fig. 4-31. Schematic of the RC constant panel.

any desired resistance setting. If desired, several fixed resistors can be used instead of the variable resistance.

As indicated by the schematic diagram, the DC potential may be obtained directly from the AC power lines by simply inserting a diode rectifier of 500-mA rating or higher in one side of the circuit. A schematic and parts list for this demonstration panel are given in Fig. 4-31. Figure 4-32 shows an RC time constant graph. It is seen that the time constant for any resistor-capacitor combination is that time needed to charge a capacitor to 63.2 percent of full capacity or to discharge it to a point 36.8 percent of complete discharge. As

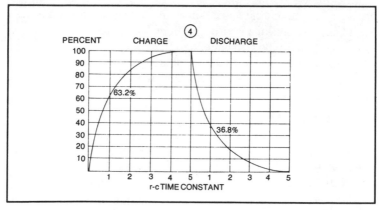

Fig. 4-32. Graph showing the growth and decay of a voltage across a capacitor.

indicated by the graph, the charge and discharge rates are not linear. It takes five time-constants to bring a capacitor to either full charge or to complete discharge.

DEMONSTRATION PANEL

This demonstrator is permanently mounted on the wall and is used by club students to solve problems involving electrical devices connected in series, parallel, or in combination. Working from the known to the unknown, the club student learns to follow a pattern that enables him to understand and analyze more complicated circuits.

Short pertinent statements, in chart form relating to the fundamentals of electrical circuits so necessary in understanding and solving electrical problems, are included on the panel. These charts may be interchanged.

In introducing circuits to club students, use incandescent lamps of the same rating, such as 60 W, designed to operate on 120V. This simplifies the arithmetic involved in the solution of early problems.

For the first demonstration, four lamps are connected in parallel, four in series, and four in combination (that is, two in parallel, two in series). When all 12 lamps are in operation, the club student easily observes the visible lighting effect of the three circuits. The four lamps in parallel are full brightness. The four in series are dim. Two of the lamps in the combination circuit are full brightness, and the other two are dim but brighter than the four lamps connected in series. See Table 4-4.

Each club student is asked to account for the various visible lighting effects of the three circuits. He begins to realize that to analyze any circuit he must know: how to read and interpret electrical diagrams or a written problem; how to use the fundamentals of arithmetic; how to understand and use Ohm's Law; how to understand and use Watt's Law; rules governing the circuit; and how to apply Ohm's Law or Watt's Law to each part of the circuit or the entire circuit.

VACUUM-TUBE DEMONSTRATOR

Introduce the subject of vacuum tubes to your radio club with this demonstrator. Easy-to-get parts and simplicity of construction make this almost a must for all radio club classes.

The detailed drawings and the description of the parts enable you to construct this model with a minimum of time and effort—about six hours for the total project. The use of the model in class is

Table 4-4. Various Combinations of 60W Lamps.

60-W Lamps in Parallel

	P Power Watts			E Pressure Volts		
	Cur. Amps. I	Press. Volts E		Cur. Amps. I	Res. Ohms R	

One Lamp

	1	2	3	4	T	Rules
R	(240)	....	....	....	(240)	Less
I	(½)	....	....	....	(½)	Sum
E	120	....	....	....	(120)	Same
P	60	....	....	....	(60)	Sum

Two Lamps

	1	2	3	4	T	Rules
R	(240)	(240)	....	....	(120)	Less
I	(½)	(½)	....	....	(1)	Sum
E	120	(120)	....	....	(120)	Same
P	60	60	....	....	(120)	Sum

Three Lamps

	1	2	3	4	T	Rules
R	(240)	(240)	(240)	....	(80)	Less
I	(½)	(½)	(½)	....	(1½)	Sum
E	120	(120)	(120)	....	(120)	Same
P	60	60	60	....	(180)	Sum

Four Lamps

	1	2	3	4	T	Rules
R	(240)	(240)	(240)	(240)	(60)	Less
I	(½)	(½)	(½)	(½)	(2)	Sum
E	120	(120)	(120)	(120)	(120)	Same
P	60	60	60	60	(240)	Sum

60-W Lamps in Series

One Lamp

	1	2	3	4	T	Rules
R	(240)	....	....	....	(240)	Sum
I	(½)	....	....	....	(½)	Same
E	120	....	....	....	(120)	Sum
P	60	....	....	....	(60)	Sum

Two Lamps

	1	2	3	4	T	Rules
R	(178)	(178)	....	....	(356)	Sum
I	(.337)	(.337)	....	....	.337	Same
E	60	60	....	....	120	Sum
P	(20.2)	(20.2)	....	....	(40.4)	Sum

Three Lamps

	1	2	3	4	T	Rules
R	(140)	(140)	(140)	....	(420)	Sum
I	(.285)	(.285)	(.285)	....	.285	Same
E	40	40	40	....	120	Sum
P	(11.4)	(11.4)	(11.4)	....	(34.2)	Sum

Four Lamps

	1	2	3	4	T	Rules
R	(128)	(128)	(128)	(128)	(128)	Sum
I	(.235)	(.235)	(.235)	(.235)	.235	Same
E	30	30	30	30	120	Sum
P	(7)	(7)	(7)	(7)	(28)	Sum

60-W Lamps in Combination

	Parallel		Series					
	1	2	3	4	1 + 2	3 + 4	T	Rules
R	(240)	(240)	(178)	(178)	(120)	(356)	(90)	Less
I	(½)	(½)	(.337)	(.337)	(1)	.337	(1.34)	Sum
E	120	(120)	60	60	(120)	120	(120)	Same
P	60	60	(20.2)	(20.2)	(120)	(40.4)	(160)	Sum

subject to many variations so that only the construction will be covered. The parts will be described in the same order that they should be made to eliminate some of the difficulties that were encountered in making the original model.

As storage is always a problem, care was taken to make a neat package to occupy a minimum of space when not in use.

The base should be constructed first, because it can then be used as an assembly jig for the elements. A piece of hardwood is best, although any piece of the proper dimensions will do. Drill all the holes ½" deep, of the proper size and spacing, as indicated in the drawings. A gray paint should be used on the base so that the base does not distract from the elements.

To make the heater element, No. 12 wire was twisted together and the bottom formed to fit the base holes. Other types of heaters can be constructed, but this is the kind used in the illustrations.

A narrow "V" was formed, also of No. 12 wire, and soldered to the top of the single rod support. Flat yellow paint was applied the full length of the No. 12 wire to represent the oxide coating on the directly heated cathode. The same base holes were used for the heater element and the directly heated cathode.

The base supports for the indirectly heated cathode were made first, using ¼" brazing rod. The supports were filed flat where they were soldered to the ½" steel tube for better mechanical support. The center two-thirds of the tube was painted yellow to indicate the oxide coating.

Eight lengths of ¼" brazing rod are needed for the grid supports. The ends should be rounded and notches filed along one side with a triangular file to accommodate the grid wires. The spacing for the notches is clearly indicated in the working drawings. See Fig. 4-33 through 4-39. The old trick of pulling wire over a screw driver does well in eliminating the kinks in the No. 14 wire that is used for the grids.

To construct the control grid, place two supports in the proper base holes, then clamp the free ends in a vise, making certain the notches are facing outward. Solder the end of the wire in the lowest

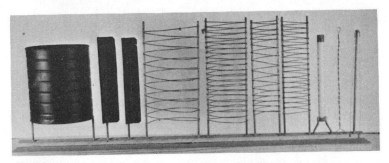

Fig. 4-33. The elements of the vacuum-tube demonstrator strung out in a row. This demonstrator can be used in many ways.

Fig. 4-34. Demonstrating the beam-forming plates in place.

notch. Form the first turn the same shape as in the working drawings, and solder only one-half turn at a time. After several turns have been formed and soldered, the base may be removed so that the remaining turns can be patterned after the first. The super-control grid is then constructed in the same two base holes, as only one of these grids will be used at one time.

Shape the beam-forming plates according to the working drawings, using any scrap tin. A tin can may be used if nothing else is available. Be sure to file the supports flat on one side for easy soldering. A flat, black paint adds realism to these plates.

Fig. 4-35. Demonstrating a beam- power pentode.

Fig. 4-36. How the demonstrator is packaged neatly for storage.

A ribbed gallon can, with the ends removed, makes a good plate. These supports are also filed flat and rounded before soldering. Use a flat, black paint on the plate, making sure not to paint the bottom of the supports that fit into the base.

This completes a model that can be used over and over again, after it has once been used to introduce the subject of vacuum tubes. The model illustrated above has been used by a number of radio club instructors.

COMPUTER DEMONSTRATOR

This demonstration unit is a small section of a large computer. It not only demonstrates binary arithmetic in addition and subtraction problems, but also contains "and" and "or" circuits in sequenced operation and is set up to show the basic principles of computer logic. Thus, it can be a useful tool in the training of the use of binary numbers, programing and the electronics of computers.

For purposes of demonstration, the multivibrators have been replaced with double-pole-double-throw switches and the readouts are simple transistor devices that light low-current dial lamps. To illustrate the use of binary arithmetic, A and B are set up as the numbers to be added (or subtracted) and C is the number carried over from addition of the previous column or the borrow in the subtraction problems.

To add a long column of numbers, the unit uses the same method as an actual computer. That is, two numbers are added—A

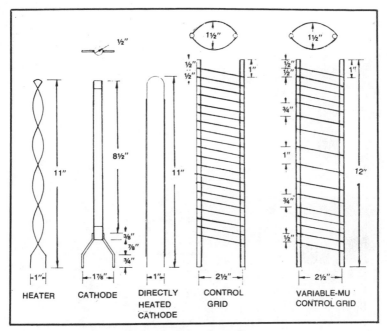

Fig. 4-37. Dimensions of the different tube elements.

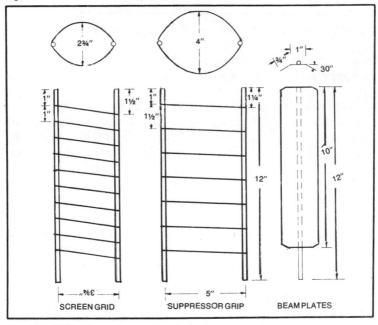

Fig. 4-38. Dimensions of other tube elements.

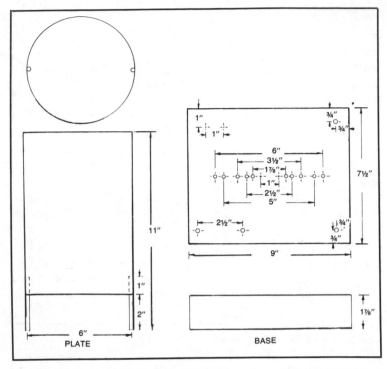

Fig. 4-39. Dimensions of the base and plate of the vacuum-tube demonstrator.

and B—then the answer is moved to the B position, the next number from the column is placed in the A position, and the process is repeated.

Pin jacks are provided to the right of the switches so that computer or radio club students can read directly the output of the switches. These represent the voltage that would be generated by the computer multivibrators on a regular computer. The lowest terminal, marked "com," is the common terminal for all voltage readings. Thus, the club student has a starting point to trace all the voltage through the system.

The "and" and "or" circuits are of the elementary diode type and are mounted so that their terminals are accessible to measure voltages. Their number and placement in the matrix allow the club student to have the feeling of working on a system and gives him the interrelation that goes with such a complex. The unit also can be related to the truth tables and Boolean algebra because the mounting board carries the block diagram and progressive formulas.

Construction

Many actual computers allocate negative 15 volts to the number zero and zero volts to the number one. In constructing this unit, this practice was followed, except that the number zero was assigned to negative 9 volts because of the availability of surplus 6-V transformers at low cost.

To begin construction, the block diagram shown in Fig. 4-40 can be drawn or silk screened onto cardboard or thin hardboard 29" × 22". If the board were much smaller than this, the individual blocks would be too small to accommodate the two diodes and one resistor required for each.

In the unit illustrated, the older square symbol for "and" and "or" circuits was used. However, the newer semicircular symbols may be used if the instructor has the necessary drawing ability.

Once the symbols were drawn on the board, holes were cut for the switches and dial lamps. Each dial lamp shines through a slot cut in the center of a large zero. The type of mounting used for the lamps in this unit was heavy copper wire soldered to the base of each.

If a lower cost unit is desired, the dial lamp readouts and the transistors feeding them may be eliminated. In that case, the voltmeter indicator is used for readout. When using it this way, the last "or" circuit must be terminated with a 5000 Ω resistor.

A small piece of plexiglass slightly larger than one of the blocks on the finished board was drilled to form a jig for making up diode units. If four holes are drilled to correspond to the feed-through points of the wires to the back of the board, these jigs can be used to hold diodes and resistors while they are being bent and soldered. They can also be used to determine mounting holes on the board. After all "and" and "or" circuits were manufactured and mounted with the wire bent to hold them, the interconnecting wires were run.

To hold the power supply, a small chassis was used (see Fig. 4-41). The transformer for this unit was from a remote-control TV unit, but because the current drain is very small, any transformer that has a 117V and a 6V winding will do.

The transistor switching unit is simple and requires so few parts that a surplus printed circuit board was used. Bending the transistor leads slightly and aligning them with the necessary jumpers in the board made this an easily-constructed unit. To complete the wiring, the power supply dial lamps, and other parts were interconnected with the switches and matrix.

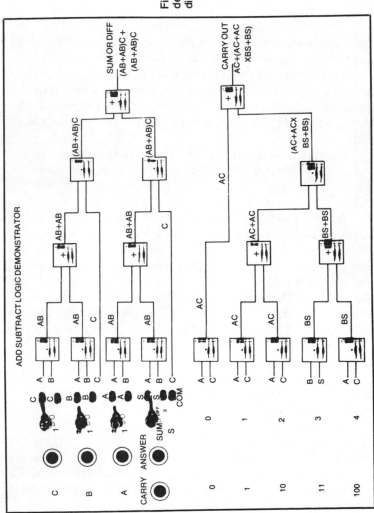

Fig. 4-40. Front of the computer demonstrator showing the block diagram and progressive formulas.

94

Because the output voltage of the matrix is low when the number is one and high when it is zero, this condition had to be reversed and the current amplified in order to work the indicator lights in the desired way. To do this, the first transistor was used to invert the voltage condition—that is, when the base voltage is high, the collector voltage is low. Directly coupled to this first transistor is a second transistor that passes current through the indicator lamp. As an example, low base voltage on the first transistor causes minimum current to its collector and a resulting higher collector voltage. This, in turn, puts a high bias on the second base and results in high current through the second transistor collector to the indicator lamp.

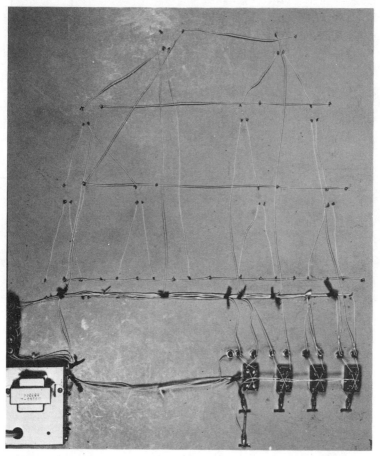

Fig. 4-41. Back of the computer demonstrator showing the power supply and wiring.

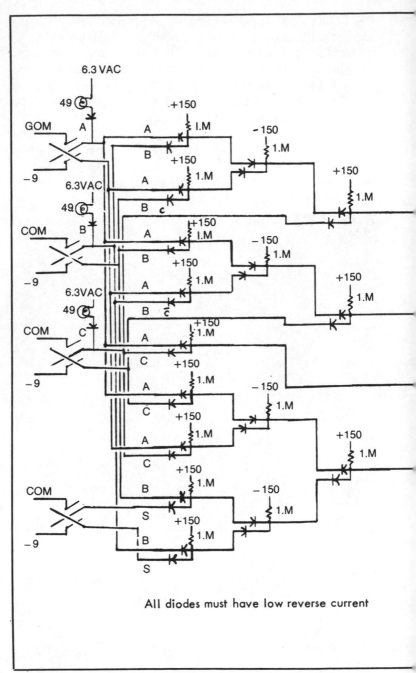

All diodes must have low reverse current

Fig. 4-42. Add-subtract logic schematic.

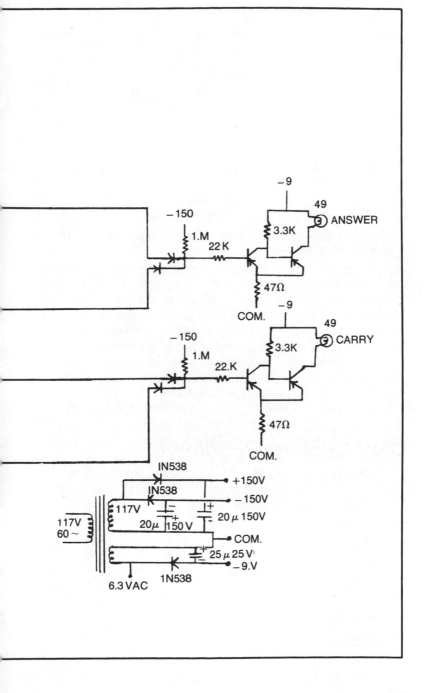

2	20 mfd, 150-v electrolytic capacitors
1	25 mfd, 25-v electrolytic capacitor
5	Dial lamps, number 49
3	Diodes, 1M538
39	Diodes, XYTAN,* Part Number 1490
	Resistors (all resistors ¼ watt)
2	47 ohm
2	3,300 ohm
2	22,000 ohm
18	1.0 meg
4	Switches, D.P.D.T.
1	Transformer, 115-v primary, 115-v and
	6 × 3-v secondary
4	Transistors, PNP, 2N586
9	Pin jacks

*XYTAN, 1755 Placentia, Costa Mesa, California. Minimum order, 40 pieces, 50 cents each.

Table 4-5 contains a list of parts. Both in this and in Figs. 4-40 through 4-42, numbers have been assigned to the diodes and transistors as a guide for building the unit. Note that the "and" and "or" diodes have only to withstand a peak inverse voltage of 9 volts and carry a current of about one milliamp. However, they must have a very low reverse current leakage. Diodes to meet these requirements can be obtained inexpensively from old printed circuit boards at most large radio parts houses.

In removing diodes from these old boards, long-nosed pliers are good to use between the diode and the solder joint. Anyone working with them must remember that a low wattage (25 watts) and fast soldering are necessary.

The diodes in the plus and minus 150V power supply must have a peak inverse voltage of about 300V and they supply very little current. A good selenium from an old tube-type portable will be quite satisfactory here. Those for the 9V negative supply must have a peak inverse voltage of 20 and carry 100 milliamps. A selenium rectifier from a tube-type portable or most any of the silicon diodes will be satisfactory here. To meet the transistor needs, almost any PNP transistor that will pass 60 milliamp when saturated at 9 volts will serve well, although not all of those that were tried were found to be satisfactory.

TROUBLESHOOTING MOCKUP

A first step in troubleshooting procedure is to check fuses. This may be done by removing the fuses from the fuse block or by checking them while they are still an integral part of the circuits of a machine or piece of equipment. Attempts to teach someone how to check the fuses while they are still an integral part of a machine's circuits are not always completely satisfactory. Danger always exists when club students are crowded around machinery and where hot wires or terminals are exposed.

For these reasons the mockup of a typical fuse block, shown in Fig. 4-43, was developed. It enables you to teach this operation under ideal conditions. The mockup is used to demonstrate how the fuses can be checked while in the machine's circuits. The demonstration insures a relatively high degree of success and safety in teaching this aspect of troubleshooting.

The troubleshooting mockup is simple and inexpensive to construct (Fig. 4-44). It consists of a panel built of half-inch plywood approximately 20″ squares. The panel is mounted into two legs approximately 3″ high and 12″ long of ¾″ lumber. The fuse-holding clips are constructed from #24 gage sheet brass. The fuses are constructed from paper towel tubes and are about 1½″ in diameter and 11″ long. The fuse ends and internal connections are made from copper foil.

Three fuses are necessary for the demonstration. Two of the three fuses have the fuse ends connected while the third one has this connection broken. Any size copper wire may be used to wire the panel and to show the power lines. A heavy wire, such as #2, was used since it is more in keeping with the size of the other parts of the mockup.

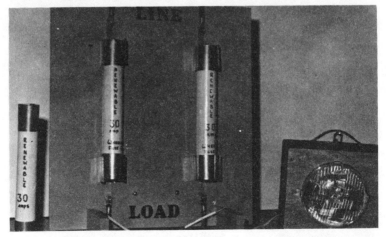

Fig. 4-43. The troubleshooting mockup provides safe, dry-run practice in the fuse-checking procedure.

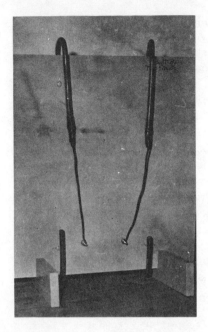

Fig. 4-44. Rear view of the trouble-shooting mockup shows its simple construction. The power supply can be a battery or a door-bell transformer.

A second part of the mockup is a light and buzzer unit to which is attached a pair of leads. The leads, when applied to the fuses (those of the mockup and never to an actual circuit), will activate the light and buzzer unit if a complete circuit is present. The lamp and buzzer unit is mounted in an $8'' \times 8'' \times 4''$ box with plugs mounted for test leads. A single-pole, double-throw slide switch is used to select the light or the buzzer aspect of the unit. This makes it possible to use a visual signal (the lamp) or an audible signal (the buzzer). A voltmeter or test light could be used but the light and buzzer unit is more dramatic and aids in maintaining interest. The power source for the mockup may be obtained from either a doorbell transformer or battery.

A CHALKBOARD SINE WAVE GENERATOR

Here is a project that will help you draw sine waves of varying amplitude and period. It will superimpose two or more sine waves with any degree of phase shift, either leading or lagging.

This teaching aid uses the simple idea of a flywheel and connecting rod. The flywheel is rotated by a rubber drive wheel that is placed in contact with the board. The connecting rod transmits this into vertical motion of the chalk holder. The whole project can be made from scrap material. The frame is constructed of wood and the flywheel is of hardboard. For theory of operation, refer to the accompanying drawing (Fig 4-45).

Flywheel A is rotated by drive wheel D which contacts the chalkboard. Cabinet hinge C attaches the drive wheel D to frame B. This drive wheel can be positioned to cause the flywheel to rotate at different speeds.

Connecting rod E operates chalk holder F which moves vertically as the machine is pulled across the board. Wheel H supports the top of the machine and serves as a visual reference for positioning it with respect to the board.

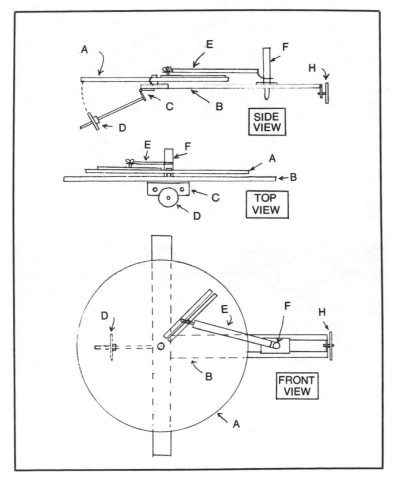

Fig. 4-45. Different views of the chalkboard sine-wave generator.

The connecting rod can be attached to the flywheel at various points by loosening the thumb screw. This allows for variation in amplitude of the waves. Since the flywheel is marked off in degrees, the wave can be started or stopped as required.

Details of construction are shown in Fig. 4-45 but feel free to use your own ingenuity. The drive and idler wheels were taken from phonograph motors and retreaded with rubber electrical tape. The chalk holder is a piece of copper tubing. The flywheel must be counterbalanced to offset the weight of the connecting rod.

AN AUDIO AMPLIFIER DYNAMIC DEMONSTRATOR

This audio amplifier dynamic demonstrator has proved to be of great value. What is unique about this demonstrator is that the troubles common to most amplifiers have been built into the circuit. Any one of 15 troubles can be put into the circuit or taken out by simply rotating one of the rotary switches. When all switches are in position No. 1, no trouble is inserted in the amplifier; as a result, the amplifier will function normally. The concepts learned with this audio amplifier are applicable to all amplifiers.

The audio amplifier whose circuit is drawn on the panel (Fig. 4-47) is typical of those found in most receivers. The circuit diagram is drawn on a 6" x 9" panel. The pin jacks on the panel are wired to the amplifier on the chassis so that the student can conveniently make voltage and resistance checks without turning

Fig. 4-46. The audio amplifier dynamic demonstrator.

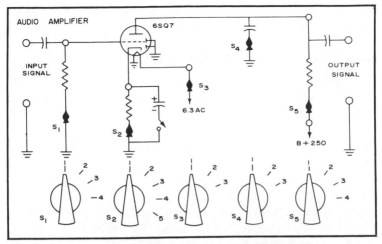

Fig. 4-47. Front panel of the audio amplifier dynamic demonstrator.

the chassis over. Five rotary switches are mounted on the front panel and one spring-return rotary switch is mounted on the top of the chassis. Referring to the schematic diagram (Fig.4-48), you will note that the front-panel mounted switches are wired to the amplifier circuit in this manner.

Switch S1 (Table 4-6) is wired in the grid circuit. S2 is wired in the cathode circuit. S3 is wired in the heater circuit. S4 is wired in the plate circuit as is S5. S6 is a spring-return rotary switch mounted on the chassis used to make the grid positive in respect to

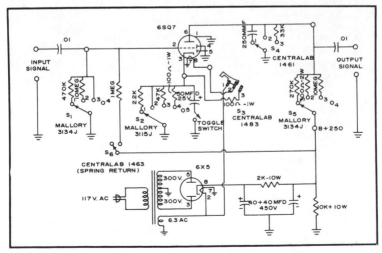

Fig. 4-48. Schematic of the audio amplifier dynamic demonstrator.

the cathode. From the schematic diagram you can determine what change is being made in each position of switches.

For Both Experimenting and Troubleshooting

This small model demonstrator can be used in two ways by a radio club student. First, he can work with the demonstrator by himself performing the following experiments:

☐ What are the three causes for "zero" plate voltage?

☐ What are the troubles that cause plate voltage to be "equal" to B + 250?

☐ What are the troubles that cause "low" plate voltage?

☐ What are the troubles that cause "zero" cathode voltage?

☐ What are the troubles that cause "high" cathode voltage?

☐ What are the troubles that cause "low" cathode voltage?

☐ What are the symptoms caused by "heater to cathode leakage"?

☐ What are the symptoms caused by an "open" cathode bypass capacitor?

Table 4-6. Parts List for Audio Amplifier Dymamic Demonstrator.

Resistors	Power Transformer
1—470k, ½ w	primary—117 v
2—10 meg, ½ w	high voltage sec.—600 v
1—1 meg, ½ w	center tapped
1—2.2, ½ w	low voltage sec.—6.3 v
1—47k, ½ w	
2—100 ohm, 1 w	
1—33k, 1 w	**Capacitors**
1—270k, ½ w	2—.01 mfg, 600 v
1—330 ohm, 2 w	1—250 mmf, 600 v
1—2k, 10 w	1—50 mfd, 25 v
1—10k, 10 w	1—40 + 40 mfd, 450 v
	Miscellaneous
Switches	2—1″ octal sockets
2—Mallory 3134J	4—banana jacks
1—Mallory 3115J	5—pin jacks
1—Centralab 1483	1—7″ × 9″ × 2″ chassis
1—Centralab 1461	1—6″ × 9″ panel
1—SPST, toggle switch	1—phase shift network consisting of:
1—Centralab 1463	3—.003 mfd 600-v
	ceramic capacitors
	2—150k, ½ w
Tubes	resistors
1—6X5	2—phone tips
1—6SQ7	1—banana plug

Secondly, after a club student has completed all the experiments and has learned the effects of these troubles on the operation of the amplifier, you can insert one of the troubles in the amplifier simply by turning one of the rotary switches. A cover can be placed over the switches to prevent the club student from seeing which switch has been turned. It will be his job to find out what trouble was put into the amplifier by you. This procedure of inserting a single trouble at a time into the amplifier is to be continued until the student finds all 15 troubles that he has studied in the experiments.

The following troubles may be inserted into this amplifier:

☐ "Shorted" plate bypass capacitor (Set S4 to position #2).

☐ "Low value" cathode resistor (set S2 to position #4).

☐ "Open" plate load resistor (set S5 to position #4).

☐ "High value" cathode resistor (set S2 to position #3).

☐ "Zero" source voltage (use 6X5 with cathode pin cut off).

☐ "Leaky" plate bypass capacitor (set S4 to position #3).

☐ "Open" heater (set S3 to position #2).

☐ "Shorted" cathode bypass capacitor (set S2 to position #2).

☐ "High value" plate load resistor (set S5 to position #3).

☐ "Open" cathode resistor (set S2 to position #5).

☐ "Low value" plate load resistor (set S5 to position #2).

☐ "Short" from grid to B minus (set S1 to position #3).

☐ "Open" cathode bypass capacitor (set toggle switch on panel to "out").

☐ "Leakage" from heater to cathode (set S3 to position #3).

☐ "Gassy" tube or "leaky" coupling (connect jumper wire across contacts on spring-return rotary switch).

Besides the foregoing, you can use the dynamic demonstrator to show amplification, amplitude distortion, phase inversion and oscillation. This audio amplifier can be made into a phase shift oscillator by connecting a phase shift network between the plate and control grid. This demonstrates that an oscillator is really an amplifier that purposely takes part of its output signal and feeds it back to its grid circuit to produce a signal.

The idea behind this demonstrator—that is, building troubles that can be switched in and out of a circuit—can be extended to other stages in a receiver. Therefore, a similar demonstrator can be built for the output stage, the power supply, etc. Since these troubles have been worked out beforehand, you can be sure that when you are ready to demonstrate them they will work without causing you any embarrassment.

TRANSISTOR BIAS DEMONSTRATOR

Proper biasing of transistors is the first prerequisite for their proper operation in a circuit. The collector voltage and current depend on the biasing. A change in room temperature may significantly change the collector voltage and current, unless the biasing arrangement incorporates an effective stabilizing action. A technician who replaces a transistor when fixing a radio may have to change the biasing if the new transistor has a very different gain.

Biasing a transistor is a more critical procedure than in the case of a vacuum tube. The simple and versatile demonstrator described here was built to accomplish the following:

☐ To show how the leakage current of a transistor varies with temperature.

☐ To show a reliable comparison of the stability provided by various types of biasing circuits.

☐ To show how a transistor can be used as the control unit of a thermostat.

The schematic of Fig. 4-49 shows how the circuit is arranged. A relay (such as Sigma GIB-51600-1 or equivalent) and an inexpensive milliammeter are connected as the load of the transistor in its collector circuit. The transistor circuit uses a small 9-V battery.

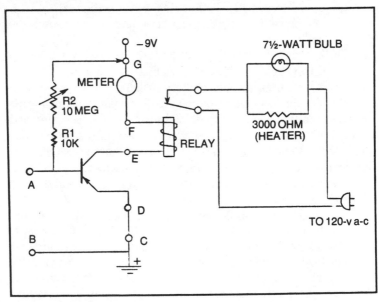

Fig. 4-49. Transistor bias demonstrator schematic.

The relay contacts control a 7½-W bulb in parallel with a 3000-Ω, 5-W resistor connected to the 120-V AC outlet.

The physical position of the components is shown in Fig. 4-50 (the relay and battery are behind the panel).

It is important to observe that the resistor is placed directly under the transistor. In that way it acts as a source of heat, to increase the transistor temperature as desired. Also, points A, B, C and D are available at the front of the demonstrator by means of Fahnestock clips, so that various bias arrangements can be quickly connected.

Points E, F and G are also available at the front by means of small connectiors or terminals; the upper end of the 10 Meg Ω potentiometer can be connected to any one of those points by means of a small alligator clip. Complete circuit flexibility is thus achieved.

Theory and Application

To demonstrate the operation of the transistor as a thermostat control, the collector current (indicated by the meter on the panel) is adjusted by means of potentiometer R1, until the relay releases its normally closed contacts and closes the 120V circuit. Now the lamp lights up, and the heating resistor begins to warm up; its heat reaches the transistor.

As the transistor gets warm its collector current increases (this can be visually observed on the meter). As the result of heat, the collector current finally reaches a value that energizes the relay, and the light and heating resistor are turned off.

As the transistor cools, the collector current decreases. Eventually, the collector current returns to the initial value, the relay is de-energized, the light and heating resistor go on again, and the cycle is repeated.

In other words, the transistor acts as a temperature sensing device and, by means of the relay, turns the light and heating resistor on and off indefinitely. If in place of the heating resistor we had a house furnace, it would be turned on and off within the temperature range for which the bias is adjusted.

In order to demonstrate how a different type of biasing minimizes the effect of temperature on collector current, the new biasing is connected by means of the Fahnestock clips. Then, with a watch, it can be ascertained that with the more stable biasing circuit the light takes longer to be turned off. In other words, the transistor has to get hotter in order for the relay to trigger.

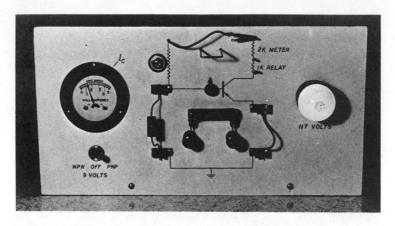

Fig. 4-50. The front panel of the transistor bias demonstrator.

More precise measurements can be made if we have a means to determine the transistor temperature. This can be done by providing the transistor with an "environmental chamber" made with an aluminum box (of the kind in which 35mm color film comes packed), as shown in Fig. 4-51.

A metal coil thermometer, of the type used for home stoves, can be obtained in most hardware or variety stores. When the thermometer (note foreground of picture) is hung in front of the aluminum box, it forms a thermal box around the transistor. The

Fig. 4-51. Placing a metal coil thermometer in front of the aluminum box will indicate the ambient temperature of the transistor.

temperature indicated by the thermometer is the ambient temperature of the transistor.

Construction Hints

The panel of the bias demonstrator was made by the author from a piece of hardboard painted dull white. The circuit can be drawn on it with India ink, and the entire panel is then given a transparent protective coating. A piece of one-inch wood stock will serve as a table base to which the panel is attached with two screws.

Provision for PNP and NPN transistors can be made, if desired, by means of a switch to reverse battery polarity and meter terminals. The transistor should be mounted in a socket, so that different transistors can be used on the panel.

AN AUDIO AMPLIFIER PRINTED CIRCUIT VISUAL AID

The audio amplifier may be the most used circuit in the radio or TV receiver, and a radio club student must learn to find his way around this circuit electronically and physically. To simplify the teaching problem, a series of easy-to-trace printed circuit boards that can easily be duplicated has been designed.

When teaching the triode audio amplifier, use the "PC" or "photocard" method. This layout is designed for use as an instructor's display panel for small group demonstrations. All of the parts required can be taken from salvaged electronic equipment.

Circuit information can be filed on 3 × 5 cards. Two cards will fit nicely on a half sheet of paper and five cards will fit on an 8½" × 11" sheet.

Figure 4-52 is the master copper pattern to duplicate the boards. Fill in the solid lines with India ink to make a resist pattern positive. Run the copy through a photocopier to make a negative resist pattern.

In Fig. 4-53, the exact placement of each part may be seen over the copper conductors. For checking the circuit, send the signal into the volume control (A). A capacitor on the input will block any dc from a previous circuit. The arm of the potentiometer is connected to the coupling capacitor (B). The grid resistor is at point (C). Make a direct connection on this board to take bias voltage readings or to use it directly from another circuit. Figure 4-54 has the schematic diagram.

On the cathode, a low-impedance line is terminated on the output side of the board. The signal at this point has a little less

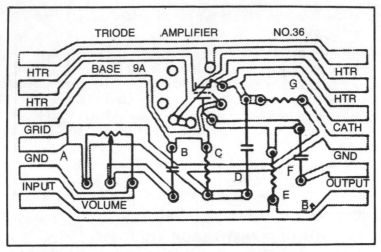

Fig. 4-52. The master copper pattern to duplicate the PC boards.

amplitude than the grid peak-to-peak voltage. However, it is useful either as a cathode follower circuit or to drive a transistor low impedance input.

If an output is not wanted, use a bypass capacitor (D). With a cathode capacitor, the plate output signal will have a slightly increased output. Ordinarily, leaving out the cathode bypass capacitor will cause degenerative feedback that increases the overall high frequency response of the amplifier.

Fig. 4-53. The exact placement of the parts of the audio amplifier printed circuit visual aid.

110

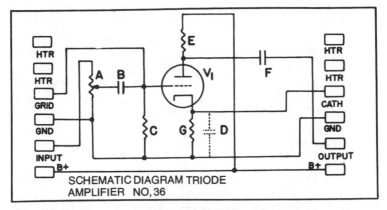

Fig. 4-54. Schematic of the triode amplifier used as a PC visual aid.

For the plate circuit, a load resistor is located at E. The DC from the B+ line is dropped to the correct plate voltage. Alternating current signal variations from the grid are also dropped across the plate load resistor. This high-impedance output is coupled to the output terminal through the capacitor at F.

Any signal distortion at this point may be corrected by using the proper choice of tubes and components. Such data may be found in laboratory manuals or articles appearing in trade magazines. Note that the signal voltage in the plate circuit is inverted by 180 degrees in phase with respect to the grid signal.

Construction Hints

Small solder tabs at the ends of each component allow uniform spacing across the printed circuit board. Cut each component wire to ⅜" and bend about 1/16" at the end for a foot to solder on the board. Bend the tabs on the volume control to meet the solder tabs on the board. A spot of solder is placed on both edges of the control to hold it down.

Use only good quality capacitors. The small mylar types such as the Elmenco "dp," Mallory "pvc" and Sprague "orange drop" are recommended in the 400 VDC tubular line. All resistors should be checked on a ohmmeter before using. For the tube socket, a standard nine pin printed circuit type is desired. However, for the tube socket, use the common Bakelite socket with a mounting ring and center sleeve. The latter two items should be removed before mounting the socket on the board.

After the board component mounting has been completed, solder the connection clips to each end. This particular type of

111

connection is a common stock hardware item and is relatively inexpensive. Use the common Fahnestock clip to hold wires; several wires can be used with effective holding power, eliminating the use of stacking wires with special clip connectors.

Each board can be connected by using the lug from an 8-pin octal tube socket. The wiggle lug is made by Amphenol, special part No. PLY 009-018-51. On the output side of this board, solder the flat end of the lug between the Fahnestock clip and the PC board island. Place it in a position where the "U" of the lug is at the edge of the PC board.

To make the input side of the PC board responsive to the grip of the wiggle lug, use the same type of Fahnestock clip. This is mounted by drilling a small hole through the PC board island near the end of the small rectangular area. The Fahnestock clip is held down by an eyelet (General Cement No. 7253-C, .087 × 7/32"). Pinch off the remaining part of the eyelet on the bottom with diagonal pliers. Put a drop of solder on the end of the clip and the eyelet. Do not allow the solder to run under the clip and the PC board copper conductor. Push the wiggle clip under this space for interboard connections.

Since this board has interconnecting clips and lugs, more boards can be connected to the system. Use a one-tube regenerative detector, FM or AM, before the amplifier. Following the amplifier is an audio output amplifier. Connect a small power supply to the end of the audio output board.

Typical Operating Data

Select tubes on the basis of base diagram 9A. This uses a number of common tubes: types 12AU7, 12AT7, 12AX7, 12BH7, 6211, 5963, 5965, 6201, 6681, and 7025. The wide selection of tubes permits this particular circuit to function as a pulse amplifier to demonstrate its use in TV sync amplifiers, oscilloscope amplifiers, and even the discharge section of a blocking oscillator. Select suitable component values for each circuit.

Table 4-7 lists the operating data on this circuit. The table indicates the list of component values used for these readings.

THE ELECTRONIC METER GOES OVERHEAD

How can a large group of radio club students be taught to read the various scales of a VTVM, VOM, or other mutimeter? The meter itself could be used. However, because of the detailed print on the meter face, the effectiveness of the instruction diminishes beyond about 6'.

Table 4-7. PC Visual Aid Tube Typical Operating Specification.

6211		12AT7		BOARD VALUES
B+	300 v.	155 v.	275 v.	155 v.	A= 1 Meg.
E_p	60 v.	30 v.	100 v.	60 v.	B= .02 uf.
I_p	2.4 ma.	1.25 ma.	2.75 ma.	.95 ma.	C= 470K
Gain	25	20	60	30	D= none
PLATE PP	100	30	90	30	E= 100K
GRID PP	4	1.5	1.5	1	G= 470 ohm

This deficiency can be overcome by transferring a full-sized replica of the meter face and front panel to an overhead transparency. The transfer can be done in four simple steps as follows:

□ Remove the face from the meter.

□ Make a photocopy of the meter face (see Fig. 4-55).

□ Outline the front of the meter enclosure around the photocopied face. Include function-and-range switches, zero-adjust knob, lettering, etc. Shade front panel with a soft lead pencil to highlight (see Fig. 4-56). If the front panel of the instrument bears detailed printed information pertinent to meter operation, it too can be photocopied as was the meter face (see Fig. 4-57). This figure shows how highlighting the front-panel portion of the copy with a soft lead pencil improves contrast between silk-screen print and panel background.

□ Make a transparency using the finished copy as a master.

When the transparency is projected on the screen, you can explain the use of the meter to the entire radio class. He can ask for specific meter readings by positioning the range-and-function switches and the needle with a few strokes of a grease pencil (see Fig. 4-58).

SUPERHETERODYNE DEMONSTRATOR

The superheterodyne is the most common type of radio receiver in use today, and yet it is extremely difficult to explain to radio club students the receiver's action in mixing two varying

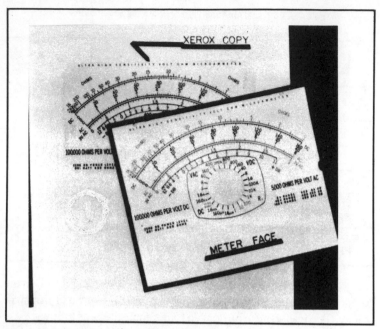

Fig. 4-55. Make a photocopy of the meter face.

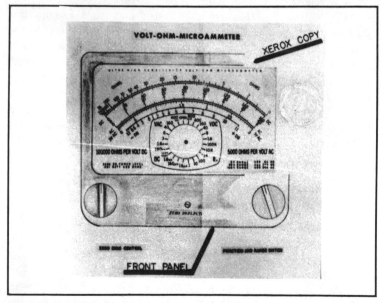

Fig. 4-56. Shade the front panel with a soft lead pencil to highlight what would be the darker areas of an actual electronic meter.

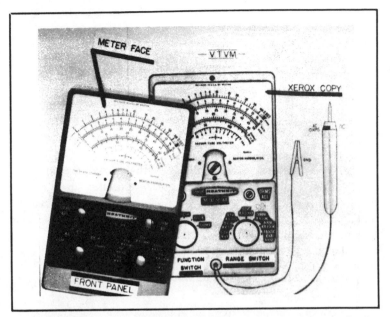

Fig. 4-57. Detailed printed information on the electronic meter can also be photocopied if necessary.

Fig. 4-58. After making a transparency of the copied electronic meter face, you can instruct your club in how to read a meter.

signals or produce an intermediate-frequency to *difference* signal. This circuit can be constructed as a demonstrator of the process of heterodyning two ultrasonic signals to produce an audio sinusoidal signal that can be seen on an oscilloscope screen and can be heard through a signal tracer or any audio amplifier.

Note that there are two inputs, (Fig. 4-59). One represents the carrier signal, while the other represents the local oscillator. By adjusting the two signals so that they beat together to produce a difference signal within the audio range, very exciting and convincing results are obtained.

The oscilloscope serves to illustrate that signals are present, displays their relative amplitudes, and reveals the shape of the modulated envelope. The audio amplifier or signal tracer serves to amplify the weak audio signal generated by heterodyning so that the student can actually hear the effect.

Begin with S1, S2, S4, and S5 open. S3 is to be closed until later. Set the oscilloscope sweep frequency to the 20-100 Hz range. Keep the signal tracer or audio amplifier either turned off or at minimum volume.

Set both audio generators for approximately 1000 Hz output. Close Sl. Observe the signal on the scope. Note its amplitude. Open S1. Close S2. Observe the signal in the scope screen as before. Note its amplitude. Adjust AF generator No. 2 output to have a higher amplitude than AF No. 1.

Leave S2 closed. Close S1. Observe the oscilloscope. Synchronize the waveform and draw a sketch of it, showing the envelope.

Vary the frequency of a-f generator No. 1. Observe how the amount of modulation of the envelope varies.

Adjust for about 70 percent modulation. Open S3, which adds a diode to the circuit, demodulating the waveform to produce half-wave rectification. Sketch the waveform.

Close S5. This connects the 0.05 μF capacitor across the output. Observe how this affects the output waveform by filtering out the higher frequency and leaving the envelope. Explain the jagged edges observed on the waveform rise and fall slopes.

Vary the frequency of AF generator No. 1. What happens to the frequency of the observed waveform?

Connect the signal tracer or AF amplifier to the circuit. Can sound be heard?

Vary the frequency of AF generator No. 1. Does the pitch or frequency of the sound heard vary? Can you achieve a very low pitch?

Open S5 and close S3. Adjust each AF generator to produce approximately 100 kHz signal. This signal is now well beyond the range of human hearing.

Observe the scope waveform. Synchronize as necessary.

Open S3. Can sound be heard? Close S5. Can sound be heard?

Vary the frequency of AF generator No. 1 as before. Open S5 and close S4.

Again vary the frequency of AF generator No. 1. What observations can you make regarding sound intensity?

Why does the 0.001 μF capacitor have a different effect than the 0.05 μF capacitor?

Vary the frequency of a-f generator No. 2, and watch its effect on the output.

Adjust both generators to exactly the same frequency. This can be accomplished by carefully adjusting the attenuator on one generator while watching the trace on the oscilloscope. When both signals are exactly identical in frequency, observe a horizontal line on the screen. This is the "zero beat" position which produces no output.

WIRING DEMONSTRATION PANEL

It has been said that a picture is worth a thousand words, but a good demonstration panel is worth many pictures *and* many words.

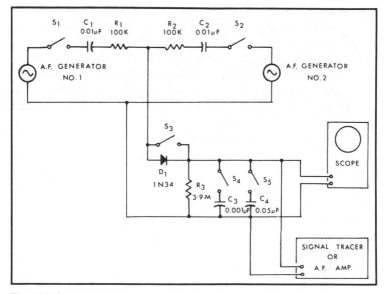

Fig. 4-59. Schematic of the superheterodyne demonstrator.

Such a wiring demonstration panel is easy to build and fairly inexpensive, especially if you have a few odds and ends of lumber and electrical supplies in stock.

The demonstration board described here is a training device that can be easily constructed and will provide the opportunity for actual wiring problems to be accomplished. The model (Fig. 4-60) can be wired for 110 or 220 volts and provides the following demonstration areas:

Panel A contains a three-way switch arrangement for a ceiling light, front doorbell button, methods of running wire with notched or drilled studs, conduit connections, auxiliary circuits from a separate fuse box, two flush-mounted wall outlets, one half of each controlled by a toggle switch.

Panel B contains a meter, obtained from the local power company, main distribution box, flooring and sub-flooring, and floor joists. The lower section has a basement light, appliance outlet 220 volts, doorbell and transformer, grounding connections and junction boxes.

Panel C has a built-up wall with insulation, vapor barrier of plastic and wood paneling. A rear doorbell is installed here and two flush-mounted outlets are controlled by a toggle switch.

The ceiling panel D contains only the light fixture (18) and ceiling joists.

Wall studs are 16" on center and it is easy to demonstrate problems that might arise when outlets or wiring are changed or added to the existing structure. Many different types of connections are used—splices, solderless, clamp type, along with rubber, friction, and plastic tapes. The demonstration panel measures only 18-½" thick by 49-½" wide and 72" tall when in the closed or storage position.

SERIES PARALLEL DEMONSTRATOR

This demonstrator was designed to impress radio club students with the inherent differences of series and parallel circuits when considering the properties of voltage, current, resistance and wattage. The obvious advantage of being able to switch between the two circuits and make almost instantaneous comparisons is enough to make the project worthy of construction.

Size and layout of the board are flexible, depending only on providing sufficient room for all the components. These can also be varied—cleat or ring receptacles may be used; and high-wattage resistors could also be used, in which case neon pilot lamps might

Fig. 4-60. The complete wiring demonstration panel.

be desirable. Resistors have the added advantage of making accurate working resistance measurements possible, something completely alien to light bulbs with their extreme range between hot and cold resistance.

The original board measured 24" × 30". The spacing of the receptacles from the side was dictated by the use of a rotary hacksaw in a drill press to cut the access holes in the Masonite.

The method of wiring the board is shown in Fig. 4-61. Number 14 TW wire was stripped and soldered to the binding posts. The screws were removed from the receptacles and a connecting link was soldered in (The binding posts did double duty in that they also held the receptables, providing a means of taking voltage measurements plus effectively insulating the terminal when not in immediate use.)An ordinary three-way tumbler switch was used as a function switch to activate either circuit. A 500-mA milliammeter measures the current flowing in the series circuit. Two separate 5-amp ammeters are used to record the parallel-circuit current—one shows the total current in the circuit and the other records the branch currents by use of a three-way switch at each receptacle which shunts the individual branch currents either through the meter labeled "branch current" or direct to the return line. Two SP tumbler switches are used to short out #2 and #3 receptacles in the series circuit for demonstration. The first receptacle is not shorted out for obvious reasons.

Through use of this board, anyone can visually appraise the varying intensity of the bulbs in the series circuit as they are cut in and out of the circuit. IR drops can be measured at any point (see photo above) and totaled up. The valve of the circuit current can be seen in an inverse ratio to the circuit resistance, while a second voltmeter records source voltage. The totally different effects can be immediately seen and compared in the parallel circuit by merely throwing the function switch to activate the circuit. Reinforcement of the properties and comparison are as immediate as the function switch.

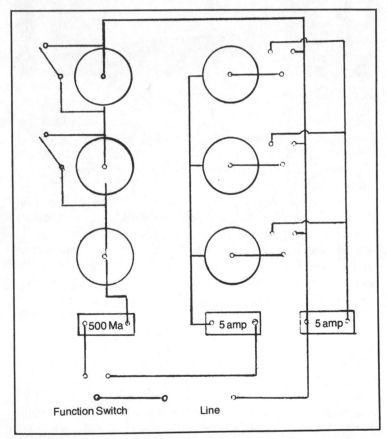

Fig. 4-61. Schematic of the series parallel demonstrator.

Here is visual and concrete proof of the laws and rules governing series and parallel circuits, as well as the abstractions presented by the basic laws of those illustrious gentlemen, Mr. Ohm, Mr. Watt, and Mr Kirchhoff.

A GENERATOR DEMONSTRATOR

Generation of electricity by using a rotating machine generator is a basic electrical concept, but one which is very difficult to teach to radio clubs. This process is not easily understood by beginning (or even some advanced) students because the immense size and power involved in power plant generating equipment make it very difficult to simulate this process in the lab.

The demonstrator described here is small and inexpensive, yet is very useful in helping club students get a good basic understanding of electrical generation and associated concepts.

The demonstrator (Fig. 4-62) consists of a variable power supply and two mechanically-linked DC motors (the same type motors commonly used in slot cars). The variable power source and first motor represent the "natural" power supply and drive the second motor—the generator—through a mechanical coupling. The generator has a resistance attached to it to either use the electricity generated or serve as a load.

Metering provisions are supplied for both the motor and generator so that measurements of voltage and current input and output can be obtained for calculating power and efficiency. In operation, the meters normally measure voltage, but give a current reading when the spring-loaded switches are depressed. The load of both motor or generator may be varied by adjusting the variable load resistor on the generator or the DC power supplied to the motor.

The efficiency of the system—the ratio of the power output to the power input—is easily calculated. The power input and the power output are simply determined by using the power formula, W(wattage) = I(current) × E(voltage). Current and voltage readings are obtained from the meters, of course. By using the appropriate conversions, power can be read in other units, such as horsepower, if desired.

It is possible to plot an efficiency curve with different input and output loads to determine at what point the total system operates at peak efficiency. Current surge in starting the motor is

Fig. 4-62. The front view of the generator demonstrator shows the separate controls for the motor (left) and the generator (right). Pressure sensitive lettering is used.

also easily demonstrated by simply switching the meter to current at the time power is applied to the motor. The current should surge at the beginning then drop back to a lower, steady value after the motor has reached its normal operating speed.

Construction Procedures

Lay out the front panel on ⅛" hardboard or sheet metal, providing holes for meters, switches, and wires, and punch and drill all necessary openings. Construct the housing (12" × 12" × 2-¾"- of ¾-inch stock. This design provided for flush mounting of the front panel and a carrying handle. Paint or otherwise finish the front panel, and apply all labels. See Table 4-8.

Mount all parts, including meters, switches and motors. The motor-generator combination must be mechanically coupled; this can be done by using strapping tape. Wire and solder all leads behind the front panel. Install the transformer at this time. *Do not install multiplier or shunt resistors at this time.*

Using the following procedure, select the 10K multiplier resistors and install them. Hook up a lab voltmeter which is known to be fairly accurate in the circuit shown in Fig. 4-63.

The voltage applied should be adjusted so that the lab meter reads 5 volts at mid-scale on a 0-10-volt scale. Now try several 10K resistors in series with the panel meter until the panel meter reads half of the full scale or 0.5 mA. This represents 5 volts. You may need to experiment with several values around 10K in order to get good results. Calibrate *both* meters (motor and generator) for voltage in this manner.

In order to change the sensitivity of the 0.001-amp meter so that it will read 1.0 amp full scale, a shunt resistor must be placed in parallel with the meter to be calibrated. The resistance needed is around 0.05 ohms, so a piece of fine copper wire or longer piece of nichrome wire will serve this purpose. Hook up the meter in the circuit shown in Fig. 4-64.

Adjust the power supply to about 5 volts and adjust the 100 ohm rheostat until the lab meter reads ½-amp on the 1.0-amp scale (center of scale). Now slide the alligator clip at either point A or B (not both) until the panel meter reads half scale (½ mA). This means that *both* meters are reading ½-amp of current and the wire between points A and B is the correct length for a shunt across the panel meter. You may wrap this wire around a dowel rod for conveneince and hook it into the circuit.

Table 4-8. Parts List for the Generator Demonstrator.

2	Meters, I MA full scale
2	DC motors, slot car type with permanent magnets
2	Two position switches, spring return
2	On-off switches, SPST
2	Binding posts
1	50-ohm potentiometer of rheostat
1	100-ohm potentiometer or rheostat
2	10-K ohm resistors, ½-W
1	4.7-ohm resistor, 2-W minimum
1	120 V to 12 V - 2A, stepdown transformer
1	Diode, 20 V, 3 amp
2	3-V light bulbs (or one 6-V bulb)
1	AC supply cord and plug
2	Knobs, ¼" shaft, pointer type
1	Carrying handle
1	Fuse and fuse holder, 3 amp

Miscellaneous wood, screws, wire, solder, etc.

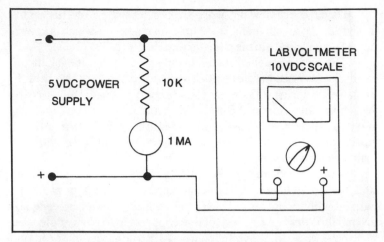

Fig. 4-63. Assembly diagram of the generator demonstrator (top).

Finally, recheck all connections, plug in power supply to AC, turn on the motor and check-out all metering and loading for proper operation. See Figs. 4-65 and 4-66.

A COMPUTER TRAINER

Teaching basic computer programming, operation, or maintenance? Why not build your own trainer?

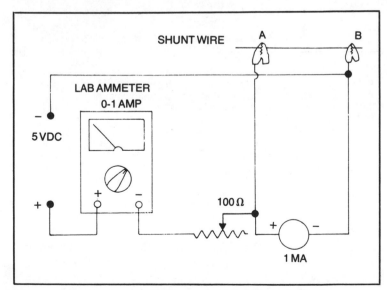

Fig. 4-64. Assembly diagram of the generator demonstrator (bottom).

Fig. 4-65. The interior of the generator demonstrator shows parts placement and wiring. The light bulb in the center is a 3V bulb in series with the 3V bulb mounted on the panel. The two act as a 6V bulb. The shunt resistors consist of nichrome wire wrapped around a dowel rod. They are located at the right and left edges of the panel. The power supply is in the lower left corner.

The portable unit described here uses simplified solid-state design. Built largely with surplus and salvaged components, the trainer cost about $200. It can be hand carried.

Design approach objectives included: to apply electronic theory and laboratory test equipment; to design and produce plug-in printed circuit assemblies related to the computer's function such as: hybrid variable frequency generator, generator buffer, low frequency schmitt trigger, etc. to design and produce a chassis to accept the circuit boards, mother circuit, power source, fuse, wiring, and operator's panel; to design and produce a special

125

power supply to furnish − 15 volts, − 3 volts, and +10 volts; and to design and produce the operator's panel with circuit drawings and logic symbols engraved on the panel, with appropriate fitted jacks, switches, fuses, and lights.

Two Common Problems

In learning any basic concept, anyone is usually confronted with two common problems: language and symbols of the specialty and regulating the timing. The design of the trainer took both these problems into consideration. For example, the operator's panel

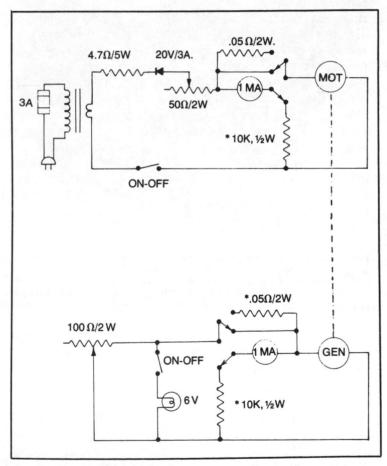

Fig. 4-66. This wiring diagram of the generator demonstrator shows meter switches in normal "E" positions.

includes both symbols and circuit diagrams so the beginner can easily assimilate the language and symbols. Each AND, DIODE, or FLIP-FLOP is labeled by symbol and circuit (Fig. 4-67). Further, the cycling, or impulse, aspect of the circuits has been slowed down or controlled at an interval of one cycle per second to give the learner ample time to observe each cycle change. The controlled impulse time contributed to an improvement in the level of understanding.

The trainer has one power supply to furnish four DC voltages: +10 volts regulated, −15 volts regulated, −3 volts regulated, and −15 volts unregulated (Figs. 4-68 and 4-69). Two transformers, a rectifier, and a heat dissipator are included. The transformer primaries T_1 and T_2 are connected in parallel for 110 volts 60 cycle AC through power switch S_1 and fuse F_1, and an indicator lamp L_1 on the front panel is in parallel with the primaries and glows when the power is on. The secondary of transformer T_1 is connected to a bridge rectifier circuit composed of CR_1 through CR_8. Diodes are connected in parallel to increase the current and handling capability of the circuit as shown in the rectifier board schematic (Fig. 4-70). Care was taken to design a power supply which was well regulated, yet heavy enough for unusually heavy loads often encountered.

While the actual physical layout and selection of individual circuits can be designed to meet the needs of the individual course content, great care should be exercised in the design of the power supply. Generally, four to six individual voltages may be required within two divisions—regulated and unregulated—with both positive and negative potential. Regulated voltage should be extremely stable to minimize undesirable operation of critical circuits when excessive loads are placed on the power supply.

Various model programs may be written and tested in the training unit, including such demonstrations as: OR and AND gate, flip-flop, logic, and decimal to binary conversion.

Operation of the individual OR gates may be demonstrated by proper connection of jumper leads. Any of the six OR gates may be used. For example, using jumper leads, make the following connections:

from	to
S-10 Black	OR 1 Black
S-9 Black	OR 1 Green
S-8 Black	OR 1 White
OR 1 Red	Lamp driver 10 Yellow

Fig. 4-67. Circuit drawings and logic symbols are engraved on the face of the operator's panel of the computer trainer, which is fitted with jacks, switches, fuses and lights.

Next place switches S-8, S-9, and S-10 in logic "0" position (down). Lamp 10 is off in this condition. By switching either S-8, S-9, or S-10 to logic "1" position (up), the lamp will light, illustrating that logic "1" input to either 8, 9, or 10 will produce a logic "1" output.

Logic circuits are used in digital computers to produce an output for a certain input condition. The OR gates of the training device have three inputs, therefore when a signal (−3 volts) is placed on one, *or more*, of the input terminals there is one output.

	S-8		S-9		S-10	=	OUT-PUT
	0	+	0	+	0	=	0
	1	+	0	+	0	=	1
or	0	+	1	+	0	=	1
or	0	+	0	+	1	=	1
also	1	+	1	+	0	=	1

A VERSATILE DEMONSTRATOR FOR RESONANT CIRCUITS

Explaining the action of resonant circuits in oscillators and radio frequency amplifiers is a chore. The demonstrator described in this article eases the task by giving an oscilloscope presentation of a tuned circuit "ringing," showing the operation of a sawtooth relaxation oscillator, graphically proving that the flywheel effect exists in radio frequency amplifiers, and demonstrating that a

Fig. 4-68. Plug-in circuit boards of the computer trainer are related to the computer's function. Chassis also accepts mother circuit, power source and other components.

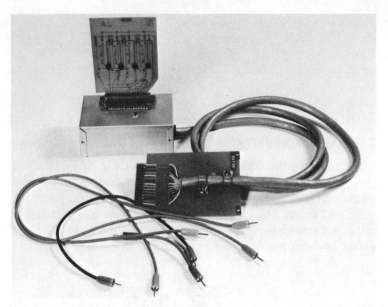

Fig. 4-69. The power supply furnishes four DC voltages and is designed to accommodate excessive loads often encountered.

resonant circuit can multiply the frequency of an audio voltage by 2× or 3×.

The entire works is built on a piece of vector board 4-¾ by 2-½ in. The two electrical connections for the 90 V battery supply also serve as mechanical support. The on-off switch is an alligator clip that connects the two batteries in series. No frills. Just simple utilitarian design that works.

The unit consists of three separate circuits—a relaxation oscillator, a tuned circuit that can be pulsed by the output of the relaxation oscillator and a tuned circuit (the same one) that can be driven by an external audio frequency source—that can be used in various combinations. Although the relaxation oscillator was intended only to pulse the resonant tuned circuit, it can be used for demonstration when lecturing on RC oscillators. A scope connected from common to test point 1 will show the sawtooth waveform of the voltage across the .1 μF capacitor.

Circuit Theory

The theory of operation of this portion of the circuit (battery, 474kΩ resistor, .1 μF capacitor, and neon bulb) is as follows: when the switch is first turned on, the current through the resistor is very high and the voltage across the capacitor is low. The capacitor charges and the voltage drop across the resistor decreases while the voltage across the capacitor increases. When the voltage across the capacitor reaches approximately 65 V the neon bulb (NE2), connected in parallel with the capacitor, fires and shorts the charged capacitor. With the capacitor discharged, the neon bulb deionizes and the entire cycle starts over again.

A sawtooth waveform of voltage change is produced across the .1 μF capacitor. The frequency of the sawtooth waveform can be changed by changing either the 474 kilohm resistor or the .1 μF capacitor. See upper trace in Fig. 4-71.

While the .1 μF capacitor is charging in the process described above, the .02 μF capacitor is also charging. When the neon bulb fires, the .1 μF capacitor is shorted and the .02 μF capacitor is discharged into the resonant tuned circuit (88 mH and 1 μF). The spike (see lower trace in Fig. 4-71) is the voltage across the 100 ohm resistor. The 100 ohm resistor was added between the common connection and TP 3 to provide a voltage for the scope presentation of the surge of current when the .02 μF capacitor is discharged into the tuned circuit.

To display the sawtooth voltage and the pulse output of the relaxation oscillator, the scope common should be connected to common connection, one channel of scope to TP 1 for sawtooth display and the other channel of scope to TP 3 for pulse display.

Ringing the Circuit

To demonstrate the resonant effect of a tuned circuit, the scope should be connected between common and TP 2. The other channel of the scope will show the sharp pulse if connected to TP 3 (see Fig. 4-72). The sharp, single pulse in the bottom trace starts the circuit "ringing." The train of oscillations shows vividly that the frequency of oscillation does not change, although the amplitude decays as the circulating current diminishes.

To demonstrate the adverse effect of parallel R, a resistor connected between TP 2 and TP 3 will cause faster decay of the voltage amplitude. If an additional capacitor is connected across TP 2 and TP 3, the frequency of oscillation will be lowered; the number of cycles between driving pulses will be decreased. A shorting strap from TP 2 to TP 6 will reduce the inductance and increase the frequency of resonance.

One of the early models of the demonstrator did not ring well. The sine wave voltage across the resonant circuit decayed very rapidly. Only about four cycles were visible on the scope. The 1 μF capacitor was a paper insulation type. It was discovered that this

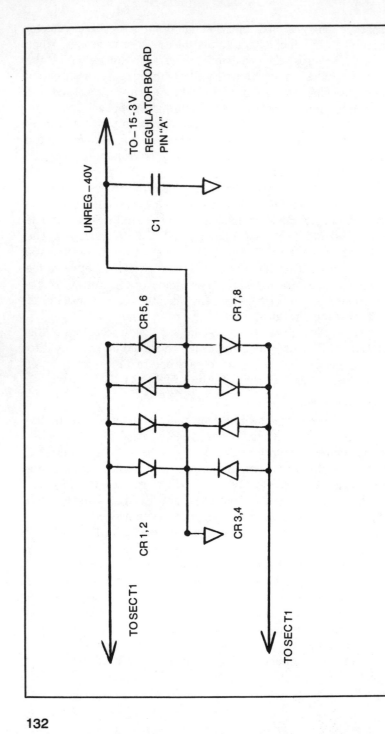

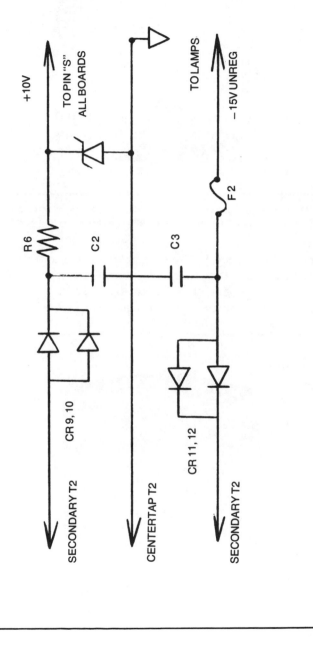

Fig. 4-70. Schematic of the rectifier board of the computer trainer.

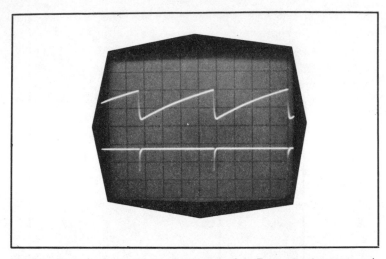

Fig. 4-71. Upper trace: sawtooth voltage across 0.1 μF capacitor (common and TP1). Lower trace: pulse of current that starts L/C circuit ringing (common and TP3).

paper capacitor had low leakage resistance and was lowering the Q of the tuned circuit. The paper capacitor was replaced by a tubular, metal-cased type with glass terminal end seals. The leakage resistance of this capacitor is very high, and with the higher Q of

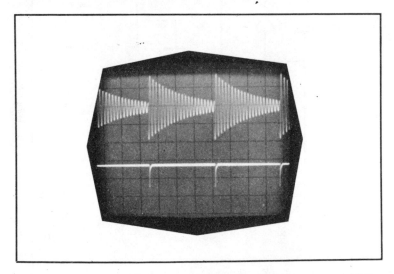

Fig. 4-72. Upper trace: voltage across resonant circuit (common and TP2). Lower trace: pulse of current that starts L/C combination ringing (common and TP3).

the tuned circuit ringing action was good. The circuit sustains oscillations for approximately 21 cycles between driving impulses.

The tuned circuit can be used to demonstrate the pulsing action of the plate current in Class B and Class C amplifiers using resonant circuits as plate load. Two additional terminals were added (TP 4 and TP 5) and a diode and resistor were installed on the board. The toroid coil was removed from the board and a primary of 20 turns of #30 enameled wire was scramble-wound over the toroid inductor. Neither the number of turns nor the size of wire is critical.

Other Uses

When using the device to demonstrate pulsing input, the battery is turned off and the relaxation oscillator is not operating. An audio generator connected between common and TP 5 will provide an audio voltage that is changed to half-cycle pulses by the diode 1N914 and the 390 ohm resistor. With the half-cycle pulses coupled into the toroid via the 20-turn primary, the circuit will oscillate vigorously and provide a scope presentation that can be related to Class B and Class C operation of vacuum tube or transistor amplifiers.

Figure 4-73 shows the resonant circuit operating with a driving pulse every cycle. The audio generator in this case is on the resonant frequency of the parallel tuned circuit, 537 Hz.

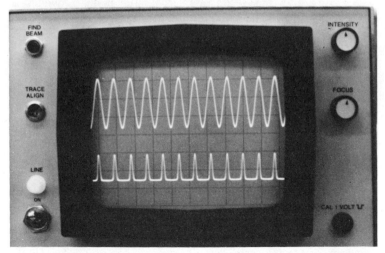

Fig. 4-73. Upper trace: voltage across resonant circuit (common and TP2). Lower trace: driving pulse from audio generator (common and TP4). Note: audio generator is connected between common and TP5.

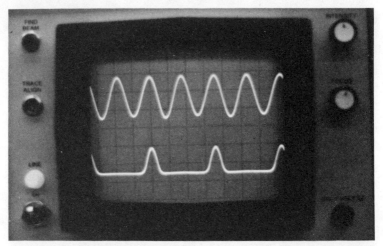

Fig. 4-74. Upper trace: 537-Hz AC voltage across 88 mH/1 μF tuned circuit (common and TP2). Lower trace: half cycle pulses at frequency of 268.5 Hz (common and TP4). Note that this pair of traces is used to demonstrate the operation of a double amplifier.

In Fig. 4-74, the output frequency remains constant at 537 Hz even though the driving voltage from the audio generator is on 268.5 Hz. This is similar to the operation of a radio frequency doubler stage and shows that the tuned circuit will produce good sine wave output even though only pulsed every other cycle.

Figure 4-75 shows the output with the input frequency at 537/3 or 175.6 Hz. The decay in amplitude between driving pulses in this 3 × multiplier is evident in the scope traces.

The parallel resonant circuit (88 mH toroid inductor and 1 μF capacitor) is the heart of the unit. The resonant frequency of the tuned circuit is 537 Hz (CPS). The 88 mH toroid inductors are inexpensive and available from electronics surplus supply houses. These very high Q inductors were originally used as loading coils in telephone lines. See Figs. 4-76 through 4-79.

RESISTOR CUBE AND ITS USE

The resistor cube covers the basic principles of direct current. It offers experience in circuit tracing, simplifying circuits with equivalent circuits, tracing electron flow from negative to positive, tracing current in a series circuit and a parallel circuit, adding voltage drops in a series circuit, adding voltage drops in a series circuit and a parallel circuit, applying Kirchhoff's and Ohm's laws and approaching an analytical problem from some known or assumed starting point.

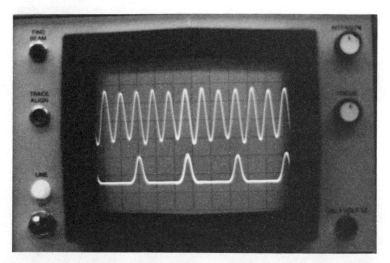

Fig. 4-75. Upper trace: output on 537 Hz. Note the slight decay in amplitude between driving pulses (common and TP2). Lower trace: driving pulses from audio generator on 537/3 Hz (175.6 Hz) (common and TP4). Note that this pair of traces shows the operation of a frequency tripler amplifier. The audio generator is connected between common and TP5.

A cube is formed from 12 resistors, all with the same resistance value. (The resistor value chosen for this project was 1 kilohm.) The problem is to calculate the total resistance between points A and B (see Fig. 4-81).

After trying to solve the problem, number the resistors (Fig. 4-82) and redraw the equivalent circuit into a two dimensional form (Fig. 4-83). Examining the equivalent circuit makes it clear that the total current will first divide evenly into R1, R2 and R3. Using Kirchhoff's laws, assume:

$$I1 = 1\,mA$$
$$I2 = 1\,mA$$
$$I3 = 1\,mA$$

Then, dividing currents, we have:

$$I4\,(I\,1A) = .5\,mA, I\,5\,(I\,1B) = .5\,mA$$
$$I6\,(I\,2A) = .5\,mA, I\,7\,(I\,2B) = .5\,mA$$
$$I8\,(I\,3A) = .5\,mA, I\,9\,(I\,3B) = .5\,mA$$

Recombining the divided currents gives us:

$$I\,10\,(I\,1A + I\,2A) = 1\,mA$$
$$I\,11\,(I\,1B + I\,3A) = 1\,mA$$
$$I\,12\,(I\,2B + I\,3B) = 1\,mA$$

Fig. 4-76. The resonance demonstrator is mounted on two 45V batteries connected in series. Clip-on capacitors in the foreground can be connected across components and change the resonant frequency of the tuned circuit or the firing frequency of the relaxation oscillator.

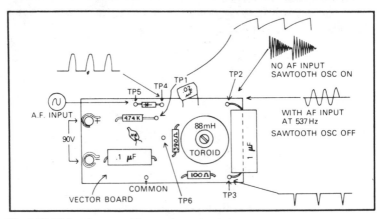

Fig. 4-77. Wiring diagram of the resonant circuit demonstrator.

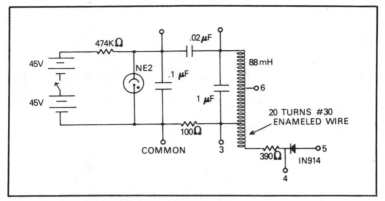

Fig. 4-78. Schematic of the resonant circuit demonstrator.

Thus, for the total current, we have:

$$IT = I1 + I2 + I3$$
$$= I10 + I11 + I12$$
$$= 1\,mA + 1\,mA + 1\,mA$$
$$= 3\,mA$$

Since all resistors have the same values, all currents will divide evenly, and the fact that any two divided currents never recombine with each other is immaterial.

Using Ohm's law, we now find the voltage drop of each resistor:

$E1$	$= I1$	$\cdot R1$	$= 1\,mA \cdot 1\,kilohm = 1V$
$E2$	$= I2$	$\cdot R2$	$= 1\,mA \cdot 1\,kilohm = 1V$
$E3$	$= I3$	$\cdot R3$	$= 1\,mA \cdot 1\,kilohm = 1V$
$E4$	$= I4$	$\cdot R4$	$= 1\,mA \cdot 1\,kilohm = .5V$
$E5$	$= I5$	$\cdot R5$	$= .5\,mA \cdot 1\,kilohm = .5V$
$E6$	$= I6$	$\cdot R6$	$= .5\,mA \cdot 1\,kilohm = .5V$
$E7$	$= I7$	$\cdot R7$	$= .5\,mA \cdot 1\,kilohm = .5V$
$E8$	$= I8$	$\cdot R8$	$= .5\,mA \cdot 1\,kilohm = .5V$

Fig. 4-79. The demonstrator removed from the batteries. Lock washers and knobs fit over the battery terminals. Note the wire leads around holes on the vector board.

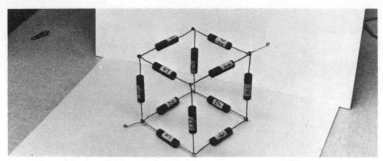

Fig. 4-80. Pieces of numbered masking tape help to identify the individual resistors on the resistor cube.

$$
\begin{aligned}
E9 &= I9 & \cdot R9 &= .5\,mA \cdot 1\,kilohm = .5V \\
E10 &= I10 & \cdot R10 &= 1\,mA \cdot 1\,kilohm = 1V \\
E11 &= I11 & \cdot R11 &= 1\,mA \cdot 1\,kilohm = 1V \\
E12 &= I12 & \cdot R12 &= 1\,mA \cdot 1\,kilohm = 1V
\end{aligned}
$$

Now, add up the voltage drops of any single path between points A and B (for example, E1 + E4 + E10). This gives us 1 V + .5 V + 1 V = 2.5 V. Since all other branches are essentially in parallel, the total voltage across points A and B is 2.5 volts.

The total resistance can now be found:

$$
\begin{aligned}
RT = ET \div IT &= (2.5\,V) \div (3\,mA) \\
&= .833\,kilohm
\end{aligned}
$$

Re-examining the equivalent circuit shows that total resistance can be found by using the resistance formulas. Because all resistors are the same value, we have:

R1, R2, R3, are in parallel
R4, R5, R6, R7, R8, R9 are in parallel
R10, R11, R12 are in parallel

Therefore, the total resistance consists of essentially three parallel branches in series, but this is difficult to prove with ohmmeter measurements. Using the previous analysis, measurements are made on the physical resistor cube to verify our calculations. Any difference between readings and calculations can be explained for the resistor tolerances. Individual resistor voltage measurements can be made by adjusting the total input current to 3 mA.

DIGITAL COUNTER-BINARY DEMONSTRATOR

Here is an ideal way to build a digital counter and get acquainted with integrated circuits. This digital counter-binary

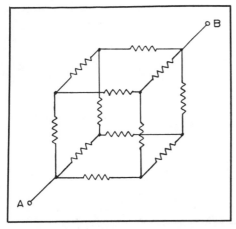

Fig. 4-81. Each resistor is 1KΩ.

demonstrator is especially good because of the many features it offers. In addition to a manual toggle built into the device, it may also be toggled externally by electronic timers or mechanical-electrical transducers. As an added feature, this counter is designed to give a binary representation of the digit displayed in each decade. The block diagram is helpful demonstrating.

The digital counter schematic (Fig. 4-84) reduces this rather complex circuit to a simple black box diagram composed of basic elements: the binary counter, the decoder driver, the readout device and a signal or pulse conditioner. This allows anyone to study the overall circuit operation instead of becoming bogged down in learning how the flip-flops operate or what makes the decoder driver work.

Fig. 4-82. Numbered resistors.

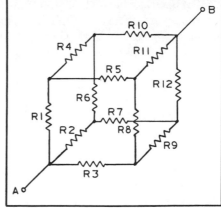

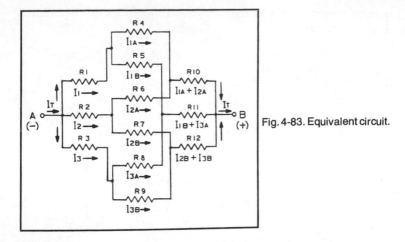

Fig. 4-83. Equivalent circuit.

How It Works

The total circuit operation can also be explained quite easily. The counting pulse first encounters the 7400 quad two-input NAND gate integrated circuit. The purpose of this IC is to provide the flip flops with a suitable counting waveform. If one tried to toggle or switch the flip flops directly mechanically, erroneous output results would be obtained because of the natural arcing of the switch contacts. Thus the 7400 is used as a bounceless switch which gives the counting pulse sharp rise and fall time.

The conditioned pulse is then fed to the 7490, a binary-Coded decimal counter. This IC contains the four flip-flops that are permuted to count only to nine and then reset. This flip-flop circuit is also supplied with an external reset terminal which allows the accumulated count to be reset to zero as the user requires. The 7447 is used as a binary to digital converter and also drives the 707 readout device. The LED (light emitting diode) readout consists of seven segments that glow when a current is passed through it. By directing current to certain LED segments, any digital number can be formed. When two of these counters are connected in such a way that a second series of flip flops and decoder receives a pulse when the binary number nine leaves the first set of flip flops, you will have a two decade digital counter (Fig. 4-85).

Neat Fit

Because the ICs contain all the circuitry except for a few resistors, the entire two decade counter can be designed to fit on one 4″ × 6″ printed circuit board. The binary lights had to be

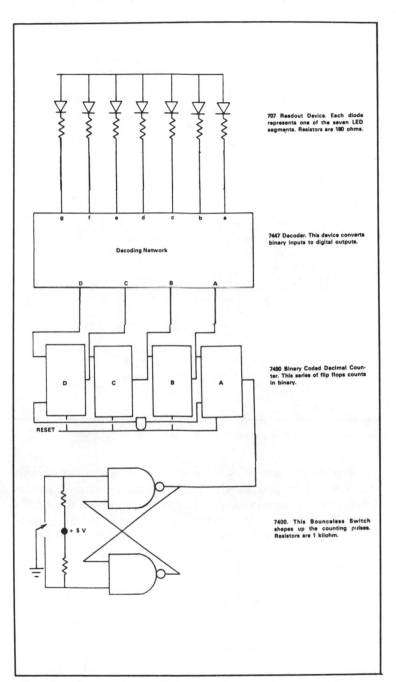

707 Readout Device. Each diode represents one of the seven LED segments. Resistors are 180 ohms.

7447 Decoder. This device converts binary inputs to digital outputs.

7490 Binary Coded Decimal Counter. This series of flip flops counts in binary.

7400. This Bounceless Switch shapes up the counting pulses. Resistors are 1 kilohm.

Fig. 4-84. One-decade digital counter-binary demonstrator.

143

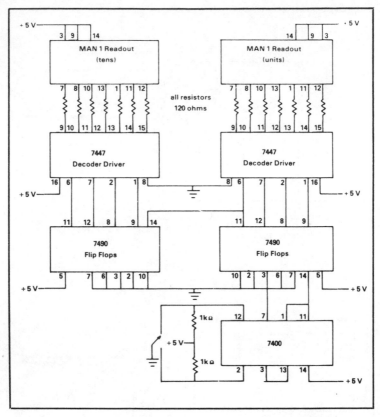

Fig. 4-85. Complete two-decade digital counter-binary demonstrator.

mounted on a separate 4″ × 1″ board. The complete circuit fits into a 5″ × 9″ × 1-½″ Bud chassis. A piece of translucent red plastic is used in the readout window to cut down the brightness of the LED and also to add color to the counter. An external 5V power supply is necessary to operate this counter, because the LEDs will draw a maximum of a half an ampere, which is too much for any battery small enough to fit in the chassis.

OHM'S LAW DEMONSTRATOR

Beginning hobbyists can visualize the Ohm's Law relationship of voltage, current and resistance in a circuit with this light board demonstrator. Jumper leads with alligator clips make the connections required by the circuit. The intensity of the lamps will vary according to the type of circuit constructed and will illustrate for the student the changing conditions in series, parallel and series-parallel circuits.

The circuit is first drawn in schematic form, checked and then wired on the light board. After inspection, the circuit is put in operation and the results noted.

How to Build the Board

Cut the tempered hardboard base to size. Prime the board and apply a finish coat of paint. Stencil or silk screen the electronic symbols. Lay out the components (refer to Fig. 4-86). Drill the hardboard and mount the components. Wire and solder the appropriate circuit connections. See Figs. 4-87 and 4-88 and Table 4-9.

BINARY-DECIMAL NUMBER DEMONSTRATOR

Beginning hobbyists often have trouble understanding the various number system used. The shop-built demonstrator described here displays output as both decimal and BCD (binary-coded decimal) and gives you an effective device for teaching both decimal and alphanumeric systems.

The truth table (Fig. 4-91) indicates the appropriate coding for both decimal and BCD outputs. The schematic which displays the interrelationships of the components. (Fig. 4-92) implements the truth table. The circuit operates in the following manner:

A 555 integrated circuit clock or astable multivibrator is used to produce a square wave clock signal adjustable to several

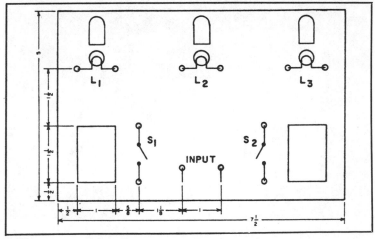

Fig. 4-86. Component layout of the Ohm's Law demonstrator.

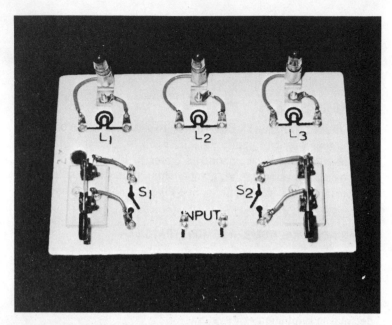

Fig. 4-87. Basic wiring of the Ohm's Law demonstrator.

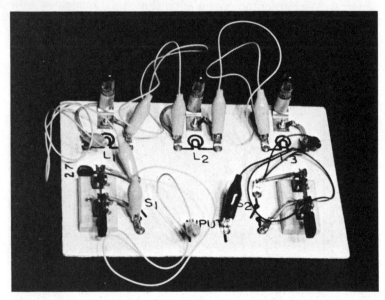

Fig. 4-88. Any number of circuits can be designed to help visualize the operation of Ohm's Law.

Table 4-9. Parts List for the Ohm's Law Demonstrator.

Qty.	Component	Description
1	Tempered hardboard	⅛ × 5 × 7-½
2	Single pole, single throw knife switch	¼ × 1 × 1-½
3	Pilot light sockets	Bayonet base
3	#47 light bulbs	Bayonet twist type
19	Solder lugs	Copper or tin
15	Metal screws	4-40 × ¼
15	Hex nuts	4-40
1 ft.	Hookup wire	Insulated
1 set	Jumper leads with alligator clips	Multi-colored

seconds. This clock will be used to automatically sequence the display circuits from number to number. See Table 4-10.

A 7490 decade counter integrated circuit is used to count the input clock pulses and index its output sequentially. The 7490 will count to 9 and then begin again at zero as long as pulses are delivered to it. The output of the 7490 is a four bit code, ABCD, which represents the binary equivalent of numbers between 0 and

Fig. 4-89. Completed unit of the binary-decimal number demonstrator.

Fig. 4-90. Rear view of the completed binary-decimal number demonstrator.

9. The 7490 can be reset to zero at any time by using a switch on pins 2 and 3. This switching must be bounceless and therefore a 7400 quad NAND integrated circuit is included as a debouncing circuit.

The BCD output of the 7490 cannot drive the 40 mA lamps which were used in this circuit. LED's could be used directly with a 200 Ω limiting resistor in series with each, but the visibility in a classroom might be poor. The use of 2N2907 PNP transistors provides enough current to activate the display lamps. The BCD output from the 7490 must, however, be inverted in order to properly drive the base of the transistors. The use of an additional 7400 quad NAND gate as a quad inverter provides the correct polarity signal to the transistors. The lamps are mounted on a printed wiring board and are appropriately labeled ABCD. In order to provide a manual display of the BCD output, switches are provided to activate each transistor. The positive connection to pin 14 of the 7400 is removed via the automatic-manual switch so that the automatic display signal may be deactivated.

The seven-segment display is developed by applying the BCD output from the 7490 to a decoder-driver, the 7447 integrated circuit. The 7447 will provide a seven-segmented display in response to a BCD input. Since the display required more current than the 7447 is designed to supply, driver transistors were used to provide an adequate current. Each segment of the display consists of three light emitting diodes. The display board is a plug-in printed wiring board. The LEDs are covered with a red translucent plastic sheet to improve display visibility. Lamps such as were used in the BCD display were not used here since the combined current requirement would have been beyond the capability of the power supply.

A regulated power supply using a LM309K 3-point regulator provides a 5V, 1A supply which is quite adequate for the entire circuit. Except for display boards and power supply, the circuit is contained on one large printed wiring board. It may be laid out using any conventional means, etched and drilled and all parts mounted. The use of sockets for the integrated circuits is desirable

BINARY CODES DECIMAL (BCD)				7 SEGMENT BINARY CODE							DECIMAL
A	B	C	D	a	b	c	d	e	f	g	
0	0	0	0	1	1	1	1	1	1	0	0
0	0	0	1	0	0	0	0	1	1	0	1
0	0	1	0	1	1	0	1	1	0	1	2
0	0	1	1	1	1	1	1	0	0	1	3
0	1	0	0	0	1	1	0	0	1	1	4
0	1	0	1	1	0	1	1	0	1	1	5
0	1	1	0	1	0	1	1	1	1	1	6
0	1	1	1	1	1	1	0	0	0	0	7
1	0	0	0	1	1	1	1	1	1	1	8
1	0	0	1	1	1	1	1	0	1	1	9

1 = HIGH = ON

0 = LOW = OFF

7-SEGMENT
DISPLAY CONFIGURATION

```
    a
 f |   | b
    g
 e |   | c
    d
```

Fig. 4-91. Binary-7-segment display truth table.

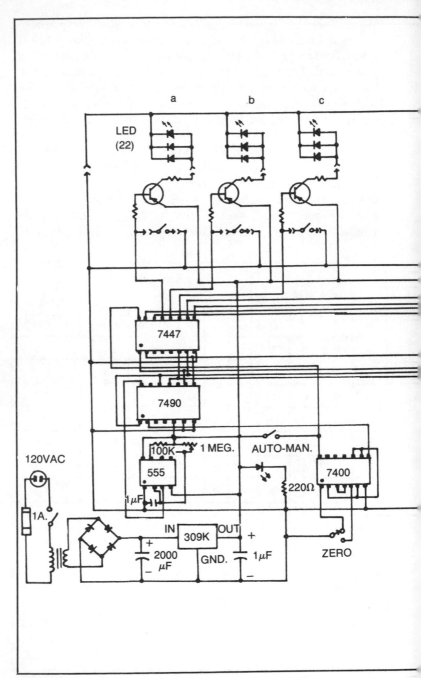

Fig. 4-92. Schematic of the binary-decimal number demonstrator.

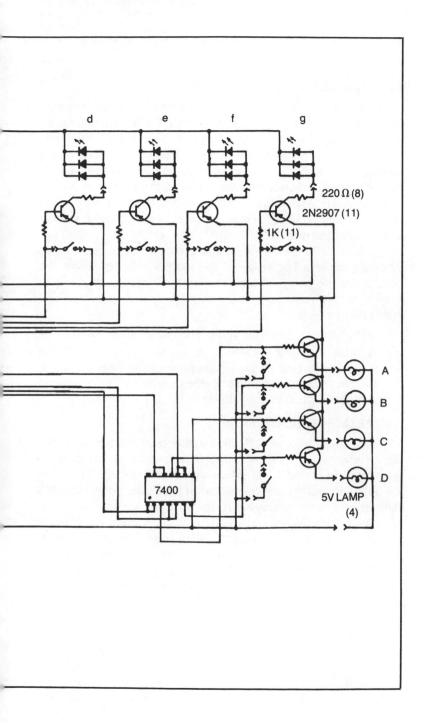

d   e   f   g

220 Ω (8)

2N2907 (11)

1K (11)

A

B

C

D

7400

5V LAMP
(4)

since the risk of damage during soldering is great. See Figs. 4-93 and 4-94.

ANOTHER GENERATOR DEMONSTRATOR

The fact is you can make ac generation understandable using a standard bicycle light generator consisting of a permanent magnet and wire coil. The device is inexpensive, practical and easily transported. Begin with a small bicycle frame salvaged from a local bike shop, and mount it on a plywood base. As you rotate the pedals at a fairly high rpm, the no-load output of the generator will be about 80 Vpp or 28.3 Vrms. The frequency at this rpm is about 400 Hz. When this output is connected to an oscilloscope, a smooth sine wave will be displayed.

Construct the meters on the board from the movements of two damaged meters. The range of the meters will depend on the generator used, the wheel diameter, and the load connected.

Applications

With *S1* open (see Fig. 4-95), you can measure the no-load output of the generator. When *S1* is closed, you can obsrve the output voltage drop as current is supplied by the generator. You will also see the current (lamp brightness) increase or decrease as the pedals are turned faster or slower.

A SOLAR ENERGY DEMONSTRATOR

Demonstrate solar collection and energy storage with this device which is easily constructed of inexpensive items from the hardware store, and a half dozen aluminum beverage cans. The system combines the principles of a flat plate solar collector with a built-in salt or saline storage unit. Here's how it works:

A flat plate solar collector consists of a reservoir to hold either a liquid or air medium, a plate to absorb solar radiation, input and output connections, and a closed air space to prevent loss of radiation from the absorbing plate. Normally, flat plate collectors are connected to an external unit to store the heated liquid or air. In this system, however, heat storage is built into the collector. The storage cells are used aluminum beverage cans filled with a super-saturated (50-50) salt solution. See Figs. 4-98 and 4-99.

Liquid circulating through this closed system is heated by the use of an absorber plate made of plexiglass. The under surface of the plate is painted silver and the top surface is covered with black sandpaper—wet or dry and coarse grit, which adds to the absorbing

Table 4-10. Parts List for the Binary-Decimal Number Demonstrator.

Qty.	Item	Qty.	Item
7	220 Ω ½ W resistor	22	LED, red
11	1500 Ω ½ W resistor (1000 Ω will do)	1	printed wiring board mount potentiometer, 1 M Ω
1	10 K Ω ½ W resistor	1	bridge rectifier, 12 V/1 A
1	1 μF/10 V electrolytic capacitor	1	555 integrated circuit
		3	dip sockets, 14 pin
2	7400 quad Nand IC	1	dip socket, 16 pin
1	7490 decade counter IC	1	transformer, 6 V, 2 A
1	7447 7-segment decoder/driver IC	1	transformer, 6 V, 2 A
		1	LM309K, 5 V regulator
11	2N2907 PNP switching transistor	1	2000 μF/10 V electrolytic capacitor
13	SPST toggle switch	1	.1 μF/10 V ceramic capacitor
1	fuse holder and 1 A fuse	-	copper clad board— approximately 12″ × 12″ wood, plastic, line cord, nuts, bolts, etc.
1	SPDT toggle switch (zero) sockets		
4	lamps, miniature 5 V, 40 MA, bi pin type		

capacity of the plate and increases its surface area. The reservoir is covered with clear plexiglass, forming an insulating airspace above the absorbing plate which reduces heat loss. The system can be enlarged by connecting several units in series with an external storage tank with larger capacity.

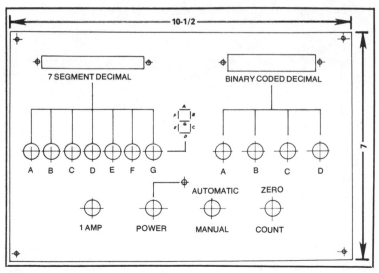

Fig. 4-93. Front panel layout.

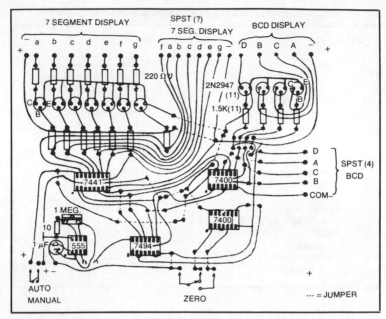

Fig. 4-94. Printed circuit board at half scale (component side dark lines, foil side light lines).

Liquid entering the system strikes a storage cell directly, creating a wave effect in the reservoir and increasing the circulation of heated liquid below the surface of the absorber plate.

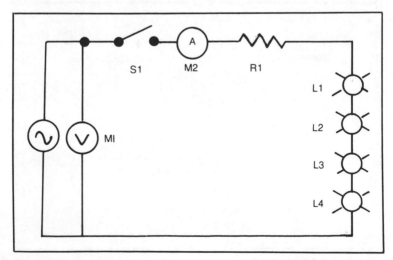

Fig. 4-95. Schematic of the generator demonstrator display board.

Fig. 4-96. A wheel and bicycle light generator is connected to the display board of the generator demonstrator.

Storage cells are mounted through holes cut in the cover and absorbing plate, and rest on the bottom of the reservoir. They are painted flat black above the surface of the absorbing plate, and

Fig. 4-97. The display board measures output and shows voltage drop and circuit current (lamp brightness).

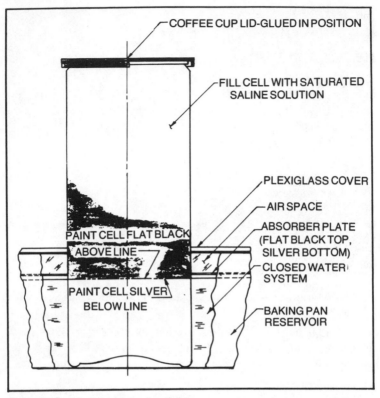

COFFEE CUP LID-GLUED IN POSITION

FILL CELL WITH SATURATED SALINE SOLUTION

PLEXIGLASS COVER

AIR SPACE

ABSORBER PLATE (FLAT BLACK TOP, SILVER BOTTOM)

CLOSED WATER SYSTEM

BAKING PAN RESERVOIR

PAINT CELL FLAT BLACK ABOVE LINE

PAINT CELL SILVER BELOW LINE

Fig. 4-98. Detail of the solar cell and plate.

silver below it. The black surface provides additional absorption capacity, the silver surface radiates additional heat into the circulating liquid. The super-saturated salt solution in the storage cells adds a heat storage capacity to the cell and prevents freezing of the solution.

To further prevent heat loss, the entire system is mounted in rigid fiberglass insulation. The surface of the insulating material that is exposed to solar radiation is painted flat black to further reduce heat loss.

By use of a thermo-syphon or a small pump, liquid heated in the solar unit can be recirculated directly through the input side or stored externally before recirculation. Specialized piping and storage tanks found in conventional saline solar systems are not needed with this self-contained unit.

The problem of salt corrosion is overcome by the "throw-away" nature of the aluminum solar cells, which are easily removed and replaced at regular intervals to prevent the salt solution from leaking into the circulating liquid.

ELECTRONICS TRAINERS FROM CASTOFFS

Surplus, castoff and junk-box parts can be converted into an inexpensive three-unit trainer for electro-mechanical switching and controls that can be used for both simple and complex circuits. Here's how:

A switch board, a relay board and a motor board comprise the trainer. To build them, you will need some surplus switches, junk-box relays, a handful of terminal or barrier strips, a couple of castoff small AC and/or DC motors, some scrap plywood and a little imagination. Use whatever switches, relays, and motors you have on hand or can find most easily. Components on the boards can be rearranged to fit your needs.

Construction

To make the switch trainer, sand and varnish a piece of ¾" plywood and mount selected switches in vertical rows, using wood screws (see Fig. 4-100). Provide plenty of terminal strips and a line cord. The lamp socket can be bought at a hardware store, if necessary. Label each switch with a number or letter for easy identification, and be sure to place a warning label on the board in a prominent position, since 100 VAC will be exposed.

For the relay board, mount three or more different types of relays: perhaps an SPDT, SPST and one with multiple contacts, some of which are normally open, and some normally closed. Make the board's dimensions to suit your needs. I used 12 V coils, but use whatever you have. For convenience, all coils should be the same voltage (see Fig. 4-101).

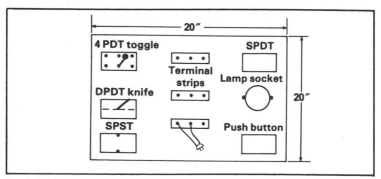

Fig. 4-100. Switch board electronics trainer layout.

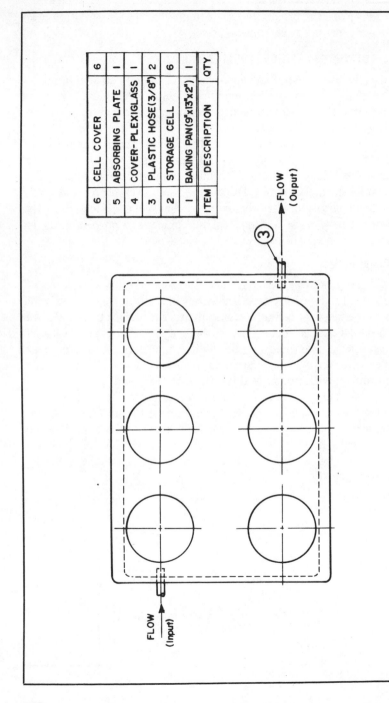

6	CELL COVER	6
5	ABSORBING PLATE	1
4	COVER-PLEXIGLASS	1
3	PLASTIC HOSE(3/8")	2
2	STORAGE CELL	6
1	BAKING PAN(9"x13"x2")	1
ITEM	DESCRIPTION	QTY

FLOW (Ouput)

FLOW (Input)

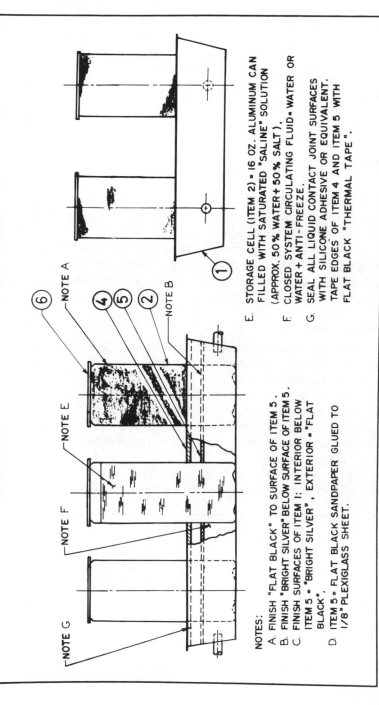

NOTES:
A. FINISH "FLAT BLACK" TO SURFACE OF ITEM 5.
B. FINISH "BRIGHT SILVER" BELOW SURFACE OF ITEM 5.
C. FINISH SURFACES OF ITEM I: INTERIOR BELOW ITEM 5 = "BRIGHT SILVER", EXTERIOR = "FLAT BLACK".
D. ITEM 5 = FLAT BLACK SANDPAPER GLUED TO 1/8" PLEXIGLASS SHEET.
E. STORAGE CELL (ITEM 2) = 16 OZ. ALUMINUM CAN FILLED WITH SATURATED "SALINE" SOLUTION (APPROX. 50% WATER + 50 % SALT),
F. CLOSED SYSTEM CIRCULATING FLUID = WATER OR WATER + ANTI-FREEZE.
G. SEAL ALL LIQUID CONTACT JOINT SURFACES WITH SILICONE ADHESIVE OR EQUIVALENT. TAPE EDGES OF ITEM 4 AND ITEM 5 WITH FLAT BLACK "THERMAL TAPE".

Fig. 4-99. Dimensions of the solar energy demonstrator.

159

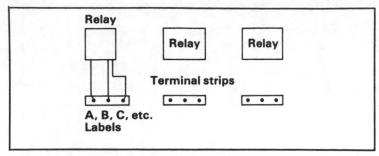

Fig. 4-101. Relay board electronics trainer layout.

Bring each relay contact and coil terminal out to separate lugs on the terminal strips, and label them. Identify on a master sheet which contacts go with which terminal.

For the motor board, use a castoff synchronous 100 VAC fan motor, an old AC-DC motor from a junked can opener, and a small surplus 24 VAC pm motor. The board is assembled in the same way as the previous units (ss Fig. 4-102).

The Possibilities

Now that you have them built, how do you use them? They can be used singly, or all three can be interconnected with clip leads for a simple motor control unit. Job sheets can be written for each experiment for individualized or self-paced instruction. Here are some sample applications to get you started:

Try to identify the types of switches and draw their schematic symbols. Next wire the lamp, using two switches, either of which will turn the lamp on or off.

The job sheet could read: "Given two SPDT switches, a 100 V lamp and socket, power source, and appropriate hookup, deter-

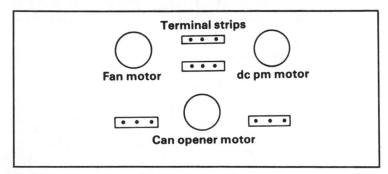

Fig. 4-102. Motor board electronics trainer layout.

mine a way to connect the light so that either of the switches will turn it on or off. Figure 4-103 shows one way to wire the board. Figure 4-104 shows one circuit design for the relay board.

MASS-PRODUCING AN ELECTRONICS PROJECT

The application of a different radio club technique converted what would have been a routine electronics project into a new and interesting experience for a basic electronics class. The purpose of this project was to provide an activity utilizing the scientific knowledge and operational processes previously covered in the course. Equally important was the implementation of club student leadership and active participation by all.

The project was a small vhf receiver designed to operate from a 12-V car battery. The mass production of the receiver was discussed early in the school year with the necessary supplies being ordered at that time. A field trip to the Motorola Company was also undertaken at that time to supply the students with essential background for planning their mass-production assembly line. In addition, several club students from the class were selected to visit Western Electric Company, Chicago, to observe the laboratories and electronic switching system.

Production planning and control as it related to our product was to be one of a continuous manufacture of a single item. All of the work necessary to manufacture this project was to be identified. Project selection and the building of a prototype must be completed well in advance by the teacher. This is necessary to determine the feasibility of the product and to prevent costly errors.

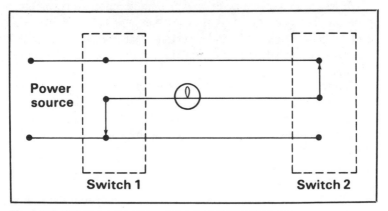

Fig. 4-103. Wiring circuit on the switch board electronics trainer.

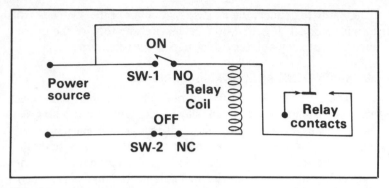

Fig. 4-104. Circuit design for the relay board electronics trainer.

In order to maintain a reasonable cost figure, the project selected could not run over six dollars per student. This meant that the instructor would have to control costs and assume the position similar to the chairman of the board of directors. From this point on, the development of production plans, controlling activities, and the preparation of reports were assumed by the students. Devising, organizing, and managing the assembly line was entrusted to a nucleus of "executives" elected by the class at the beginning of the project.

All fabrication of the chassis and case was completed in the metal shop. These metal parts were then brought to where an assembly line was set up along two long work tables. The "personnel manager," with the assistance of the "methods engineer," wrote out a set of instructions for each worker along the line describing the operations he was to perform.

It was necessary for additional grommets, metal screws, etc. to be ordered once the design of the set was established. The "stock-supply engineer" made certain that these parts were ordered and that all parts were at their proper places along the assembly line.

To try out the assembly line and determine if the radio set design and manufacture would be satisfactory, a prototype was built according to the mass-production plans. Most of the production problems were encountered during the construction of this prototype. These problems were solved by modifying construction plans and tightening the organization and discipline.

As the assembly line then went into production, it made a working example of student ingenuity and cooperation. Parts and working tools stretched along the worktables where the product gradually progressed from a simple metal chassis at the beginning to a completed, compact *vhf receiver* at the end. One student was attaching tube sockets and tuners while others were wiring, soldering, winding coils, and testing.

Adding to the experience was the fact that the assembly line was set up and organized by the club students, with their instructor acting in only an advisory capacity after introducing the concept. Setting up an assembly line proved to be a challenging and interesting problem.

LAUNCHING AN AEROSPACE PROGRAM

This project is particularly effective since it uses many of our normal metal and electric working skills and techniques. It is also an excellent project on which further electronic components can be built incorporating advanced electronic concepts.

Layout and Design

The switch box (Fig. 4-105) layout and design has been made for either a 12 V lantern battery or a 12 V gel cel. The gel cel will need some balsa shimming. It is a little smaller but can be recharged through the voltmeter jacks.

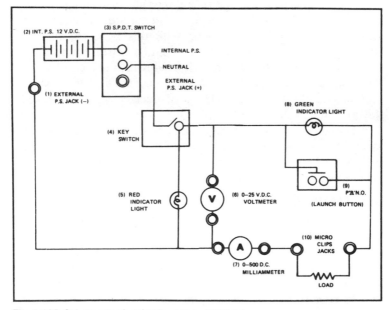

Fig. 4-105. Schematic of rocket launcher switch box.

The electronic components, jacks, switches and lamps should be purchased before the holes are drilled in the 18 ga. aluminum chassis. Drill all holes before bending in a box and pan brake.

Project Function

The design provides for several learning situations. The internal power supply (gel cel) can be recharged without removing it from the box. The charger can be plugged into the voltmeter jacks and slow charged at 300 mA.

Should the battery fail in the field, the external power jacks can be attached to an automobile cigar lighter. The voltmeter will monitor the battery and show an applicable and readable deflection upon launching.

The milliammeter will likewise show change when the launch button is depressed. The reason for these deflections can be important lessons in understanding the electronics of rocket engine ignition. The safety features include the key switch and a mechanical cover over the launch button.

Operation of the Switch Box

When the rocket launcher switch box is ready to use, turn on the key switch and the red indicator light will show that power (internal or external by the S.P.D.T. switch) is ready. Then connect the micro clips to the rocket ignition wire and plug them into the appropriate jacks. If there is continuity to the engine, the green indicator light will go on. With this light on, there is insufficient amperage to heat the ignition wire. Remove the mechanical cover of the launch button following the clearing of the area around the rocket. Make the final countdown and press the buttom to ignite the engine. While the launch button is depressed, the green light will go out.

Both meters should be attached and demonstrated with the switch box (without engines) in the classroom. They can be left in the shop when the unit is used in the field. A simple shunt will substitute for the missing milliammeter.

MAKE SOLAR ENERGY SHINE TWO WAYS

Solar energy is coming into its own as an alternative energy source with applications for home, industry and schools, and it has a lot to offer as subject matter in radio club classes. Solar energy research gets its big boost when a club student developes a sun-tracking solar energy parabolic collector system.

Energy from the sun falls on the earth's surface as both direct and diffuse radiation; direct radiation permits economical collection, concentration, absorption, conversion, and utilization in the form of heat energy. In theory, the intensity of the sun's radiant energy on the earth's surface is 1353 W/m², or enough to power a 20 cu. ft. deep freezer, assuming no losses. In actual practice, however, a 3 by 8 ft. solar collector panel can absorb only up to 85 percent of the sun's energy. In units of heat, this converts to about 5000 BTU's on a bright day; in temperature gradients, such a panel would produce over 185 °F at solar noon in a very few minutes.

Using the Sun

Radiation from the sun can be put to practical use in three ways: (1) heliochemical, (2) helioelectrical and (3) heliothermal. The heliochemical process—photosynthesis—enables sun rays of certain frequencies to cause organic substances such as carbon to turn into carbon dioxide and water, which unite with soil nutrients and produce plant life and oxygen. Man has not yet been able to make much use of nature's heliochemical process on his own.

The helioelectrical process, on the other hand, is devised entirely by man and is useable; an example is the silicon solar cell which converts solar energy directly into electricity. Unfortunately, solar cell production is expensive, so this approach is not widely used except in specialized high-cost applications such as space satellites, remote buoys, etc. Storing the produced electricity is also a major drawback of the helioelectrical process.

Currently, the most practical way to harness the sun's energy is with heliothermal devices—low-cost units which convert solar radiation directly into heat energy and which have been used for centuries to generate temperatures as high as 6000 °F. They have been employed for solar distillation, cooking, crop drying, water and air heater systems and sundials. Heliothermal systems hold the most promise for industry. Two methods will be discussed for fabricating a heliothermal device: the flat-surface method and the parabolic method.

Flat-surface Approach

The principle of the solar energy flat-surface system is as simple as its construction (see Fig. 4-106), and almost any common materials can be used in making it; all that is needed is a flat-plate aluminum or copper solar collector panel connected to a thermal energy storage system. The panel is made up of a number of copper or aluminum tubes mounted on flat metal plates in a windowed "dead air" space. The surface of the flat metal is specially treated to make it more receptive to the infrared rays of the sun.

165

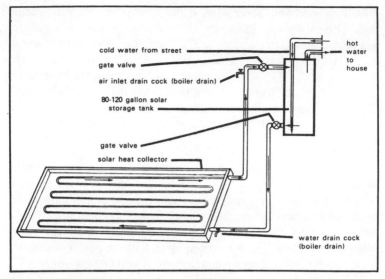

Fig. 4-106. Thermosyphon solar water heating system. (Drawing courtesy Environmental Information Center, Florida Conservation Foundation, Inc.)

The tubing is attached to the surface of the flat plate so that thermosyphon circulation is developed, causing water to move toward the area of lowest temperature. The heated circulating water is generally stored in an insulated tank for later use. The entire system functions much like a greenhouse and offers the advantages of being reliable, durable, and low in maintenance cost.

Making a heliothermal device is relatively easy; the low-cost, fully demonstrable device is fabricated from readily obtainable materials. The heat transfer section is a concentric length of ⅜ in. id plastic tubing mounted in a frame with an acrylic or glass cover. The flat plate forming the bottom of the frame is painted black or very dark green to hold the heat. The glass or acrylic cover allows the radiant energy to pass through but permits very little of the heat to escape from the dead air space between the cover plate and the heat absorber plate.

One end of the plastic tube carrying the heated water is connected to a small circulating pump operated by a 1.5V electric motor powered by a solar cell on sunny days, or by a 1.5V flashlight battery on cloudy days. The heated water is then stored in an insulated and vented hot water tank for later use. The circulatory system is completed by the clear plastic tube which links the hot water tank back to the solar collector panel. If the unit is placed in the sun, temperatures as high as 150 °F can be attained by this simple device. On cloudy days or in a classroom setting, a heat lamp can be used to simualte the sun's infrared rays.

The Parabolic Method

The second approach for demonstrating the high-temperature heliothermal characteristics of the sun's rays involves the use of a parabolic reflector. By definition, a parabolic surface is a plane curve formed by the intersection of a right circular cone with a plane parallel to a generator of the cone, resulting in a curve similar to the one shown in Fig. 4-107. This surface design allows the sun's rays to be collected and focused on a very small area called the focal point.

A 6 ft. parabolic collector has an area of 27,000 cm² and receives 38,000 calories/min. with a normal solar radiation of 1.4 langley. At 65 percent efficiency, it can deliver up to 25,000 calories/min. (or 1.5 million calories/hr.) to the focal point, or approximately the equivalent of 1.8 kW of delivered heat power. This represents a sizable heat source by almost any standard.

A CLUB COMPUTER SYSTEM

This system is designed to give the computer club a broad exposure to the state-of-the-art techniques and new technologies found in industry today.

All club students can be on-line at once, and each student has a minimum delay "hands-on" contact with the computer. Three languages are available, depending on the problem.

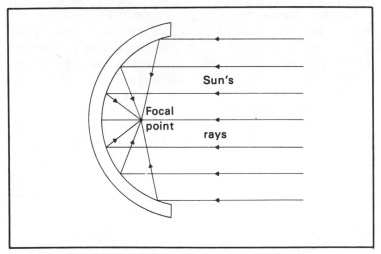

Fig. 4-107. Geometry of parabolic curve.

ADD ETCHED CIRCUITS TO YOUR RADIO CLUB

Etched circuitry is an electronic development that is of fairly recent origin. Some of the earliest forms of this technique were found in the proximity fuse of World War II. Certain advantages of this technique, to be discussed later, have led to its increased use in the manufacture of electronic devices such as test equipment, radios, and television receivers. Since etched circuits are rapidly replacing conventional hand wiring in the manufacture of electornic equipment, it seems that this technique should be demonstrated and used in radio clubs.

What Are Etched Circuits?

The etched circuit eliminates the need for punched and formed metal chassis, numerous tie-points and terminal boards which serve to hold electronic parts, and numerous hand-placed and soldered wires. The etched circuit consists of an insulating board which has bonded to it a thin metal foil circuit pattern, usually copper. This metal-foil provides for the connections to the various electronic parts that comprise the completed circuit. The leads for the various parts are inserted through holes drilled or punched in the insulating board and soldered to the copper-foil circuit connections.

The insulating board is made by saturating paper or cloth sheets with a thermosetting resin such as phenolic resin. Stacks of the impregnated sheets are then placed in a heated press with a sheet of copper foil and subjected to a pressure of approximately one ton per square inch at a temperature of 350°F. The pressure and heat form the sandwich into a hard board with a copper-foil layer bonded to it. Copper-faced laminate is available commercially in thicknesses from 1/16" to 3/32" with copper foil bonded to either one or both surfaces. The copper foil used is electrolytically deposited and is usually either .001" or .003" thick. The copper foil is then transferred from the electroplating bath to a large, slowly rotating lead drum from which it is stripped. Rolls of copper, approximately 2,000' long and 60" wide and of uniform thickness, are produced by this method.

Several different materials, each having particular advantages, are used as the laminate to which copper foil is bonded. Among these materials are XXXP phenolic, melamine-fiberglass, epoxy-fiberglass, silicone-fiberglass, and teflon-fiberglass. All of the laminate materials, with the exception of XXXP phenolic, have excellent high-temperature characteristics which means that temperatures in the range of 300° to 400°F, such as encountered in soldering, will not cause the copper foil to lift from the insulating board. XXXP phenolic is least expensive and has rather poor temperature characteristics but, with proper soldering technique, is entirely satisfactory for school-shop use.

How Are Etched Circuits Made?

Etched circuits are produced by placing a resist on the copper-foil side of the laminate in the form of a circuit configuration. The board is then placed in an etching solution which removes all of the copper not protected by the resist. After etching, the board is punched or drilled to receive the leads of the circuit components. After the leads of the various components have been inserted in the holes, they are soldered either by hand or by dip soldering.

In producing etched-circuit boards commercially, the engineer works from a schematic diagram and prepares a drawing of the conductor pattern desired for a particular piece of electronic equipment. This drawing is inked and photographed to produce a negative of the circuit. The copper-laminate is coated with a light-sensitive resist material called cold-top enamel. The negative is placed on top of the treated copper and exposed to an arc light. After exposure, the board is etched in a ferri chloride solution. Areas of the copper, those that were exposed to the light, are etched away, while the areas of copper protected by the opaque lines on the negative, remain. Other methods of producing etched-circuit boards are also used, such as offset printing and slik screen. These are shown elsewhere in this book.

What Are the Advantages of Etched Circuits?

Etched circuits eliminate wiring errors. The circuit can be reproduced indefinitely without error. Tedious hand-wiring operations are eliminated, thus reducing the time spent in numerous inspection operations. Etched circuits lend themselves to automated assembly techniques. Circuit variations due to wiring differences are eliminated.

CLUB OPERATED SATELLITE TRACKING STATION

Man's curiosity about the far reaches of outer space has led him to explore it with man-made satellites of every type and description. Some are designed to take pictures of the earth and its cloud cover, others to measure radiation, magnetic field, particle bombardment and micrometeoroids—to name just a few.

Space information is transmitted back to earth stations via radio waves on a number of different frequencies ranging from 1.5 to 8000 MHz where it is picked up, recorded, and analyzed.

Many of these satellites transmit their information on frequencies within the range of amateur shortwave receivers, and, with an appropriate antenna, the satellites can be readily tracked as they pass overhead. Although such a station may not be able to decode or make an analysis of the particular information sent, there are a number of interesting facts it can provide for observation and study. The operator may be able to indentify the satellite by its direction, orbital period, and frequency. Utilizing the Doppler Effect, he can determine its height and speed and, occasionally, determine ionization effects on radio propagation.

Perhaps the most gratifying instructional result with our school's tracking station has been the great interest and enthusiasm shown by club members and visitors who have observed our electronic apparatus operations in conjunction with the National Space Program.

Station Requirements and Specifications

The construction of a club-operated satellite receiving and tracking station is relatively uncomplicated. A basic Yagi type of antenna tuned to the center of the 136 to 137 MHz band, a TV antenna rotor, and a radio receiver capable of tuning in on the 136 to 137 MHz band serve as the core of the project.

After checking through the numerous references, the following general specifications and components for the station were selected and utilized in the project.

Antenna
    10 decibel gain
    50 ohm impedance
    TV type rotors for positioning control

Converter and Pre-Amplifier
    Input frequency range 136 to 137 MHz
    Output frequency 20 to 21 MHz
    Output impedance 50 ohms
    Gain 20 decibels
    Noise figure 4.5 decibels maximum

Radio Receivers
    Frequency range 20 to 21 MHz

Input sensitivity 8 micro volts or better

Predetection bandwidth 30 MHz

Audio output impedance 500 ohms

Two receivers required, one designed for frequency modulated signals and the other for amplitude modulated or CW type signals.

Other Miscellaneous Equipment

Regular school-shop equipment such as frequency meters, oscilloscopes, tape recorders to be used to carry out the various tests.

In construction of the antenna, many variations of the system may be made; however, the following requirements should be met: a 10dB gain over a simple dipole and some degree of pointing ability.

The antenna shown in the photo is a Winegard Model 2K1006, Channel 6 with each element cut down in size by the ratio of 1.6 to 1. (This is the ratio of the Channel 6 frequency to 136.5 MHz, which is the center of the satellite band.) For example, one of the elements of the TV antenna is 63.5 inches long:

$$\frac{1.6}{1} = \frac{63.5}{X}$$

=39.68 inches the length of this element for 136.5 MHz. Spacing between elements was considered to be adequate except between the reflector and the driven element, which was also shortened by the same ratio.

An 18" length of 1" iron pipe was welded across the top of a 10' length of pipe to form a tee shape, on which a rotor was mounted to carry the horizontal boom for the elevation control. This boom, in turn, supported the two vertically polarized antennas, which were one wave length or 81.2" apart.

The vertical mast passes down through the two lower rotors to a thrust bearing. (The two rotors were used to give greater turning power, but it was discovered later that one was ample for the job.)

The little box mounted just below the top rotor is the junction box for the various cables that make up the impedance matching network. A coaxial balun is used to match the RG-8, 52-ohm transmission line to the phasing network between the two antennas (Fig. 4-108).

The entire antenna system is mounted on a 2" pipe protuding up through the roof. The length of the mast was just long enough so that, when the antenna was pointing straight up, the lower end was just a little over a wave length or 80″ above the roof.

Because the transmission line had to be 100′ long, and had to pass through a long length of pipe to reach the radio room, RG-8 was used to reduce losses as much as possible.

For converter and radio receiver an Ameco Model CN144 converter was retuned for the 136-137 MHz band and its output of 20 to 21 MHz fed into either of two BC-683 surplus receivers.

Although the BC-683 is of World War II vintage, it does a fine job for this application. One of these is used without modification for the reception of FM type of signals. The other BC-683 was modified for AM and CW type of signals.

The operation of the station is not difficult. Briefly, the operator turns on the equipment, points the antenna up into the southern sky, and tunes the receiver back and forth over the band. In time, usually not more than an hour, signals will be heard.

These signals are received in various forms: steady CW, beeps, tones, etc. They will last from approximately a minute on up to 15 or 20 minutes and longer.

By operating the rotor controls, signals can be held to the highest level as the satellite moves across the sky. Most satellites will again pass overhead about 90 to 110 minutes later.

ACCEPTING THE SOLAR ENERGY CHALLENGE

Learn about heating water quickly and economically, about temperature, rates of heat loss, radiant energy, conduction and insulation, and convection and interference to flow! Learn to compare and present results, and about measurement in general.

Here are some tips were learned the hard way. The first step should be to enlist the support and enthusiasm of anyone who might become involved in the project.

The second factor to settle early in your planning is selection of a site for the solar collectors. Your choice will require approval of the city building people, depending on local ordinances and who owns the building. We got our city officials to help by suggesting that our findings might prove to be the prototype for solar water heaters for the city buildings.

A must before you start is a separate area for constructing, assembling and painting the collectors and other solar devices. A lockable room is essential to avoid damage or loss of projects.

Questions to Consider

How will you protect the collectors from freezing? (Possible

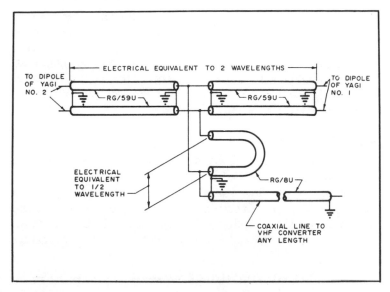

Fig. 4-108. Impedance-matching network for satellite antenna.

answers are antifreeze plus heat exchanges, low-temperature alarm or thermostatically-actuated drain, collectors mounted behind sky-lights inside the building, etc.)

Will you use tubing, roll-bonded sheet, open trickle-plate system, or something else?

What materials will you use for the collector and plumbing? Will you use a tube at the focus of a parabola? If so, how will you position it?

How will you induce flow from collector to storage? (thermosyphon with elevated storage tank, or pumps?) If you use pumps, how many, and how large?

If you have several collectors, how will you provide for controlling flow in series, in parallel, or in a combination of both?

How will you determine the rate of flow in each loop? (Drain and measure the volume discharged in a given time interval, or flowmeter?)

Where will you place thermometers to determine instantaneous temperatures? What will you record about outside weather conditions, and how?

What compass direction should the collectors face for best results during the *school* day? (This is a tough one!) Should all the collectors face the same compass direction?

Will you want to vary the slope of the collectors? How? When?

Design your assemblies with generous clearances to reduce the need for accurate work. For instance, if you use saw slots in wooden frames for the glass, fiberglass, or collector sheets, make the slots wide enough so that the sheets will slide in freely, and deep enough so that they will remain captive despite warping or inaccurate spacing at assembly.

Assembling the Collectors

We bought four roll-bonded, aluminum collectors. Each one was 3′ × 8′, for a total of 96 sq. ft. of collector surface. Our insulation was two layers of 1″ closed-cell polystyrene, chosen because it would not absorb moisture. We mounted these rigid sheets on a panel of ½″ exterior-grade plywood, and between two 2 × 4s. These 2 × 4 side rails were grooved on a table saw to accept

Table 4-11. Parts List of the Rocket Launcher Switch Box.

Part No.	Description
1	External power supply— standard phone jacks and plugs #3265139
2	Internal power supply— 12 V dc battery #32R47772
3	Single pole double throw switch #34R33695
4	Key switch #33R64015
5	Red indicator light #34R52620
6	0-25 dc voltmeter #99R5109 with jacks
7	0-500 dc milliammeter #99R51104 with jacks
8	Green indicator light #35R52646
9	Push button switch, normally off (launch button) #34R34594
10	Microclips #32R37330 with jacks and 20′ line cord
	Hook-up wire
	Cigar lighter plug for external power with 20′ line cord and banana plugs

Note: numbers are for Lafayette stock numbers used as reference.

the aluminum collector sheet just above the insulation, and two sheets of translucent cover material were spaced about ½″ above the collector and about ⅜″ apart. Both layers of insulation and both sheets of cover material were adhered and sealed with non-asphaltic, adhesive caulking. Horizontal battens to hold the cover material flat and to reduce internal convection losses were fastened every foot or so over the collector and between the cover sheets. The ends, top, and bottom were closed with 1″ boards, fastened with screws.

We had some difficulty making a connection to the stubby tube at one end of the collectors, while the other was long enough to protude through a hole in the end board. Screw-type hose clamps collapsed the thin-walled tubing, but we solved this by inserting a short length of ½″ CPVC tubing to keep the tubing round. It was my aim to seal the top and side reasonably well, but to allow condensation and leakage to drain through the uncaulked joints in the lower end.

All wooden parts were sealed, primed and painted. All our rigid plumbing was high-temperature polyvinyl chloride, CPVE, which assembles very easily. Flexible couplings were provided with rubber hose and screw-type clamps. Automotive heater hose is not suitable for line pressures, so you must use good garden hose. I had to glue up and turn down CPVC fittings to match the CPVC tubing to the ¾″ id hose dictated by the collector panels. After checking for leaks, we insulated all the tubing.

Position and Mounting

We faced the collectors due south, and designed and built projecting supports of 3″ steel angle, which pivot the upper ends of the collectors at 45 degrees to a southeast-facing wall. The lower end of our collectors can be swung out from the wall, and locked at the selected angle, thanks to another pair of simple weldments. You may want to examine the best compass direction for your location to get the maximum solar heating, which maximized the heating over the whole day. For our location, an angle of 9° east of south would face the sun at about 11 a.m., and as the sun continued to rise, the heating would continue in a broad peak.

Our circulating pump is 1/25 hp, and does the job. A regular, home-style, circulating pump would probably be too high in capacity, and would require cycled, on-off operation to give the water enough time to heat up in the collectors. The temperature gain must be high enough to offset heat losses in the piping. We brought our four lines together through plastic glazing in a window. These lines, the control valves, and a seven-day outside temperature recorder are all mounted on a ¾" sheet of plywood on the wall. We stored the heated water in a discarded, insulated, 40-gal. water heater, supported on a welded steel stand, as high as possible over a sink.

Chapter 5
Computers

Computers have finally invaded the home, and personal computers can be seen in almost every neighborhood. These projects include demonstrators of computer fundamentals, as well as an actual microcomputer and video displays.

AN N/C DEMONSTRATOR

Look familiar? This demonstrator (Fig. 5-1) for teaching numerical control (N/C) had a former life as an "Etch-a-Sketch" toy. Its two axes help you visualize the X and Y inputs from the tape to the table of a piece of real N/C equipment.

The table will move in the ±X or ±Y direction when either the X or Y knobs are turned clockwise or counterclockwise. To make a 45-degreee traverse of the table, both knobs are turned simultaneously at the same speed.

To make this demonstrator, remove and discard the glass and powder from the original toy, taking care not to distort the wires or pulleys in the four corners of the machine. Simulate the N/C table by gluing two hardboard sheets together, and then cutting a conical hole in the lower sheet to match the shape of the stylus. Use epoxy glue to fasten the table to the stylus. Carefully wax the rails to prevent excess epoxy from adhering to the mechanism.

Simulate a typical N/C machined part from hardboard and mount it on the table with mock fixture stops to represent an actual production situation. The arbor support is a piece of ⅛″ polycarbo-

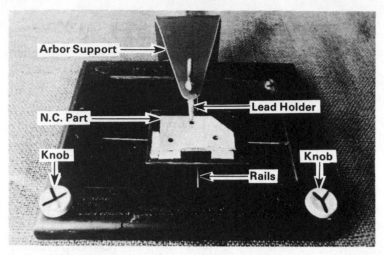

Fig. 5-1. N/C demonstrator.

nate plastic material bent 90 degrees with the help of a heated rod, and pop riveted to the base. The vertical arbor is an old drafting compass lead holder attached to an eyebolt and mounted with wing nuts to permit adjustment. The simulated N/C part can be replaced by a sheet of white paper for demonstrations that will leave a pencil trace line to show the path of the cutter.

ROM UNIT

This unit was designed to demonstrate and explain the more difficult concepts of time division multiplex, scanned displays, and $5-- \times 7--$ or $7-- \times 9--$ LED matrix displays. $5'' \times 7''$ or $7'' \times 9''$. As a side benefit, you will also gain knowledge of the USA standard Code for Information Interchange (ASCII) and binary-coded decimal (BCD) systems.

The heart of the unit is a National Semiconductor MOS read only memory (ROM) MM5241 (see Table 5-1) programed for character generator use. It has the capability of upper case letters, numbers and various punctuation.

Input switches *S1* through *S6* select a particular character by applying either +5V or ground to pins 5 through 10. The scan speed is controlled by the frequency of the external clock, which can be practically any sine or square wave generator. With a suitable input frequency, the scan can be slowed to the point that a student can take voltage measurements at the 1C pins and develop a chart of operation for each of the clock states.

Particulars

The input clock is fed to *1C3 SN 7490* decode counter; the output is a BCD representation of the number of clock cycles. The code is sent to *LO*, *L1*, and *L2* (pins 11,13,14) of *1C1*, to select the dot pattern for one of the five column. *1C4* and *1C5* decode the information to control column drivers *Q9* through *Q13* (Fig. 5-2).

The dot pattern is sent from pins *15* through *22*, *1C1*, to transistors *Q1* through *Q8* (Fig. 5-3). These transistors provide voltage level interfacing and isolation. Drive capacility for the LED matrix is furnished by *1C2*, *DM 8863*.

When the output of *1C3* is *010* at the input of *1C1*, it selects the dot pattern for column three. At the same time, the BCD is fed to *1C4*, which gives a pin output of pin 2-H, pin 4-L, pin 6-H, pin 8-H, pin 10-H, and pin 12-H. These three high states to pins 9, 10, and 11 of *1C5* cause activation of *Q11* to apply supply voltage to column three. *1C2* provides a ground to the respective LEDs for the dot pattern selected by *1C1* (Fig. 5-4).

Some alterations to the circuit can be made, if desired, to reduce the number of components; or, if for an advanced project *1C4* and *1C5* can be replaced by a decimal decoder, such as an SN 7448. The transistors can be replaced by 1C drivers, such as SN 75491 or SN 75492 (Figs. 5-5 and 5-6).

TWO VIDEO DISPLAY UNITS

One of the most important devices in the field of computer technology, and many other areas of visual information transfer is the video display unit. Used not only as interaction devices by computer programers, they also perform unilateral information transfer at such places as airports and race tracks. The following two projects used PC board construction techniques (Figs. 5-7 and 5-8).

| R1-R8: 10 K, .25 W |
| R9-R16: 100 K, .25 W |
| R17-R21: 1 K, .25 W |
| IC1: MM 5241 |
| IC2: DM 8863 |
| IC3: SN 7490 |
| IC4: SN 7404 |
| IC5, IC6: SN7415 |
| Q1-Q13: Any NPN med. signal |
| S1-S6: Single pole, double throw |

Table 5-1. Parts List for the ROM Unit.

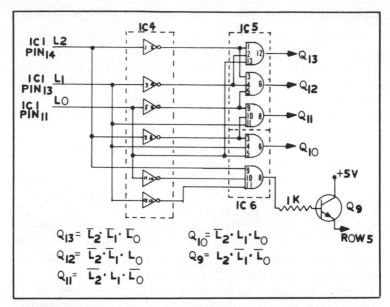

Fig. 5-2. IC4 and IC5 decode the information to control column drivers Q9 through Q13.

The first unit (Fig. 5-9) is a single-character display using the 5×7 dot format. The demonstrator has an external clock input to allow any speed of operation desired. With the use of a pushbutton

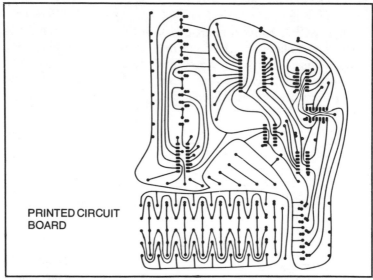

Fig. 5-3. Printed circuit board for the ROM Unit.

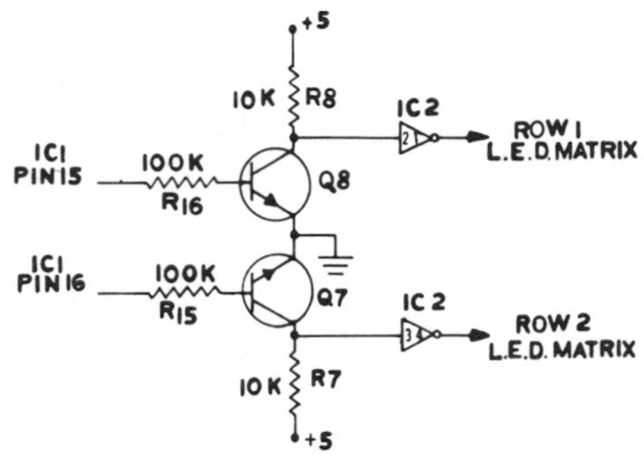

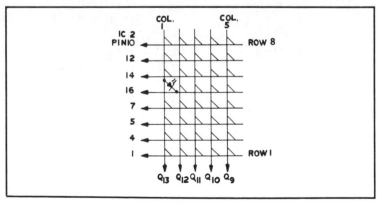

Fig. 5-4. The dot pattern is sent from pins 15 through 22 of IC1 to transistors Q1 through Q8.

and a debounce circuit at the clock input, you can examine the circuit for each individual state of operation and prepare state diagrams or timing charts. Additional topics which can be demonstrated are intensity modulation, the American Standard Code for Information Interchange (ASCII), sweep circuits, logic gate operation, digital counters and parallel-to-serial conversion.

The second unit produces a single line of 16 characters. It provides additional topics over the single character display and is used as a second step to a typical display unit. The additional topics demonstrated are bus systems, tri-state bus drivers, multiplexing, random access memory integrated circuits and the timing of signals required to operate the memory chips. In operation, the student

Fig. 5-5. IC2 provides a ground to the respective LEDs for the dot pattern selected by IC1.

181

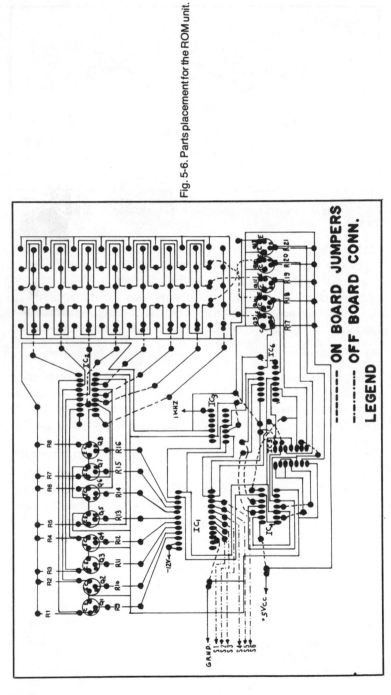

Fig. 5-6. Parts placement for the ROM unit.

ON BOARD JUMPERS

OFF BOARD CONN.

LEGEND

182

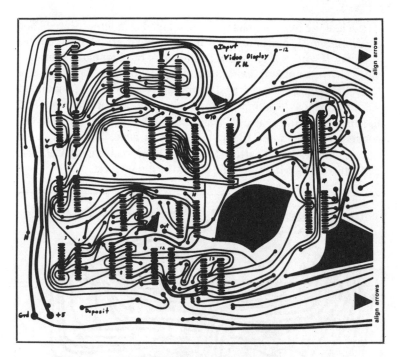

Fig. 5-7. PC board for the 16-character unit. Match the arrows on the plan.

selects a character position (*S1-S4*), an ASCII character code (*S5-S10*), and presses the deposit switch (*S11*). He is allowed to write into any position selected by the switches which address locations in the RAM. Again, an external clock input is used to allow complete control of speed of operation.

Build both demonstrators and use them to lead into the operation of an actual unit. The increasing complexity from single character to 16 character is easily made, as much of the circuitry is identical. A word of caution: these two units are the best motivators and interest builders we have developed and, as such, can turn apathetic students into ravenous digital enthusiasts.

Single-Character Display

Refer to Figs. 5-9 through 5-13 for the following discussion of circuit operation of the single-character demonstration unit. For a discussion of individual chip operation, refer to the manuals from any of the manufacturers of SN7400 series integrated circuits. The external clock signal is applied to 1C1, a divide-by-16 counter, and to 1C9, the parallel-to-serial converter. 1C1 provides a dot count for each of the five columns in the displayed character. Sixteen dot time intervals are counted, but the first eight are blanked by the output of 1C4 pin 2 (Fig. 5-10). At the count of 15 for each column, a vertical reset pulse is generated at pin 12 of 1C2. This pulse is logically 'ored' with the signal of pin 2 1C4 by 1C7. The combined output provides the complete blanking pulse for vertical reset and the unused dot intervals. 1C 5 and 1C 4 provide a load pulse to the converter (1C9) when the pulse count is zero. The load pulse causes the parallel bit pattern from 1C 8 to be loaded into 1C 9. The pattern is then serially stepped out by the input clock.

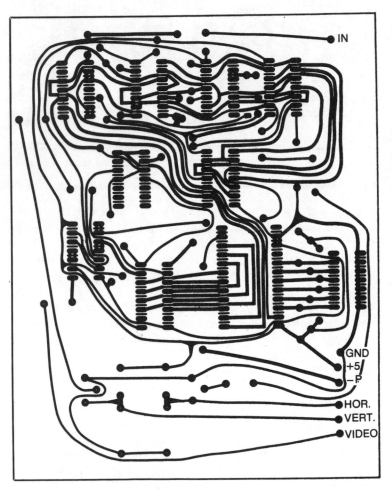

Fig. 5-8. PC board for the single-character display.

184

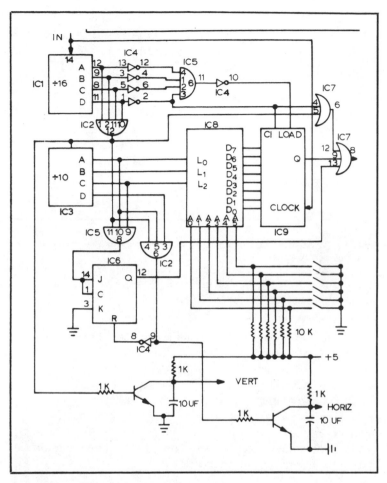

Fig. 5-9. Schematic for single-character unit.

1C3, a divide-by-10 counter, counts the vertical reset pulses and feeds this line number in binary coded decimal to the line inputs (L0, L1, L2) of 1C8 (Fig. 5-11). Flip-flop 1C6 is set by the pulse from pin 8 of 1C5 each time the line count reaches six. The flip-flop produces blanking of lines six through nine. A pulse from 1C2 pin 6 resets the flip-flop on completion of the ninth line. This reset pulse also produces reset of the horizontal sweep generator. The sequence of timing is now complete and the unit will retrace the character.

Figure 5-12 illustrates the two blanking inputs to 1C7 and the composite video pattern with dot information and blanking com-

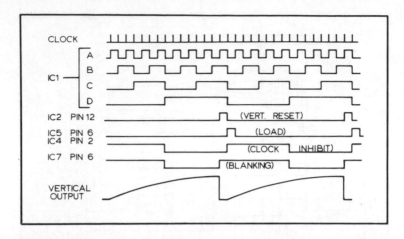

Fig. 5-10. Sixteen dot time intervals are counted, but the first eight are blanked by the output of IC4 pin 2.

bined. The pattern illustrated in Fig. 5-13 is the serial information and resulting dot pattern for the character "A".

16-Character Display

Figures 5-14 through 5-16 illustrate the operation of the single-line-16-character display unit. Dot count and line count are provided by 1C1 and 1C2, respectively. Character count is supplied by 1C3 and pin 11 of 1C2. 1C8 is a quad tri-state bus driver and passes the character address to two SN7489 (16 × 4) memory chips. The character count selects a location in the memory, the memory outputs the stored character code to the character generator, the dot pattern for each line is loaded into the interface and output in serial fashion to the CRT.

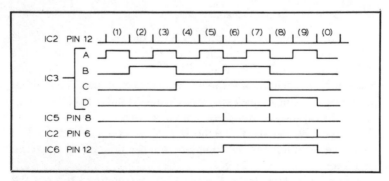

Fig. 5-11. IC3, a divide-by-10 counter, counts the vertical reset pulses and feeds this line number in binary-coded decimal to the line inputs of IC8.

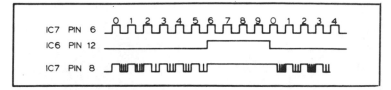

Fig. 5-12. The two blanking inputs to IC7 and the composite video pattern with dot information and blanking combined.

To write a location in memory, switches *S1* through *S4* select a position and switches *S5* through *S10* select the character. The sequence of operation is illustrated in Fig. 5-16. After switches *S1* through *S10* are set, the deposit switch is pressed. This signal sets flip-flop *1C12* and output pin 5 goes on high, enabling an and gate in *IC5*. Pin 8 of the flip-flop goes low, which disables the tri-state drivers in 1C8. When the line count reaches seven, an output pulse from pin 6 of 1C5 enables the tri-state drivers in 1C13 and also produces a write pulse for the memories. When the line count returns to zero, a reset pulse is applied to the flip-flop and control is returned to the character counter 1C3.

The vertical and horizontal sweep circuits are not included in the schematic for the 16-character display unit as they are identical to the ones for the single character unit. If a wider range of adjustment is needed for the larger display, a series combination of potentiometer and limiting resistor can be used in place of the 1KΩ load resistors.

The video output of both units is fed to the Z-axis (beam intensity) modulation input of the oscilloscope. We use a Tektronix T-912, but any scope with blanking on positive will do.

A MICROCOMPUTER

The microcomputer constructed here consists of four separate printed circuit boards. They are the central processing unit

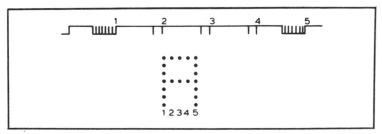

Fig. 5-13. The serial information and resulting dot pattern for the character "A."

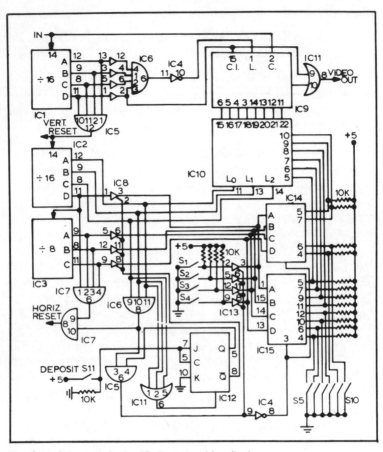

Fig. 5-14. Schematic for the 16-character video display.

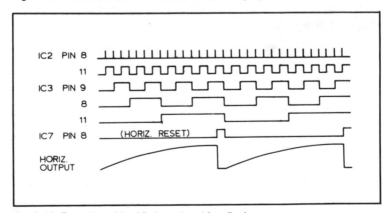

Fig. 5-15. Operation of the 16-character video display.

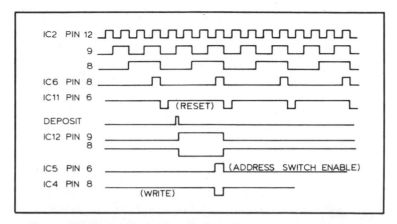

Fig. 5-16. Operation of the 16-character video display.

(CPU) (Fig. 5-17) and its support chips: the random access memory (RAM) (Fig. 5-17), the erasable programable read only memory (EPROM) (Fig. 5-8) and the input-output (I/O) unit (Fig. 5-18). All boards are single-sided for ease of construction and laid out for ease of signal observation. A jumper-component placement guide is included for each.

The CPU board contains an 8080 microprocessor, an 8238 bus controller and driver, an 8224 clock generator and four SN 74125 three-state bus drivers. A 7805 5V regulator is included on all boards. If the HOLD capability of the microprocessor is desired, the jumper connecting the bus enable control line must be removed. The complete bus signal assignment is shown in Table 5-2. Some provision should be made for temporarily connecting the reset line to ground. This will allow interruption of any program that goes awry and return control to the monitor program.

The EPROM board has sockets for five 2708 1Cs. The monitor program given with this computer design occupies about one-half of one 2708 chip. It will be fully discussed later. For reader convenience, however, Fig. 5-19 gives a hexadecimal listing of the monitor routine. If desired, one of the many 4K versions of BASIC could be adapted and placed on 2708 chips. When used with the monitor, it will allow an individual to start with machine language and move into the high level language of BASIC.

The chips must be properly placed to assure sequential access (Figs. 5-20 and 5-21). Viewed from the component side, the first chip must be in the right-hand socket. Successive chips should be placed on the board moving to the left.

Many units are available which will program the 2708 EPROM (Figs. 5-22 and 5-23). Some are available constructed or in kit form. Plans have been published for suitable units in most of the more popular electronics magazines.

The working memory board contains two 2114 RAM chips (Figs. 5-24 and 5-25). They provide 1K (1024) of memory locations. This amount of memory has proved adequate for machine language programing, but when using BASIC more memory will be needed. Any number of boards can be used on one computer by proper selectin of each board's starting address. The addresses are selected in 1K increments by on-board jumpers.

Input and output from the computer is through the I/O board (Figs. 5-26 and 5-27) which contains an 8251 1C, a software programable universal synchronous asynchronous receiver transmitter (USART). Programing is accomplished by the first part of the resident monitor program. The format is one start bit, seven data bits, no parity, and two stop bits transmitted at 110 baud. Output from the board will drive a standard teletype current loop.

This computer was originally designed as a training unit for an electronics technology program. Two other dedicated peripheral boards have been designed and constructed for the unit; a digital to analog conversion board and an intelligent graphics board.

Terminals are one of the major problems. The cost of a teletype is prohibitive, so an oscilloscope is used. These serve as display devices for CRT terminals. The following video display

Table 5-2. Bus Signal Assignment.

PIN SIGNAL	17 not used	34 Phase 2 (Clock)
1 ground	18 not used	35 READY
2 A13	19 A1	*36 RESET
3 A14	20 A2	37 D0
4 A11	21 A0	38 D2
5 A12	22 A3	39 D3
6 A9	23 WAIT	40 D4
7 A15	24 A10	41 D7
8 A8	25 not used	42 D1
9 HOLD REQ	*26 BUS ENABLE	43 D5
10 A7	27 not used	44 D6
11 A6	28 −5V	*45 I/O WRITE
12 A4	29 not used	*46 MEM WRITE
13 A5	30 not used	*47 I/O READ
14 INT REQ	31 not used	*48 MEM READ
15 +12V	32 not used	*49 INT ACK
16 not used	33 HOLD ACK	50 +8V

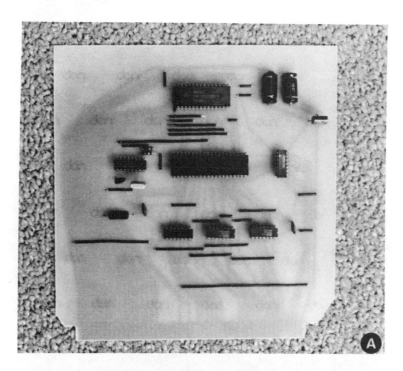

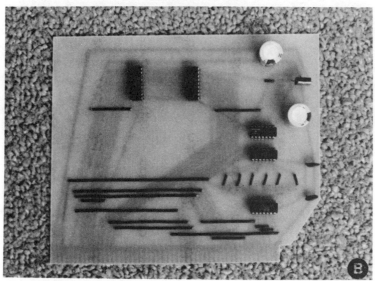

Fig. 5-17. The CPU PC board at A, and the RAM PC board at B.

191

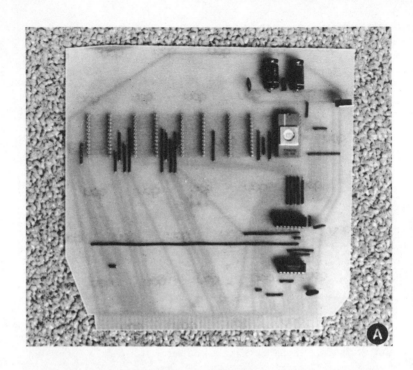

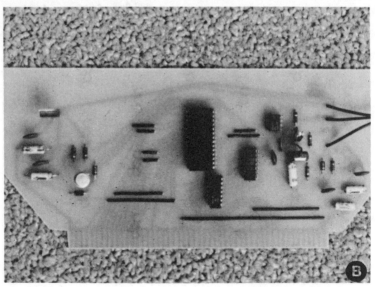

Fig. 5-18. The EPROM PC board at A, and the IN/OUT PC board at B.

192

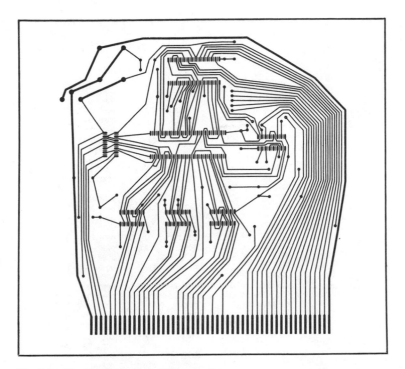

Fig. 5-20. The CPU PC board at 50 percent.

unit design has been not only a terminal but also a training device on character generators and CRT displays. The keyboard used is available from many surplus houses. It must produce ASCII code output along with a strobe pulse. Any oscilloscope with a z axis input can be used as the display device. In operation, seven rows with 32 characters in each row will be seen.

Circuit Operation

The video display unit consists of three sections: a transmitting unit, a receiving unit and a display unit. Figure 5-28 is a schematic of the transmitting unit. It takes the ASCII character code from the keyboard, adds the required framing bits and transmits the byte in serial mode at 110 bits per second (110 baud).

Input from the keyboard is fed to 1C1 in parallel. 1C1 is a parallel-to-serial conversion integrated circuit (SN 74165). As the ASCII input only requires seven bits, the least-significant bit of the 8-bit converter (pin 6) is grounded. This bit is shifted out first and serves as the start bit for the transmission.

Address		Address		Address		Address		Address	
0000-31	LXI SP	004D-CD	CALL	0036-F3		0083-F3			
0001-FF		004E-EB		0037-00		0084-00			
0002-07		004F-00		0038-CD	CALL	0085-C3	JMP		
0003-3E	MVI A	0050-3E	MVI A	0039-EB		0086-1B			
0004-00		0051-2D		003A-00		0087-00			
0005-D3	OUT	0052-D3	OUT	003B-3E	MVI A	0088-CD	CALL		
0006-01		0053-00		003C-4C		0089-90			
0007-D3	OUT	0054-CD	CALL	003D-D3	OUT	008A-00			
0008-01		0055-D4		003E-00		008B-77	MOV M,A		
0009-D3	OUT	0056-00		003F-CD	CALL	008C-23	INX H		
000A-01		0057-CD	CALL	0040-EB		008D-C3	JMP		
000B-3E	MVI A	0058-90		0041-00		008E-62			
000C-40		0059-00		0042-3E	MVI A	008F-00			
000D-D3	OUT	005A-67	MOV H,A	0043-4F		0090-FE	CPI		
000E-01		005B-CD	CALL	0044-D3	OUT	0091-40			
000F-3E	MVI A	005C-D4		0045-00		0092-D2	JNC		
0010-CB		005D-00		0046-CD	CALL	0093-A8			
0011-D3	OUT	005E-CD	CALL	0047-EB		0094-00			
0012-01		005F-90		0048-00		0095-E6	ANI		
0013-3E	MVI A	0060-00		0049-3E	MVI A	0096-0F			
0014-05		0061-6F	MOV L,A	004A-43		0097-07	RLC		
0015-D3	OUT	0062-CD	CALL	004B-D3	OUT	0098-07	RLC		
0016-01		0063-F3		004C-00		0099-07	RLC		
0017-CD	CALL	0064-00		009A-07	RLC	00EA-C9	RET		
0018-90		0065-CD	CALL	009B-47	MOV B,A	00EB-DB	IN		
0019-01		0066-FE		009C-CD	CALL	00EC-01			
001A-00	NOP	0067-00		009D-D4		00ED-E6	ANI		
001B-3E	MVI A	0068-CD	CALL	009E-00		00EE-01			
001C-2A		0069-D4		009F-FE	CPI	00EF-CA	JZ		
001D-CD	CALL	006A-00		00A0-40		00F0-EB			
001E-D9		006B-FE	CPI	00A1-D2	JNC	00F1-00			
001F-00		006C-4E		00A2-B1		00F2-C9	RET		
0020-CD	CALL	006D-C2	JNZ	00A3-00		00F3-3E	MVI A		
0021-D4		006E-7D		00A4-E6	ANI	00F4-0D			
0022-00		006F-00		00A5-0F		00F5-CD	CALL		
0023-FE	CPI	0070-CD	CALL	00A6-80	ADD B	00F6-D9			
0024-45		0071-F3		00A7-C9	RET	00F7-00			
0025-CA	JZ	0072-00		00A8-D6	SUI	00F8-3E	MIV A		
0026-35		0073-23	INX H	00A9-37		00F9-0A			
0027-00		0074-CD	CALL	00AA-07	RLC	00FA-CD	CALL		
0028-FE	CPI	0075-FE		00AB-07	RLC	00FB-D9			
0029-44		0076-00		00AC-07	RLC	00FC-00			
002A-CA	JZ	0077-CD	CALL	00AD-07	RLC	00FD-C9	RET		
002B-15		0078-D4		00AE-C3	JMP	00FE-54	MOV D,H		
002C-01		0079-00		00AF-9B		00FF-CD	CALL		
002D-FE	CPI	007A-C3	JMP	00B0-00		0100-B6			
002E-53		007B-6B		00B1-D6	SUI	0101-00			
002F-CA	JZ	007C-00		00B2-37		0102-55	MOV D,L		
0030-55		007D-FE	CPI	00B3-C3	JMP	0103-CD	CALL		
0031-01		007E-0D		00B4-A6		0104-B6			
0032-C3	JMP	007F-C2	JNZ	00B5-00		0105-00			
0033-1B		0080-88		00B6-7A	MOV A,D	0106-3E	MVI A		
0034-00		0081-00		00B7-E6	ANI	0107-2D			
0035-CD	CALL	0082-CD	CALL						

Fig. 5-19. Hexadecimal listing of the monitor routine.

194

Addr	Op	Addr	Op	Addr	Op	Addr	Op	Addr	Op
00B8-F0		0108-CD	CALL	013F-00		017B-CD	CALL		
00B9-0F	RRC	0109-D9		0140-CD	CALL	017C-D9			
00BA-0F	RRC	010A-00		0141-90		017D-00			
00BB-0F	RRC	010B-56	MOV D,M	0142-00		017E-CD	CALL		
00BC-0F	RRC	010C-CD	CALL	0143-4F	MOV C,A	017F-D4			
00BD-CD	CALL	010D-B6		0144-CD	CALL	0180-00			
00BE-C3		010E-00		0145-F3		0181-CD	CALL		
00BF-00		010F-3E	MVI A	0146-00		0182-90			
00C0-7A	MOV A,D	0110-20		0147-CD	CALL	0183-00			
00C1-E6	ANI	0111-CD	CALL	0148-FE		0184-67	MOV H,A		
00C2-0F		0112-D9		0149-00		0185-CD	CALL		
00C3-FE	CPI	0113-00		014A-23	INX H	0186-D4			
00C4-0A		0114-C9	RET	014B-CD	CALL	0187-00			
00C5-D2	JNC	0115-CD	CALL	014C-F3		0188-CD	CALL		
00C6-CE		0116-F3		014D-00		0189-90			
00C7-00		0117-00		014E-0D	DCR C	018A-00			
00C8-C6	ADI	0118-3E	MVI A	014F-C2	JNZ	018B-6F	MOV L,A		
00C9-30		0119-53		0150-47		018C-CD	CALL		
00CA-CD	CALL	011A-CD	CALL	0151-01		018D-F3			
00CB-D9		011B-D9		0152-C3	JMP	018E-00			
00CC-00		011C-00		0153-1B		018F-E9	PCHL		
00CD-C9	RET	011D-3E	MVI A	0154-00		0190-06	MVI B		
00CE-C6	ADI	011E-2D		0155-CD	CALL	0191-25			
00CF-37		011F-CD	CALL	0156-F3		0192-21	LXI H		
00D0-CD	CALL	0120-D9		0157-00		0193-A2			
00D1-D9		0121-00		0158-3E	MVI A	0194-01			
00D2-00		0122-CD	CALL	0159-45		0195-7E	MOV A,M		
00D3-C9	RET	0123-D4		015A-CD	CALL	0196-CD	CALL		
00D4-CD	CALL	0124-00		015B-D9		0197-D9			
00D5-E3		0125-CD	CALL	015C-00		0198-00			
00D6-00		0126-90		015D-3E	MVI A	0199-23	INX H		
00D7-DB	IN	0127-00		015E-58		019A-05	DCR B		
00D8-00		0128-67	MOV H,A	015F-CD	CALL	019B-C2	JNZ		
00D9-D5	PUSH D	0129-CD	CALL	0160-D9		019C-95			
00DA-5F	MOV E,A	012A-D4		0161-00		019D-01		01B2-57	
00DB-CD	CALL	012B-00		0162-3E	MVI A	019E-CD	CALL	01B3-52	
00DC-EB		012C-CD	CALL	0163-20		019F-F3		01B4-49	
00DD-00		012D-90		0164-CD	CALL	01A0-00		01B5-54	
00DE-7B	MOV A,E	012E-00		0165-D9		01A1-C9	RET	01B6-54	
00DF-D3	OUT	012F-6F	MOV L,A	0166-00		01A2-4D		01B7-45	
00E0-00		0130-CD	CALL	0167-3E	MVI A	01A3-4F		01B8-4E	
00E1-D1	POP D	0131-F3		0168-41		01A4-4E		01B9-20	
00E2-C9	RET	0132-00		0169-CD	CALL	01A5-49		01BA-42	
00E3-DB	IN	0133-3E	MVI A	016A-D9		01A6-54		01BB-59	
00E4-01		0134-23		016B-00		01A7-4F		01BC-20	
00E5-E6	ANI	0135-CD	CALL	016C-3E	MVI A	01A8-52		01BD-46	
00E6-02		0136-D9		016D-44		01A9-20		01BE-52	
00E7-CA	JZ	0137-00		016E-CD	CALL	01AA-52		01BF-45	
00E8-E3		0138-3E	MVI A	016F-D9		01AB-4F		01C0-44	
00E9-00		0139-2D		0170-00		01AC-55		01C1-20	
013A-CD	CALL	0176-CD	CALL	0171-CD	CALL	01AD-54		01C2-48	
013B-D9		0177-D9		0172-D9		01AE-49		01C3-49	
013C-00		0178-00		0173-00		01AF-4E		01C4-4E	
013D-CD	CALL	0179-3E	MVI A	0174-3E	MVI A	01B0-45		01C5-45	
013E-D4		017A-2D		0175-52		01B1-20		01C6-53	

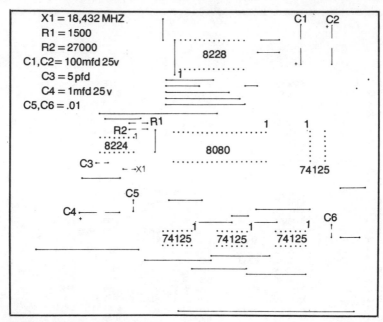

Fig. 5-21. CPU parts placement.

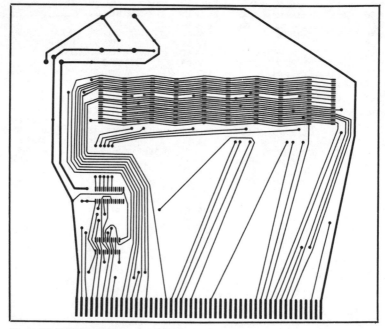

Fig. 5-22. EPROM PC board at 47.5 percent.

196

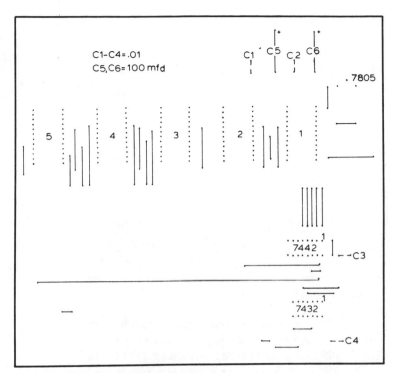

Fig. 5-23. EPROM parts placement.

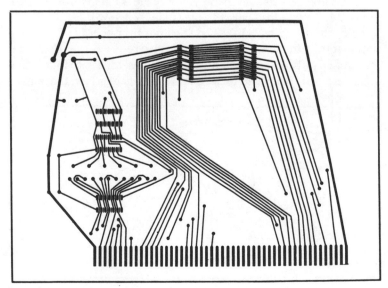

Fig. 5-24. RAM PC board at 50 percent.

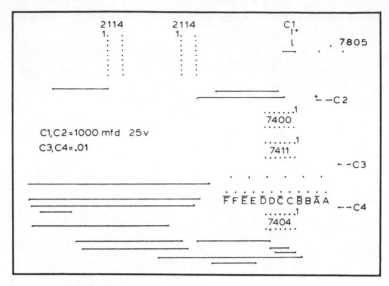

Fig. 5-25. RAM parts placement.

The load input is the strobe output from the keyboard which tells the unit the ASCII code is stable and can be loaded into the IC. The leading edge of the negative-going load pulse causes one flip-flop (pins 5-7) in 1C5 to set. The Q output of the flip-flop goes low, causing 1C1 to load the keyboard code. One inverter of 1C6 (pins 3-4) inverts load pulse so the trailing edge will set the other flip-flop in 1C5 (pins 1-14). The Q output from this flip-flop goes low resetting the first FF. This output also releases the AND gate comprised of 1C4 (pins 1, 2, 3) and 1C6 (pins 13, 12) to follow the input on pin 1. If the input to pin 2 of the AND gate is low, the output

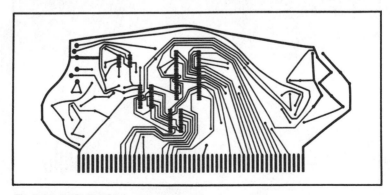

Fig. 5-26. IN/OUT PC board at 44 percent.

198

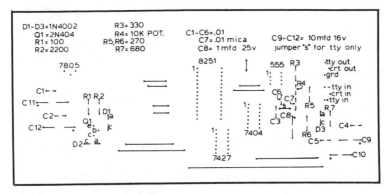

Fig. 5-27. IN/OUT parts placement.

will remain low regardless of the signal on pin 1. This prevents transmission of false bits during the load sequence. The Q output going low also releases the reset lines of the two divide-by-16 counters, 1C2 and 1C3. Pin 11 of 1C2 produces 110 pulses per second (1760 divided by 16). Each pulse causes the next bit of the ASCII code to be shifted out of 1C1. 1C3 counts the number of shift pulses and at the count of 11 causes a low to be output from pin 8 of 1C4 which resets FF 2 of 1C5. The reset inputs of 1C2 and 1C3 are taken high by the Q output of FF 2. This prevents their counting the input 1760 clock pulses. Also the reset line of FF 1 is released so it

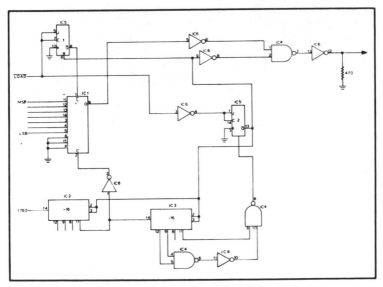

Fig. 5-28. Schematic of the transmitting unit of the microcomputer.

199

may respond to the next strobe pulse from the keyboard. The transmission cycle is completed, the output from the unit is disabled, and no action takes place until the next strobe pulse.

Decoder and Counter

Figure 5-29 contains not only the serial-to-parallel receiving unit but also the control character decoder and character position counter used by the video display section. When no data is being transmitted, the data input line to 1C10 will be low. The first pulse received will be the positive start pulse. The leading edge of this pulse, after inversion by inverter 1, will cause FF 1 to set. The Q output of FF 1 goes low, releasing the reset inputs of 1C 5 and resetting FF 2, which, in turn, releases the reset inputs of 1C7. On each positive going edge from pin 11 of 1C5, 1C1 will load the pulse value at its input into register A. As each successive shift pulse occurs, the value in the A register will be moved to the B register, the B register will have moved to the C register, etc. Even though the start bit value is initially loaded into the register A, after nine shift pulses it will have been shifted past the H register and lost. Figure 5-30 and Table 5-3 illustrate the load/shift sequence.

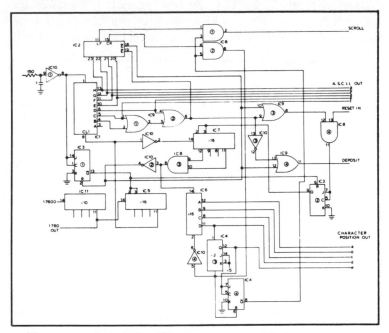

Fig. 5-29. Serial-to-parallel receiving unit of the microcomputer.

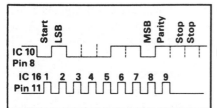

Fig. 5-30. The load/shift sequence.

1C5 produces the shift pulses and 1C7 counts them. When the ninth pulse is counted, 1C3 is reset by pin 4 of 1C10. The high output from pin 13 of 1C3 prevents 1C5 from producing any more shift pulses. The serial ASCII code has been caught and changed from serial to parallel form by 1C1. The least significant bit is output on pin 13 and the most significant bit used is output on pin 4. The same low pulse which resets 1C3 is 'ORed' with a signal from inverter 3 to produce a deposit pulse (negative-going edge) for the video display unit. The deposit pulse signals that a valid ASCII code is present and ready to be written into the video display's memory. The deposit pulse will occur unless pins 4, 5, and 6 of 1C1 are all low (three most significant bits). If low, inverter 3 will produce a high level at pin 13 of OR gate 4 and prevent the output from going low. This sequence of events prevents depositing control characters into the display's memory. The only two control characters used, carriage return and line feed, have their most

Table 5-3. The Load/Shift Sequence.

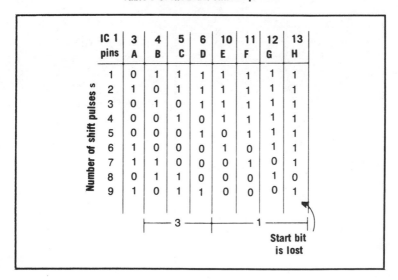

IC 1 pins	3 A	4 B	5 C	6 D	10 E	11 F	12 G	13 H
1	0	1	1	1	1	1	1	1
2	1	0	1	1	1	1	1	1
3	0	1	0	1	1	1	1	1
4	0	0	1	0	1	1	1	1
5	0	0	0	1	0	1	1	1
6	1	0	0	0	1	0	1	1
7	1	1	0	0	0	1	0	1
8	0	1	1	0	0	0	1	0
9	1	0	1	1	0	0	0	1

Number of shift pulses s

|— 3 —|— 1 —|

Start bit
is lost

201

significant 4 bits equal to 0000. For a number or a letter, at least one of the three most significant bits will be high. This will allow the deposit pulse to pass through OR gate 4. The relevant ASCII code is shown in Table 5-4.

The video display unit will determine in which line of the display a deposited character will appear, but the character position in that line will be determined by the outputs of 1C6 and 1C4. The reset pulse which is generated from a nine count of 1C7 is also counted by 1C6 and 1C4. They form a 0 to 31 counter chain. As each character is deposited, this counter chain is incremented and causes the next character to be deposited at the next position on the line. The operation of this counter chain with the line counter in

Hex	ASCII	HEX	ASCII
0A	line feed	3F	?
0D	Carr. ret.	40	@
20	Space	41	A
21	!	42	B
22	"	43	C
23	#	44	D
24	$	45	E
25	%	46	F
26		47	G
27	'	48	H
28	(	49	I
29	)	4A	J
2A	*	4B	K
2B	×	4C	L
2C	'	4D	M
2D	–	4E	N
2E	.	4F	O
2F	/	50	P
30	0	51	Q
31	1	52	R
32	2	53	S
33	3	54	T
34	4	55	U
35	5	56	V
36	6	57	W
37	7	58	X
38	8	59	Y
39	9	5A	Z
3A	:	5B	[
3B	;	5C	/
3C	‹	5D	]
3D	=	5E	Δ
3E	>	5F	

Table 5-4. ASCII Code.

the video display unit will become more apparent in the video display discussion.

1C2 is a 4-line to 16-line decoder (SN 74154). For each of the 16 possible combinations at the four input lines, a respective output line will go low. Pins 18 and 19 are active low enables. Both pins must be low for the chip to decode the input lines and produce a corresponding output. If the control character-carriage return (OD on the hex code) is received, output pin 15 will go low when a low pulse is received from inverter 2, signifying nine pulses have been counted, and a low pulse from OR gate 2, signifying no high values in the most significant three bits. A carriage return causes the video display to return to the first position of the line it is currently printing. If more characters are received, they will write over the initial ones displayed on that line. When a carriage return is decoded, the AND gate 2 of 1C8 produces a low pulse to reset the character position counter.

When a character is written into the last position on a line, FF 4 pin 8 goes low, generating both a carriage return and a line feed. The carriage return resets the character position counter and the line feed is fed to the display unit as a scroll pulse. The scroll function causes all lines on the screen to move up one position. The top-most line disappears and a new blank line appears at the bottom. Any more characters received will be displayed on the new line starting at the first position.

1C3 serves to reset the pulse counter 1C7 if a deposit is performed and also if a control character is received. If a regular character is received, the deposit pulse will set a flip-flop in the video display unit which will feed back a reset pulse to pin 13 of AND gate 4. A high will be received on pin 12 of the AND gate for a regular character and the negative pulse on pin 13 will be fed to 1C3. This negative-going edge will set 1C3 and cause counter 1C7 to be disabled. If a control character is received, the deposit line stays high, the reset in line to AND gate 4 therefore stays high, but a low is received from OR gate 3 to reset IC3.

The low is produced by OR gate 3 because a control character has all lows for its most significant three bit location.

Video Display

Figure 5-31 is the schematic for the video display section. The vertical and horizontal output are one μs pulses which reset the RC oscillators shown in Fig. 5-32. The horizontal frequency is 4000 Hz and the vertical frequency is 50 Hz. Together they produce a screen raster on the oscilloscope of 80 lines, starting in the upper left hand corner and terminating in the lower right hand corner. The complete raster is displayed 50 times each second to prevent flicker.

The ASCII code for each of the 32 characters in each of the 7 displayed rows is stored in memory chips 1C21 through 1C26. As the first line of dots of each character is displayed, the address is changed 32 times. Each address selects one of 32 possible locations in the memory chips. At each location is stored the code for the desired character at that row position. The memory outputs the code to the character generator chip. It, in turn, selects a dot pattern based on the input code and the line desired. The line is determined by the number of pulses that have occured at pin 6. These pulses trigger an internal line counter. Figure 5-33 illustrates the line count sequence.

In Fig. 5-31, 1C1 and 1C2 form a part of a counter chain. They produce the character position count which is fed to the memory. At the end of one line, IC10 is set to produce a pulse for the internal line counter of the character generator. When the character position counter starts over, the same memory location will be selected as for the previous line within the row, but now the next line of dots will be output by the character generator.

Figure 5-34 shows the character position count and also the row count. The row count is produced in part by IC4. The other value is supplied by the scroll counter, IC6. Both values are added by IC11. Their composite output determines the row count sequence.

When the scroll counter is zero, the normal count sequence is produced by IC4 on the screen. When a line feed is decoded by the circuitry of Fig. 5-28, a pulse will be fed to the scroll counter. The scroll counter output will give a constant offset to the count from IC4. This offset causes the rows of characters to move up the screen one row position with each line feed input. Table 5-5 illustrates this sequence. As the rows of characters are scrolled up the screen, the top row disappears and a new blank row appears at the bottom. The blank row is produced as follows: the top row with its characters is moved to the bottom or eighth position, but the entire row is blanked from the screen. IC17 decodes the row count from IC4 and indicates by a low on pin 9 when the bottom row is being scanned. This signal causes IC 13 and IC14 (quad 2-line to 1-line data selectors) to select input pins 3, 6, and 10.

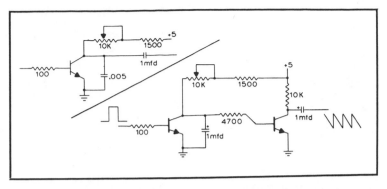

Fig. 5-32. RC oscillators used with the video displays of the microcomputer.

These pins are wired to produce ASCII code 20 (hex), which is the code for a blank or space. This same positive output from pin 10 of IC16 causes a positive to appear at the video output and, through the OR gate of IC19, causes the read/write line to be driven high to signify a write operation. The memory chips will thereby have the ASCII code for a blank written into each of the 32 positions for whichever line happens to be in the bottom position. When that line is scrolled up to the seventh position, it will appear as a totally blank line.

Characters that are being fed in will appear on the row that is in the seventh position. They will be displayed on the screen until that row has been shifted past the top row position. IC17 produces a low on pin 7 whenever the display is scanning row position 7. This low signal is one input of an OR gate in IC19, the other input comes from FF 4. The flip-flop serves as a deposit flag to the unit; when a deposit pulse is received, the $\overline{Q}$ output goes low. If the output from IC17 pin 7 is low (unit is scanning row position 7) and the $\overline{Q}$ output from FF 4 is low (a deposit has been requested), a low will appear at the OR gate output. The low, after inversion, causes IC8 and IC9 to switch from the normal counter input to the input from the

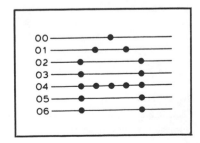

Fig. 5-33. Line count sequence.

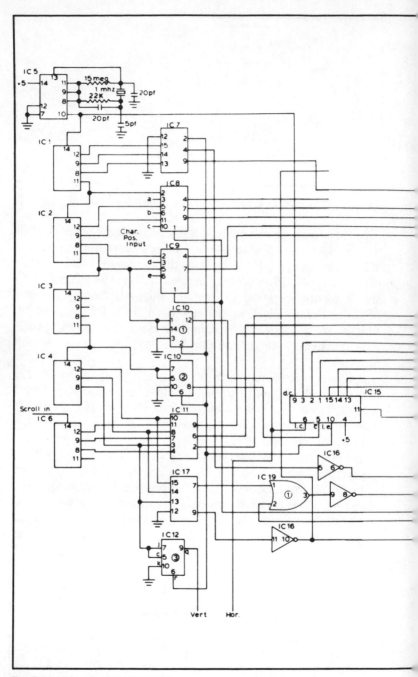

Fig. 5-31. Video display schematic.

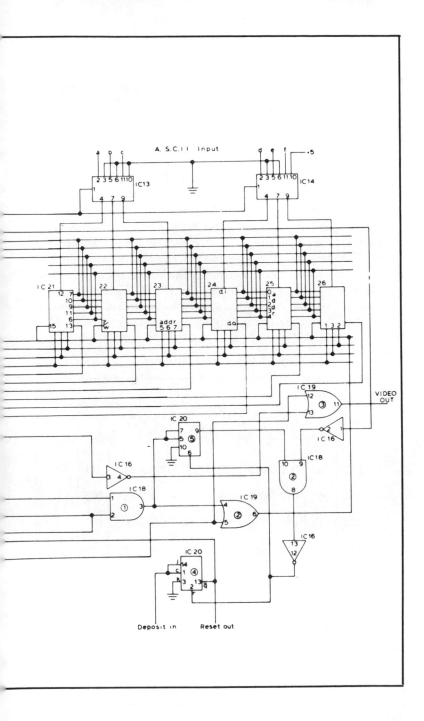

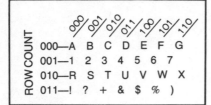

ROW COUNT	000	001	010	011	100	101	110
000—	A	B	C	D	E	F	G
001—	1	2	3	4	5	6	7
010—	R	S	T	U	V	W	X
011—	!	?	+	&	$	%	)

Fig. 5-34. Character position count.

character position counter. The character position input will be the memory address for the next character position.

The same pulse which causes IC8 and IC9 to switch inputs is fed to AND gate 1. When dot count three is reached, signified by the input at pin 1 of AND gate 1, the output of the gate will go high, producing through OR gate 2 a positive write pulse for the memory chips. The memory address for the write operation is supplied by IC8 and IC9, character position input. The bit pattern to be stored, ASCII code input, is supplied by IC13 and IC14. When the input to AND gate 1—pin 1 terminates, signifying the end of dot position three, the output goes low causing FF 5 to set. Position 8 of the dot counter is indicated by the input at pin 9 of AND gate 2. This pulse,

Table 5-5. Line Feed Input Sequence.

Scroll counter value	000	001	010	011	100	101	110	111
000	000	001	010	011	100	101	110	111
001	001	010	011	100	101	110	111	000
010	010	011	100	101	110	111	000	001
011	011	100	101	110	111	000	001	010
100	100	101	110	111	000	001	010	011
101	101	110	111	000	001	010	011	100
110	110	111	000	001	010	011	100	101
111	111	000	001	010	011	100	101	110

IC 4 output (row axis) / IC 11 output (column axis)

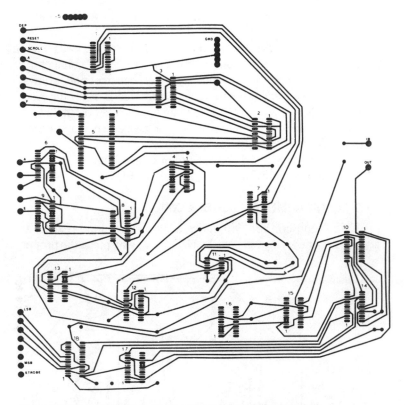

Fig. 5-35. Video R-T PC board showing foil side at 100 percent.

combined with output of $\overline{FF}$ 5, produces a reset pulse for both flip-flops in IC20. The $\overline{Q}$ output of FF 4 goes high, which terminates the deposit sequence.

Adjustments

For proper operation, the baud rate oscillators must be adjusted on the computer I/O board and on the video interface board. The computer I/O board oscillator is set to 17,600 Hz while the video interface oscillator is set to 1760 Hz. This ratio of 10 must be maintained within 1%. All commercial units use crystal oscillators to stay within the tolerance, but RC oscillators were chosen for the trainer to require adjustment. If both oscillators are varied, still maintaining the factor of 10, the baud rate can be altered. The video interface should be connected to the transmit out and the positive receive in terminals on the computer I/O board (see the crt out and cut in on the I/O board part placement of Fig. 5-27.

For maximum flexibility, the computer trainer was constructed utilizing a motherboard and PC card edge connectors (Figs. 5-35 through 5-39). This type of construction allows memory expansion and repositioning of the cards to allow for ease of signal viewing when studying specific operations. By including extra sockets, special computer cards can be designed and used. These can include digital-to-analog conversion, graphics display (multiplexed memory) and specialized interface. Others might be light pen graphics and a Z-80 microprocessor CPU card.

CRACKING THE NUMERICAL CONTROL CODE

This "code cracker" simulates a mechanical tape reader. It rewards anyone trying to learn the N/C codes who picks the proper

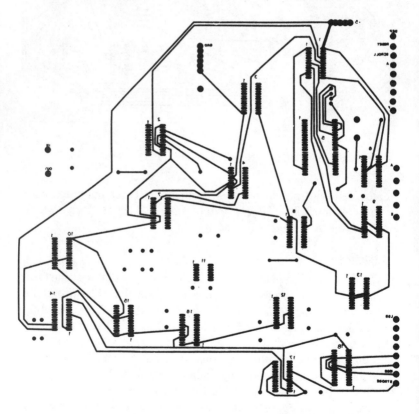

Fig. 5-36. Video R-T PC board showing component side at 100 percent.

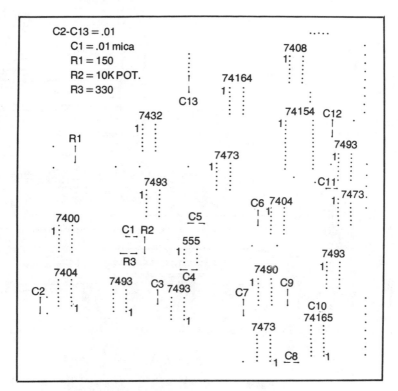

Fig. 5-37. Video R-T parts placement.

combination of holes corresponding to numbers in the binary decimal system.

The unit consists of a board on which are mounted four switches and nine bulbs. Throwing the correct switch or switches causes the appropriate light to come on. Learning the binary decimal code becomes a challenging game.

On tapes for N/C machines, the first four channels of a standard paper or magnetic tape are reserved for holes which represent numbers built on a base of two—hence, the *binary* system. Where mechanical tape readers are used, fingers drop into holes. The resultant motion is converted into electrical impulses which the machine control interprets as numbers.

The hole combinations are as follows:

☐ A hole in channel one is worth 1
☐ A hole in channel two is worth 2
☐ A hole in channel three is worth 4
☐ A hole in channel four is worth 8.

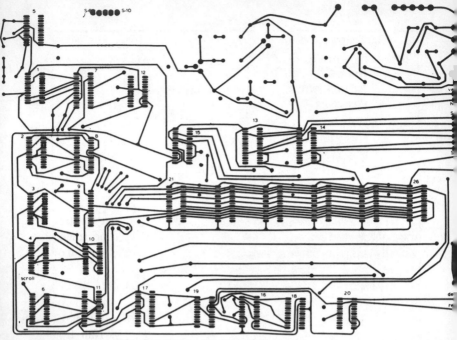

Fig. 5-38. Video display PC board showing foil side at 100 percent.

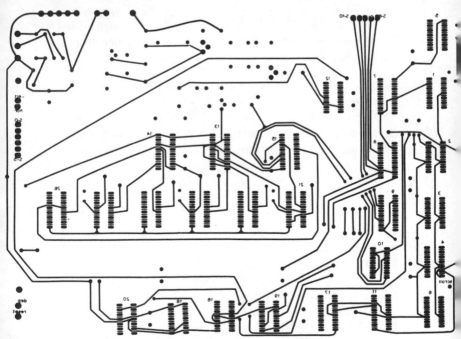

Fig. 5-39. Video display PC board showing component side at 100 percent.

212

The following combinations are used to express numbers 1 through 9:

- ☐ Hole in 1st channel = value of 1
- ☐ Hole in 2nd channel = value of 2
- ☐ Holes in 1st and 2nd channel = value of 3 (1 + 2)
- ☐ Hole in 3rd channel = value of 4
- ☐ Holes in 1st and 3rd channels = value of 5 (1 + 4)
- ☐ Holes in 2nd and 3rd channels = value of 6 (2 + 4)
- ☐ Holes in 1st, 2nd, and 3rd channels = value of 7 (1 + 2 + 4)
- ☐ Holes in 4th channel = value of 8
- ☐ Holes in 1st and 4th channels = value of 9 (1 + 8)

This device demonstrates the ease with which an electro-mechanical system selects numbers. The circuit can be traced visually and it becomes apparent that all other circuits are open or inoperative. All the mystery of tape reading disappears.

The unit can be operated with or without electricity. In the latter case, all circuits and lamps are painted on the board. Materials for its construction are readily available. Details of construction, as shown in Fig 5-40, are as follows: Nail five plywood strips of varying lengths to a larger piece of plywood. These strips provide channels for long dowels which serve as trip rods for the relays. Secure eight shorter dowel pins through drilled holes to the trip rods (Fig. 5-41). A "U" shaped copper strip is

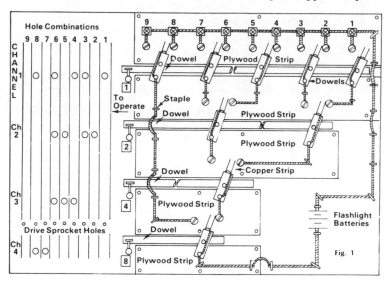

Fig. 5-40. Construction details of the numberical control simulator.

213

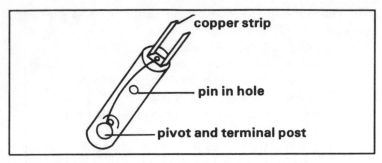

Fig. 5-41. Detail of switch used in the numberical control simulator.

nailed to the end of each short dowel for relay points. Figure 5-42 shows construction of the four manual switches, which are analogous to tape reader fingers.

A safe and cheap power source comes from two flashlight dry cells held in a clip. Wire can be salvaged from an old drop cord. Use nine small receptacles and bulbs from a Christmas tree light string to complete the bill of material.

FLIP-FLOP COMPUTER DEMONSTRATOR

Computers are now part of the daily scene in science, business, home and industry; their use will find ever-increasing applications in the coming years. The general public regards the computer as some sort of "magic box" which automatically comes up with all the answers amidst the furious blinking of lights and the revolutions of tape reels. This design was developed, therefore, to help dispel the air of mystery which, to the layman, envelops the computer.

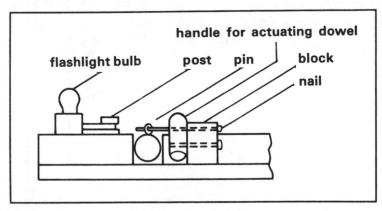

Fig. 5-42. End view of one of the four manual switches.

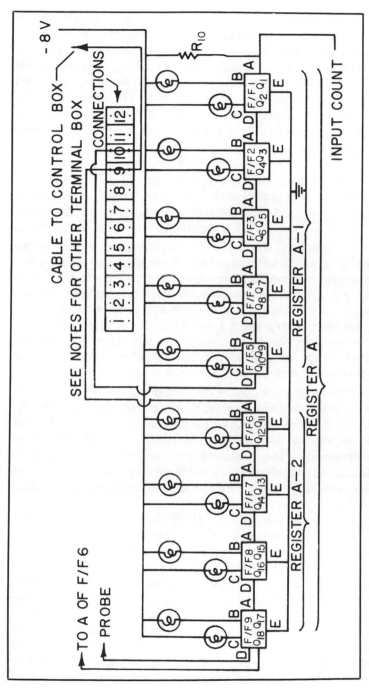

Fig. 5-43. Nine flip-flop circuits are connected in series. These give a discrete count of 2^8.

215

This computer demonstrator is a device by which you can trace the basic circuits and their operation. For this reason, most of the wiring and the components are exposed to view. To illustrate the fact that a computer flip-flop (F-F) circuit can differentiate *only* red lights are installed to represent "zeros" between "ones" and "zeros", and white lights represent "ones." The computer can be manipulated manually with a probe, or through the remotely controlled input dial. This computer may be hung up for easy viewing.

The basic F-F was originally designed to count serially. This computer demonstrator has a basic F-F modified not only to count serially but in a parallel mode, and to perform other arithmetic functions, e.g., addition, subtraction, multiplication and division.

Nine F-F circuits are connected in series, which will give a discrete count of 2^8 (Fig. 5-43). (Note: Any number of F-F circuits may be used, depending on needs.) These nine circuits are broken up into two separate circuits, which allow the use of all or part of the count. This innovation gives two distinct registers to use for division: e.g., input would be the divisor, register A-2 would store the quotient, and register A-1 would store the dividend. The dial is capable of using the additive/subtractive technique for division by adding the divisor to the stored dividend and recording the cycles in the quotient register simultaneously.

The input control box (Figs. 5-44 through 5-47) has a switch on the side which will allow for the separation of the registers A, A-1 and A-2. There are also six push buttons (Fig 5-44) on the top of the control box (two are now unused to be available for future programming). The remaining four pushbuttons are used to illustrate circuits allowing one to add not only in parallel mode, but also to multiply. The illustration of octal arithmetic is also done with ease.

This equipment used 8V DC broken down from 110V AC through a battery eliminator. Conventional batteries would also be a satisfactory power source.

The most unique feature of this design is the simplicity of the one basic F-F circuit used to accommodate the arithmetic functions and to illustrate the other computer operations. Ordinarily, these operations have been accomplished through the use of complex circuitry. A contributing factor which gave these advantages to the design was the ability to initiate (set or reset) any F-F circuit as an input for the next F-F. This is done by simply shorting any emitter to the base or collector of the desired transistor. The repeated shorting of one transistor to the other transistor of an F-F will serve as an input pulse for arithmetic functions or computer operations.

The computer demonstrator is capable of performing and illustrating the following arithmetic functions and computer operations:

☐ Binary. B.C.D., octal and hexi-decimal arithmetic:
   addition
   subtraction
   multiplication
   division

☐ Complement arithmetic:
   binary—2's complement, 1's complement
   octal—8's complement, 7's complement
   hexi-decimal—16's complement, 15's complement

☐ Representing positive and negative numbers
☐ Overflow
   additive
   subtractive

☐ Exponential arithmetic
   positive and negative exponents
   positive and negative coefficients
   conversion and translation

☐ Extracting roots
   additive/subtractive method
☐ Linear interpolation in a single plane

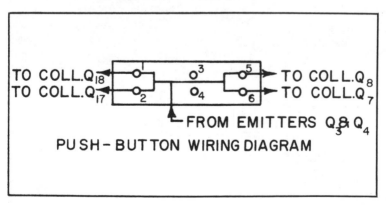

Fig. 5-44. Pushbutton wiring diagram for use in the flip-flop computer demonstrator.

217

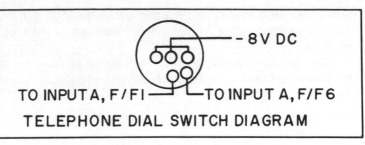

TELEPHONE DIAL SWITCH DIAGRAM

Fig. 5-45. Telephone dial switch wiring diagram for use in flip-flop computer demonstrator.

Theory of Operation

The numerical quantities and their functions to be illustrated will be translated from machine code (true binary) to the following bases: binary coded decimal, octal and hexi-decimal. The number bases that are used lend themselves to easy conversions merely by substitution. Considering the design characteristics, the largest quantities which can be represented (on this model) in their respective bases are shown below:

True binary	111111111	511
B.C.D.	1,1001,1001	199
Octal	111,111,111	777
Hexi-decimal	1,1111,1111	1FF

A digital computer operates internally by binary operations. Serial binary addition is illustrated using register A-1, a five-bit register. All logic is cleared by loading register A-1 with a count of

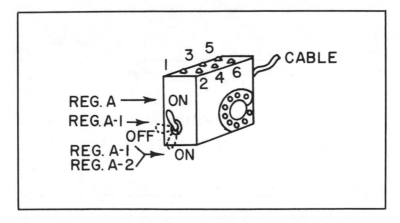

Fig. 5-46. The input control box used in the flip-flop computer demonstrator.

218

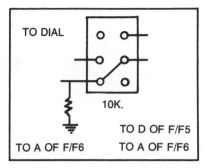

TO DIAL

10K.

TO A OF F/F6

TO D OF F/F5

TO A OF F/F6

Fig. 5-47. elector switch diagram for use in the flip-flop computer demonstrator.

31 plus 1. Logic may also be cleared by using the shorting probe or loading every three circuits with a count of seven. A single switch can also be incorporated to clear all logic.

Sample Problem: 10 + 10 = ?

Register A, Logic cleared:

5 white lights off, representing 5 "not ones"

5 red lights on, representing 5 "zeros"

ooooo (White lights, all OFF)

●●●●● (Red lights, all ON)

(All F-Fs are reset.)

The first input pulse to F-F #1 changes its state from reset to set. (Red light goes off and white light goes on.) This state now represents a binary "1." Any subsequent input pulse to F-F No. 1 changes its state. When F-F No. 1 is set and an input pulse is introduced, the set side of the F-F serves as an input for the next F-F, thereby changing its state. Only when an F-F is set, can it potentially change the state of the next F-F.

To add 10 + 10:

Dial 10 to an all-cleared register, and dial 10 again, giving a total of 20 input pulses.

The solution is indicated by the lights as follows:

(Solid dots represent lights which are ON)

$$2^4\ 2^3\ 2^2\ 2^1\ 2^0$$

●o●oo   (White lights)

o●o●●   (Red lights)

$10100 = 20$

$$2^0 = \quad 1 \times 0 = 0 \qquad 00000$$
$$2^1 = \quad 2 \times 0 = 0 \qquad 00000$$
$$2^2 = \quad 4 \times 1 = 4 \qquad 00100$$
$$2^3 = \quad 8 \times 0 = 0 \qquad 00000$$
$$2^4 = 16 \times 1 = 16 \qquad \underline{10000}$$
$$10100 = 20$$

219

```
C₁, C₂—0.05μf. 200V
D₁, D₂—1N34A diode
"1", "0"—#48 panel lamps
R₁, R₇—68Ω resistors
R₂, R₃, R₅, R₆—1.2 K resistors
R₄—10Ω resistor
R₈, R₉, R₁₀—10 K resistors
Q₁, Q₂—2N 554 transistors
```

Table 5-6. Parts List for the Flip-Flop Computer Demonstrator.

Table 5-7. Terminal Box Connections for the Flip-Flop Compute Demonstrator.

Wire Color	Terminal Number	
White	1	F-F 4 collector Q_7 to control box
White	2	F-F 4 collector Q_8 to control box
Red	3	F-F 7 collector Q_{13} to control box
Red	4	F-F 7 collector Q_{14} to control box
Blue	5	F-F 2 emitters to input control box
Blue	6	Open
Orange	7	Open
Orange	8	Ground from power supply common to F-F circuits and to selector switch
Black	9	Input by-pass from F-F 5 to F-F 6 via selector switch
Black	10	Output by-pass from F-F 5 to selector switch
Green	11	Input count from dial to F-F 1
Green	12	Power supply -8 v d.c. common to F-F 1 and to dial

Notes:
1. A probe connected to common emitters 17 and 18 (or any emitters) is capable of changing state of flip-flops by shorting from emitter to collector or base. To re-set, short even-numbered transistor. To set, short odd-numbered transistor. This will serve as an input pulse for subsequent F-F.
2. To illustrate E.A.C. (End Around Carry), an output is connected from F-F 9 to Input A of F-F 6.

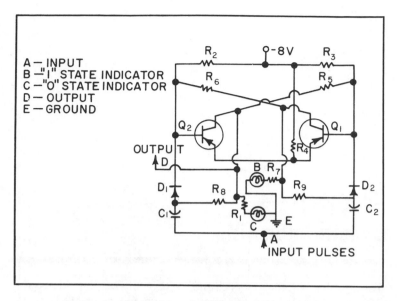

Fig. 5-48. Flip-flop circuit diagram.

By expanding the number of F-F circuits and registers, other computer operations can be illustrated, e.g., linear interpolation in two planes (X and Y), circular interpolation, parabolic interpolation, etc.

The computer demonstrator is board-mounted. Individual design modifications would depend on the needs of the user, that is, variation in the number of F-F circuits, mounting specifications, and optional equipment.

The approximate dimensions of the mounting board illustrated are 38″ wide, 7¾″ high, and ¾″ thick. A wooden strip 38″ × 2″ × ¾″ is used to hold the F-F circuits. This strip is anchored to the board with wood screws. The board and strip may be covered or coated as desired (Table 5-6).

To the extreme right, at the top of the wood strip, is a terminal block containing 12 terminal mounts (Table 5-7). Refer to Fig. 5-48 for a detailed description. From the terminal block to the remote control box is connected a cable (which may be of any convenient length) containing six pairs of color-coded stranded wire. Refer to the circuit diagram of Fig. 5-48 for a detailed description of the control box and related circuit diagrams. This design has merit, not only because of its simplicity and the visual advantages incorporated, but because of the economy of construction.

221

Chapter 6
Construction Aids

You'll probably refer to this chapter often while reading this book. After all, the projects in this chapter—from making printed circuit boards to building your own soldering gun—will come in handy when building all the other projects in other chapters.

PHOTOMECHANICAL PRODUCTION OF PRINTED CIRCUITS

The method of developing a printed circuit board described here is popular in industry so hobbyists should experience this technique.

Step-by-Step Procedure

Start with a schematic diagram of the circuit to be developed as a printed circuit. Then determine the physical limitations of the components to be used.

Make a possible parts layout to size on poster board, actually mounting parts to get a concept of what is to be mounted and where. Make a rough pencil layout or the printed circuit while the parts are mounted on the poster board.

Check and recheck the schematic and rough pencil layout to be certain you have not left anything out or made connections in the wrong places. Revise the rough pencil layout for better arrangement of parts or better runs for the printed circuit.

Mount parts on the revised layout, make temporary connections, and check the operation of the circuit. This is a very important step, as you are now going to finalize the circuit layout.

Finalize your pencil layout on graph paper, poster board, or some other good white surface. Number and label values and locations of parts for easier reference at a future time.

Make a tape layout. Make a master pattern of your tape layout. Place tape layout on camera lens (f/16 lens setting, camera and copyboard set at 100 percent; see Fig. 6-1); set timer (28 sec. at f/16). Shoot to make negative. Put exposed film in developer for approximately two and a half min.; in stop bath for approximately 15 sec.; in fixer for two to three min.; in wash for two min.; put film on light table and squeegee dry; place film in dryer on rack—approximately two min. if dryer is heated. If scratches appear or places are underdeveloped, use opaquing to touch up negative (see Fig. 6-2).

Clean copper-clad board with steel wool. Coat copper-clad board with photo resist (see Fig. 6-3). *This is one of the most important steps*. You should get a good even coat. Keep away from water.

Heat photo-resist-coated board on a hot plate or air dry it for approximately one min. This creates a bond of the photo resist. If you overdo this process it will take longer to do the etching process.

Put coated board on light table and lay master pattern (negative on top of board, emulsion side down) and expose for seven min. (see Fig. 6-4).

Remove board and negative from light table. (Save your negative to make additional boards.)

Put exposed copper-clad board in photo-resist developer for two min. (see Fig. 6-5). If you leave the board in developer too long you will wash away the entire image and you will have to start all over.

Fig. 6-1. Set lens at f/16 and camera and copyboard at 100 percent.

Fig. 6-2. Use an opaquing kit to touch up the negative.

Fig. 6-3. Coat copper-clad board with photo resist.

Rinse off the developer with water. Do not rub or wipe off exposed board.

Prepare etching solution. There are many solutions that can be used but I have found that ferric chloride is the easiest to handle, safest to use, and produces the best results. It is also popular in industry for the same work. You get only a minimum amount of fumes from it. If you get it on your hands it will stain them yellow, but you can wash it off safely with soap and water.

Place copper-clad board in etching solution (see Fig. 6-6). Agitate constantly; maintain at even temperature. Ferric chloride will take from 15 to 45 minutes to dissolve the unwanted copper. All the solutions will become weak after using and will need to have fresh etchant added periodically.

Watch for the phenolite board to show beneath the copper clad. When all unwanted copper is removed, take the board out of the solution.

Fig. 6-4. Expose pattern on board on light table for 7 minutes.

Fig. 6-5. Put exposed board in photo-resist developer for 2 minutes.

Fig. 6-6. Place copper-clad board in etching solution.

Rinse thoroughly with water to wash off the acid. Wipe dry. Use steel wool to remove the photo resist from the circuit for ease in soldering.

Drill holes in the board for component leads. Mount parts on the opposite side of the printed circuit on bare phenolite side as planned in pencil layout.

Leave the leads of the components long until after soldering. Bend the leads out after inserting through holes so as to hold them in place.

Solder the parts. Use a low-wattage soldering pencil. Check the operation of the circuit.

Packaged sets of components which can be adapted to printed-circuit work can be purchased from various firms. This is cheaper and easier than trying to stock a large inventory of parts. Chassis-type projects can also be adapted nicely to printed-circuit work.

Figure 6-7 shows one of the kits that can be used.

TENSILE TESTER AND ITS USE

A tensile tester is an excellent piece of equipment for the radio hobbyist, especially the ham operator. Wire antennas are very commonly used, and this tensil tester will tell you how strong your wire really is.

The following materials are called for, though there is some leeway for substitutions evolving from a given builder's ingenuity:

1—9″ × 12″ × ¼″ steel plate
75″—⅞″ crs round stock
8—½″ #13 nuts
3—5″ channel iron, 11″ long

225

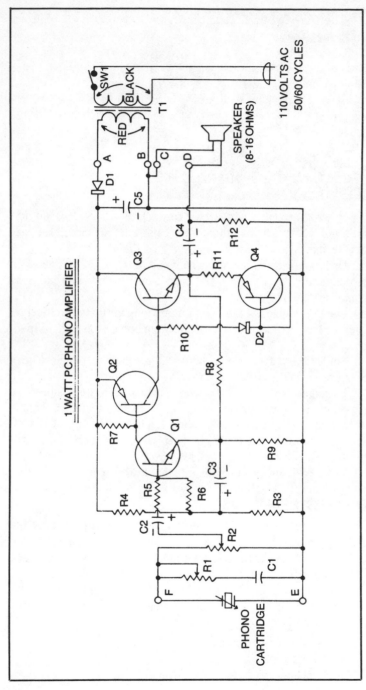

Fig. 6-7. 1W phono amplifier project PC board is shown going through the steps in Figs. 6-1 through 6-6.

There are two basic components of the tester: a hydraulic jack and a gage. The rating of the hydraulic jack is the main determinant of the capacity of the tester. The jack used with this tester has a 5000-lb capacity. It is suggested that there should be gages available for use with the tester having the following capacities: 0-500 lbs, 0-2000 lbs, and 0-5000 lbs. This will provide more accuracy on the tests at lower or higher ranges, depending on the material tested, although the tester may be utilized with a single gage.

The first step in the construction of the tester necessitates removing the fluid plug from the jack and, with the proper fittings, mounting the gage in its place. Next, make a flat steel platform approximately 9″ × 12″ × ¼″ for mounting the jack. Raise the platform about an inch with either angle bar or feet. Six inches on center, drill a ½″ hole on each side of the jack for mounting of the main support arms, which are of ⅞″-d crs. The two main support arms are 26″ long and the short arms are 11½″ long. The main support arms are turned down to ½″ at one end for a distance of 1″. The same is done to both ends of the short support arms. See Fig. 6-8.

Next, cut three pieces of 5″ channel iron 11″ long. On two of the channel irons, drill, on the center line, a ½″ hole 2½″ from each end. On the third piece, drill ⅞″ hole 2½″ from each end and a 1″ hole 1″ from each end.

At a point of 18½″ up from the threaded end of the main support arms, weld the channel (with the ⅞″ hole) to the main support arms, having the angle of the channel face down. These welds are made on the inside (⅞″) holes and are on both sides of the channel. Place one of the other channels on the jack head, with the angle down, and insert the main support arms through the appropriate holes; fasten them to the base, using the nuts previously fitted to the ends. Fasten the short support arms to the lower channel having them pass through the welded channel. Fasten the third channel with the angle up, on the top of the short arms, with the main support arm coming up through the inside holes.

ANOTHER TENSILE STRENGTH AND DUCTILITY TESTER AND ITS USE

This testing device, which you can build, measures two important properties of materials: tensil strength and ductility. The tester, although not as accurate as commercial machines, provides an inexpensive means of indicating the value of the these physical properties with minimum expense, time and labor.

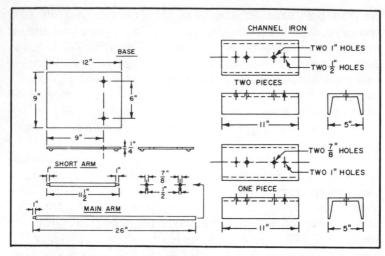

Fig. 6-8. Dimensions of the tensile tester.

Since tensile strength is one of the most important properties of engineering materials, tensile tests are frequently made and are among the simplest of all mechanical tests to conduct. It is also common practice to determine ductility of the specimen by measuring the elongation of a portion of the sample. These two properties are extremely important when considering what materials to use in manufacturing a product. Without prior knowledge of these and other properties, a product may easily fail during use.

Tensile strength may be defined as the resistance of a material to being separated by two forces acting in opposite directions. The measurement of a material's maximum resistance is usually expressed in terms of the strength of a one-inch-square cross section of material, or "pounds per square inch" (psi).

Ductility is the ability of a material to be deformed physically without fracturing. The term "plasticity" has much the same meaning, but ductitity is commonly used in reference to metals. Two common measures of ductility are percentage of elongation and percentage of area reduction.

Method of Construction

Obtain a hydraulic jack of the capacity desired. Remove the bottom drain plug and replace it with threaded pipe to which a hydraulic pressure gauge, available at any plumbing supply company, is attached. The price of the gauge depends on the capacity you wish to measure. On some jacks, a small plug inside the grain hole must be removed in order for the fluid to reach enough pressure to register on the gauge.

Mild steel, easily machined, was used to construct the metal frame. Note that on the model in the photograph another piece of mild steel has been welded to the top of the frame for additional strength. The strength and capacity of the machine can be further improved by constructing the top of the frame as one solid piece, as shown in the drawing. The hinged part of the frame is fastened by a steel pin, press fit, but a nut, bolt, and washer will perform equally well.

A guide must be constructed to prevent the frame from slipping sideways off the jack when force is applied. Drill two holes into the top of the jack to accept two 2" machine screws. Because the top piece rebounds when the specimen is broken, a brace is attached across the top of the hinged piece between the screws.

Notch the ends of the top and bottom horizontal pieces of the tester on a vertical mill to provide the space necessary for insertion of the jaw assembly.

Medium carbon steel is used to make the jaws. This permits heat treating them after the desired configuration has been obtained. In each of the upper and lower jaw assemblies one jaw is welded to the jaw bracket, while the other is left free to slide on the pin, allowing the jaws to adjust to different material thicknesses. Upper and lower jaw assemblies are inserted into the notches on the upper and lower arms of the frame into which a steel pin is pressed. Again, a nut, bolt, and washer can be used instead of the pin.

Specimen thickness capacity of this tester is ⅛" flat material. To fasten the specimen in the jaw, drill and tap two holes on one side of each of the upper and lower jaw brackets and insert hexagon-head bolts. An Allen wrench enables the operator to apply more torque in tightening each bolt against the blank end of the specimen. Other types of bolts, requiring different types of wrenches, can be used with similar results.

Bolt the hydraulic jack and frame unit to a maple base 1" thick. Although other wood can be used, maple makes a sturdy testing unit because of its hardness, strength, and weight. See Fig. 6-9.

Preparing the Specimen

The specimens, if flat in design should be made to the general shape shown in Fig 6-10. Overall length can vary, depending on jaw capacity. Cut the material to shape on a bandsaw. Specimens must be made as accurately as possible with no scratches, nicks, or irregularities in the metal. This will also insure a more consistent comparison among the different materials being tested. After the specimens are made, open the upper and lower jaws and insert the blank ends. Tighten the jaws firmly to prevent slippage.

To perform the test, raise and lower the jack handle and observe the reading on the pressure gauge. Continue this procedure until "necking," or a narrowing where the specimen is about to fracture, begins. At this time the pressure reading may start dropping back, rather than increase. This is to be expected with specimens having considerable elongaton or stretching capacity.

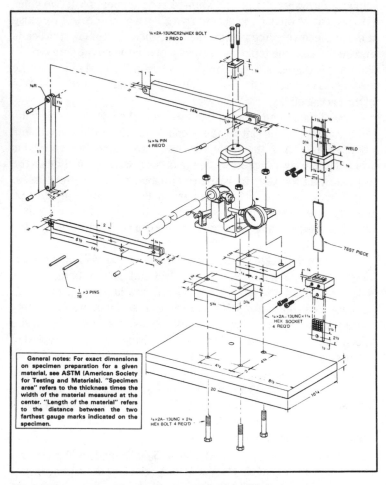

General notes: For exact dimensions on specimen preparation for a given material, see ASTM (American Society for Testing and Materials). "Specimen area" refers to the thickness times the width of the material measured at the center. "Length of the material" refers to the distance between the two farthest gauge marks indicated on the specimen.

Fig. 6-9. Dimensions of the tensile strength and ductility tester.

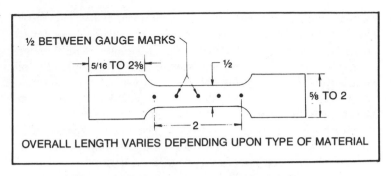

Fig. 6-10. A typical specimen for testing with the tensile strength and ductility tester.

To determine ultimate tensile strength of the specimen, record the highest point at which the pressure gauge registered. Insert that number in the formula below, and compute:

$$\text{Ultimate tensile strength} = \frac{\text{Maximum load, lbs.}}{\text{Specimen area, sq. in .}}$$

If the percentage of elongation is to be determined after the fracture has occurred, press the ends of the broken specimen together and measure the distance between the two farthest gauge marks. Insert this figure in the formula below, and calculate:

$$\text{Percentage of elongation} = \frac{\text{Final length} - \text{Original length}}{\text{Original length}} \times 100$$

A VISCOMETER AND ITS USE

Construct this inexpensive viscometer for use in automobile maintenance or ham radio dummy load oil testing. With it you can measure static oil viscosity, which is imply timing the rate of flow of a certain amount of oil through a small hole. The time will depend upon the hole size and oil quantity. Although oil viscosity (thickness), perhaps the single most important characteristic of engine motor oil, is generally difficult to determine, this technique is simple and interesting.

Industrial Measuring

Oil viscosity for industrial application is generally measured either by dynamic viscosity measured in centipoise (cp), or static

231

viscosity measured in Saybolt Universal Seconds (SUS or SSU). The Society of Automotive Engineers (SAE) designations for oil viscosity have been established to obtain simple, readily understandable values. To be marketed with an SAE number, the oil must be tested in a Saybolt viscometer at various temperatures. This unit is basically a tube with a volume of about 80 cm³ which can be heated in an oil bath. A tiny hole in the bottom of the tube permits the oil to drop out into a beaker. The time it takes to drip 60 cm³ is a measure of the oil's viscosity at the test temperature. This time, in seconds, is referred to as the SUS viscosity.

A commercially available Saybolt viscometer is quite expensive, but this viscometer costs almost nothing. Here's what you need: simple stand; plastic bottle (about 6 oz.) with a small hole drilled in the cap (ours is 0.070" and slightly countersunk on the inside), the bottom cut out to facilitate filling and reading, and two lines painted close to the top and 1-½" apart; container to catch dripping oil; method for heating small quantities of oil; thermometer; stopwatch (or wristwatch with second hand); and two-cycle semi-logarithmic graph paper.

The semi-log graph paper isn't a necessity, but it makes graphical construction quite simple. Over the temperature ranges encountered in this test, the viscosity is pretty close to a straight line when plotted on semi-log graph paper. Using standard graph paper would involve French curves and deciding how to draw the final curve.

Test Procedure

At room temperature, fill the bottle with oil to a level above both lines. Measure the oil temperature.

Uncover the cap hole and let the oil drip out. Begin timing when the oil level reaches the first line. End timing when it passes the second line (typical time is one to five minutes).

Add a small amount of heated oil to the container, refill it and repeat the previous steps for several different oil temperatures.

The result is a list of oil temperatures and corresponding times (Fig. 6-11). Graphing one against the other will result in almost a straight line on semi-log graph paper. Typically, the data will appear as a shotgun pattern. Draw a straight line which fits most of the points; this is called constructing a curve of best fit.

Test only "straight-weight" oils. Multi-weight oils are quite sophisticated; it's difficult to get good results. If you test multiple-viscosity oil, results will indicate only a general trend; they should not be interpreted as absolute.

DISPLAY CASE

Many hobbyist need a portable display case for their electronic projects. An old television cabinet seems to be a perfect housing for such a showcase.

First clean both cabinet and glass. Then fit a piece of plywood over the bottom of the cabinet to cover the holes. Next mount a 2-rpm synchronous motor and fasten a turntable to it. Two 60-W light bulbs are then installed at a slight angle between the top and sides of the cabinet, so that they focus on the turntable.

Use two single-pole, single-throw rotary switches, mounted in the original holes for the television's controls, to operate the motor and the lights (Fig. 6-12).

After attaching a plywood back and a hinged door, paint the interior and affix a metal support with casters to the finished product. Your display case can be set up in any room of the house to show off your work.

SOLDER TIPS

The average hobbyist has many and varied calls for special soft solder jobs. Some require coverage of surface, others, pin-point contact, so it is advisable to use a solder iron with replaceable points. Each tip is then shaped for its specific purpose.

It is wise to drill a well slightly above the point, then cut a feed channel to the tip. In that way a quantity of solder may be stored

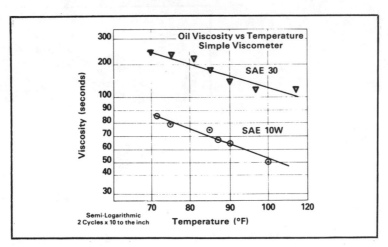

Fig. 6-11. Results of graphing using the curve of best fit based on data obtained with the simple viscometer.

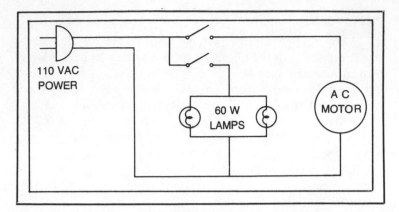

Fig. 6-12. Wiring diagram for the display case.

above and allowed to flow toward the point of contact as needed, by tilting. Round, blunt, flat, and pointed are all shapes into which a solder tip may be ground depending on the requirements.

Very great demand is made of the solder iron to reach remote points in out-of-the-way places. Radio technicians, art-metal workers, refrigerator mechanics and motor repairmen all require a tip which will reach some difficult-to-get-to spot and land on a pin point. If you thread a piece of #8 copper wire through a hole drilled into a solder-copper, then twist it tightly over the point (Fig. 6-13)

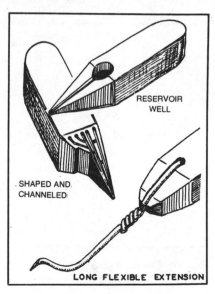

Fig. 6-13. Various solder tips you can make.

RESERVOIR WELL

SHAPED AND CHANNELED

LONG FLEXIBLE EXTENSION

you will have the answer to that problem. File a small guide channel on each face of the solder-copper from the hole to the point to keep the copper wire from slipping about, and allow for a snug joint. One end of this wire is cut off at the end of the twist while the other extends as long as is needed. The end of this extension is filed to a suitable point and tinned.

This wire may be bent to circumvent any obstacle in order to reach the place which needs the solder. It may be alternately bent whenever needed for other job requirements.

"NEW JUNK"—AN ELECTRONICS CHALLENGE

Though electronics hobbyists use a variety of methods and disagree on many issues, there is one way in which they are very much alike: they collect and save junk. There once was a time when old tube-type radios and televisions were hoarded for their junk parts, and many worthwhile projects were constructed from those collections.

Then a wonderful (yet terrible) thing appeared: the transistor. Suddenly, the junk looked much different. Component lead dress was shorter, parts were mounted on printed circuit boards, everything was small and salvaging became more difficult.

Hobbyists were ready and eager to use the new technology. But our junk piles slowly became obsolete. It was difficult to determine what junk was usable because components were harder to identify and retrieve. We stopped collecting—and using—junk. Our shops became neater looking, but we spent more money for components and missed the valuable learning experiences which the junk had provided. The following are some solutions to the unique problems teachers encounter when they try to help students use the new junk.

Identifying Parts

Many transistors, diodes and other parts are not marked at all, or they are marked with special codes used by their manufacturers. Looking up these markings in cross-reference manuals is a difficult, and at times fruitless, chore. But the same technology that gave us the new junk has also given us new tools. It is easy to check a transistor with an in-circuit transistor tester to determine whether it is NPN or PNP, as well as its beta and pin configuration.

In project circuits requiring medium-power audio transistors, this information is usually sufficient to determine whether or not the transistor can be used. These tests can often be conducted without removing the transistor from the junk circuit. Other componenets—resistors, diodes, and capacitors—can be easily checked if one lead is disconnected.

The second problem is to remove the needed items from the junk without destroying them. This was once a very difficult problem—most of us can recall our first frustrating attempts to extract transistors from printed circuit boards. A desoldering tool or solder wick has greatly increased the success rate for salvaging semiconductors.

The Advantages of New Junk

Among the benefits of using new junk is that it is much smaller than the old junk. Except for television picture tubes, the amount of junk which used to fill an entire storage room can now be neatly tucked away in a few drawers.

New junk components, as with old junk, cost less to salvage than to buy new. Some projects can be built entirely from the parts contained in one small transistor radio. Other projects which could be constructed from transistor radio and television junk are oscillators, amplifiers, sirens, power supplies and even small transmitters.

The best thing about using junk, however is that you learn more than when you use new parts. When students buy a kit and complete it, they learn little more than how to solder. Most students are not challenged much by kit instructions which tell them to connect a brown-red-red resistor between points A and R. Even when you build a project from new discrete parts, you are not forced to learn how to identify and evaluate components, because you can very easily ask a store clerk for a 1200Ω resistor, and you will readily obtain one which is in a marked container and is known to be good.

Junk can solve these problems by creating new ones. To build a project from junk, you must be able to:

☐ Identify components by their color codes or other markings;

☐ Use a transistor tester;

☐ Use a capacitor checker;

☐ Use an ohmmeter;

☐ Lay out a printed circuit board to fit the actual components they find, regardless of differences in size or pin configuration;

☐ Use appropriate cross-reference manuals and catalogs to determine component types and substitutes;

□ Use desoldering aids and appropriate heat-sinking methods to successfully remove semiconductors and reuse them;

□ Construct or adapt a suitable case for the project;

□ Build the project without pictorial diagrams or other simplifying aids.

Problems and Joys

One old problem of using junk is still with us—craftsmanship does suffer. You will be absolutely dismayed at the horrible looking pile of junk your finished projects turn out to be. The projects will not always be the customary brushed aluminum or black plastic boxes with neatly arranged switches and knobs. Component leads will be twisted, bent and otherwise mangled beyond recognition.

PLUG AND JACK

Although many different kinds of electrical connections are available, none have the strength, rigidity, adaptability, and fool-proof construction that is required for their use in a hobbyist's shop.

The plug described here is designed for use with a heavy-duty, ¼" diameter, solid-rubber, insulated, stranded, 20-gage wire. One of the chief advantages of the plug is its ability to be attached securely to its wire without requiring removal of the insulation. After the insertion of the end of the wire into the ¼" hole, a pointed screw pierces the rubbers and locks the wire firmly in place (Fig. 6-14). At the same time, it insures positive electrical contact. Should the wire become damaged, the end need only be cut and reinserted.

The jack is designed to be used on instrument panels and on all types of electrical apparatus. The base is threated for an 8-32 screw. This permits it to be used also for the terminals of a standard dry cell. For this purpose, a drilled piece of ⅛" thick black fiber is first placed on the cell terminals to prevent flexing or breakage.

The 6-32 contact screw is turned to length and pointed in a lathe by first inserting in into a holder made from 3/16" diameter threaded piece of brass 3/16" long.

The spring is made of piano wire by forming it around the jaws of specially prepared round-nosed pliers. Part of one jaw is removed for clearance. The second jaw is softened so that a 1/16" diameter hole can be drilled ⅛" deep in the center of the jaw at the point where it engages the shorter upper jaw.

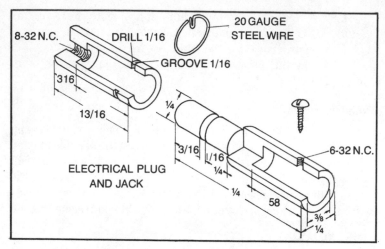

Fig. 6-14. Plug and jack dimensions and assembly.

To make the spring, cut wire to a convenient length, bend and flatten a 180° turn about ½" from the end. Insert this bend into the hole in the jaw, and close the jaw tightly so that it spreads apart the two upright wires. Then make a complete loop about the lower jaw with the long end of the wire. The excess may now be cut. A bit of trial and error might be required to secure the optimum diameter and length. Once determined, the springs can be made rapidly and in quantity.

For ease of identification, the wires can be made of two colors, red and black, the longer ones of one color and the shorter ones of another. The jacks can also be attached to a metal strip to house all the lead wires in use. This should permit a quick check of missing lead wires.

PRINTED CIRCUITS VIA ELECTRICAL TAPE

Many hobbyists find the electrical tape resist method an effective way of providing printed circuits. Minimum expense can also be added to the unique qualities of this process with only the following materials being needed:

Copper laminate board
Paper punch
Roll of plastic electrical tape
One photographic tray
Solution of ferric chloride
Scissors
Wax paper.

238

The methods of constructing printed circuits appropriate for project work include photographic and tape resist. Tape resist gives the good results. The photographic method is fine where many identical copies are needed, but for the individual hobbyist who is only going to make one copy this method proves to be a waste of time and materials when the end result and methods of basic construction are not changed.

If examination of a commercial printed circuit is made, it is seen that the layout is composed of two parts—the *tie-on points* for the components and the *tracks* connecting the respective items. When electrical tape is used as a resist, the tie-on points and tracks can be cut from a roll of tape and then placed on the copper laminate in the pattern of the layout (Fig. 6-15). On a circuit produced by the electrical tape method, all tie-on points will be circles as they are cut with a paper punch, and all the tracks will be made by cutting the tape into strips 3/32" wide. Obviously, the tape may be punched and cut before the actual layout procedure, but steps must be taken to keep the items from sticking to each other. One of the best ways found to eliminate this problem was storing them on wax paper so they will peel off easily as needed.

After the printed circuit layout is completed and the tape resist is made to adhere firmly to the copper laminate, it is time to do the etching. Here a photographic tray and the ferric chloride are needed. The tray is filled with the acid solution ¾" deep. The board is placed in the solution and the tray is agitated until the unprotected copper has disappeared. Etching time will vary on the strength of the acid. For best results, used ferric chloride should be placed in a new container and only used one more time.

The *make-ready* for the mounting of components is a simple process. Take the etched board from the acid solution, clean it in clear water and remove the tape. In the middle of every circle, drill a hole to allow the wire of a component to pass through. The hole size will vary according to need. Holes are also drilled to accommodate electrical parts that required mechanical fastening. This would include variable capacitors, potentiometers, power resistors and transformers. After all necessary drilling has been done, the board is cleaned with steel wool so that the copper is free of all grease to allow for easy soldering.

Due to the popularity of printed circuits today, the materials are readily available, particularly in the large metropolitan districts. Copper laminate can be obtained from most electrical suppliers. The names of local distributers can be found in your telephone book.

Ferric chloride is a chemical obtained from local suppliers of chemicals. Ferric chloride can be obtained in either solid form or premixed. The solid form requires dilution in water, whereas the premixed is ready to use. However, no matter which way it is purchased the end result will not change.

For those who live too far from local distributors, several mail order firms with excellent catalogs provide another convenient means of obtaining these materials.

THE BIG TRANSISTOR AND ITS USE

Nothing is more frustrating to the beginning electronics hobbyist than to complete a transistorized project that fails to work. Although hardly the only causes, a defective transistor or an unsuitable substitution of transistors are common sources of such failures. Yet substitution is often desirable, even in a working project for experimental purposes.

Construction and Application

This simple little transistor substitution box consists of a transistor socket, a three-lug terminal strip with a separate mounting lug, and three leads with small insulated alligator clips. Using any convenient small metal box, this Big Transistor is wired according to the accompanying schematic (Fig. 6-16).

To use, mount all components on a circuit board or terminal strip—depending on the construction method—making connec-

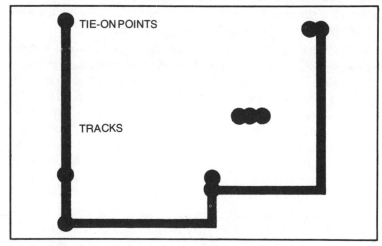

Fig. 6-15. A circuit using electrical tape as a resist.

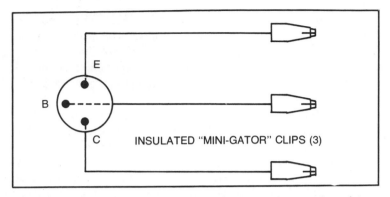

Fig. 6-16. The Big Transistor makes it possible to plug in a variety of transistors with short leads.

tions with the substitution box instead of the specific transistor. Transistors are then plugged in until you are satisfied with the performance of the project. If transistor substitution and project performance have been headaches in your beginning projects, the building of a Big Transistor may provide the relief you've been looking for. See Fig. 6-17.

THIRD HAND FOR SOLDERING SMALL PARTS

How many times have you wished for three hands when soldering connections for electrical projects? This rather simple fixture will provide you and your students with a free hand when doing soldering on individual project circuits or in mass-production assembly. A piece of plywood is used for the base. The frame is ½" angle iron, notched, bent and attached to the base. A strap is attached to the top with a machine screw and a wing nut. The solder pencil or iron is clamped to the bracket to provide you with a very useful "third hand." See Fig. 6-18.

SOLDERING PISTOL

This soldering pistol is functional, inexpensive and requires only the skill and knowledge of any newcomer. It provides opportunity to exercise a little originality in the final design of the handle.

Construction

The wood used in the project should range in thickness from 1" to 1¼". If necessary, two pieces can be joined together with glue to gain the desired thickness. The handle can be roughed out on a band saw or with a coping saw. Boring the hole is best accomplished with a speed bit, using the drill press and drill press vise. The boring can be done with a brace and bit. Then, the final form can be worked by filing and sanding. The handle can be finished in one of a number of ways. The practice of dipping as used in industry works well with this project. A section of ⅜" dowel forced up into the handle works well during dipping operations. See Fig. 6-19.

In assembling the electrical parts to the handle, the cord should be fed up through the handle and out the barrel. If a candelabra socket is used, a force fit is usually enough to hold the socket firmly in place. The screw cap should be forced down tightly on the candelabra socket. It will then be necessary to file or saw off a portion of the cap to provide clearance for the line cord. The socket will then be forced back into the handle. If this force fit is not satisfactory, a set-screw arrangement can be used. Once the socket is installed, only installation of a soldering tip remains. Then the soldering pistol is ready to be tinned and used.

Here are a few additional tips that might be useful. Materials other than wood can be used to make the soldering pistol, but wood

Fig. 6-17. The Big Transistor.

Fig. 6-18. A typical third hand for soldering small parts.

is easy to work. Heat transfer from the tip to the wood handle is minimal, even for long periods of use. Heat transfer from the tip to the table top might pose a problem. If so, a stand similar to the stand in the drawing can be easily made.

Soldering tips of different wattages and shapes are easily available from any major electronics supplier. They are of relatively low cost, especially if purchased in quantity.

The soldering tip should not be subjected to shock or quick cooling, as in water. Also, care should be exercised to avoid over-tightening the tip into the socket, especially if candelabra sockets are used. With normal care and use, the soldering pistol will give good service for a long time.

PROJECT BOX

Every electronics hobbyist shop needs a good, inexpensive project box. Metal containers cost more, present safety problems with higher voltages, and hide fine wiring workmanship. Less expensive alternatives also often fall short, leaving components exposed and subject to breakage, electric shock, and make for an unattractive final product. Regrettably, an electronics project is often judged by its case.

This inexpensive box, on the other hand, is a safe, attractive, and low-cost container that can be assembled by the student in one

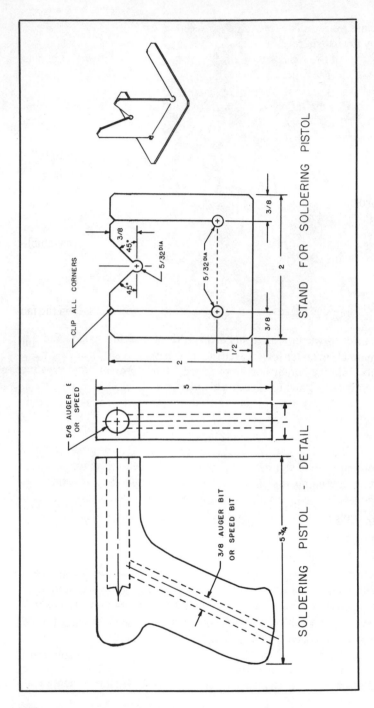

STAND FOR SOLDERING PISTOL

CLIP ALL CORNERS

3/8
45°
45°
5/32 DIA
5/32 DIA

3/8
2
3/8
1/2
2

5/8 AUGER &
OR SPEED

5

SOLDERING PISTOL - DETAIL

3/8 AUGER BIT
OR SPEED BIT

5 3/4

SOLDERING PISTOL

Fig. 6-19. Dimensions for the soldering pistol and the stand for the pistol.

class period. The key to its economy is the precutting of the parts by the instructor.

Cut the top and base from ⅜" one-side-good plywood, making 7-¾" × 3-½" pieces. On a table saw, cut a groove, 3/16" deep and ⅜" from the front edge.

Cut the sides from 1" × 4" (nominal) furring strips, making 2-½" long pieces. Groove one of the surfaces as above. The grooves are used to mount the face plate of clear acrylic.

Cut the acrylic panels on a band saw, making 6-9/16" × 2-1/16" pieces. Dimensions can vary, depending on the projects, but the suggested size is more than ample for such classroom projects as small transformers or speakers.

Box Construction

With all parts precut, the box is easily modified and assembled in seven steps:

☐ Drill holes in the face plate for all project controls and exposed instrumentation.

☐ Use epoxy cement or screws to mount controls on the face plate.

☐ Use glue and ¾" wire brads to assemble the box while fitting the plastic face plate into the precut grooves in the wood blanks.

☐ Glue or screw the circuit board on the inside base of the box.

☐ Sand the edges of the plywood thoroughly and clean all outside surfaces.

☐ Use clear spray varnish or wax to finish the outside of the box.

☐ Staple carboard or cloth over the exposed back of the box.

The box also can be used for a breadboard. Its attractiveness encourages any hobbyist to keep his workmanship visible for others to see.

A FLEXIBLE ENCLOSURE

Electronic projects should be housed in suitable attractive enclosures which may be bought or built. Commercially-made boxes, however, are expensive. On the other hand, it normally is not worthwhile to spend a disproportionate share of your time on any given project in fabricating just the enclosure.

The "shadow box" shown (Fig. 6-20) is a simple solution to this dilemma. Suitable for a wide variety of projects, it is also easy to make—a feature which your beginning students will appreciate.

This is a general layout for the box. Metal as thin as 28 gage may be used for construction of the box with satisfactory results. Simply "plug in" the desired dimension and select the straight or inclined (indicated by the dashed lines) cover style.

The following sequence of operations is suggested:

☐ Lay out or cement full-size pattern to material.

☐ Cut material to size.

☐ Drill holes necessary for component mounting.

☐ Form the base.

☐ Form the cover.

☐ In assembly, drill holes A and C, and B and D for root diameter size of screw.

☐ Drill or taper ream to enlarge holes A and B to sheetmetal screw-clearance size.

☐ Finish base and cover with matching or contrasting colors.

To add a special touch, common adhesive-backed wood or pattern-grained paper may be used to advantage in finishing.

EASY MOUNTING OF COMPONENTS

This method of mounting components of small electronic projects was worked out by the writer after seeing how much difficulty newcomers have with this type of work.

The method requires a nonconducting panel about 1/16" thick (Formica or similar material) and the riveting type of solder lug. A hole is drilled the size of the solder lug and the lug is inserted in the hole with the part to be soldered facing the back side of the panel. A couple of homemade punches head over the front side of the lug as an eyelet. The components are then mounted on the front of the panel and the leads soldered on the back along with any connecting wires of the circuit. See Fig. 6-21.

This method leaves nothing suspended in air and makes a neat looking assembly. While not as up to date as the printed circuit board, this method is similar to mounting techniques used in commerical products and kits.

SOLDERING GUN

This soldering-gun project uses a number of different processes. These processes are: woodworking for the handle; electricity for wiring the plug and switch as well as winding the coil; metalworking for forming the buss bar for the secondary coil and securing the covers and laminations; and electroplating of the covers, tips and the buss bar. See Fig. 6-22.

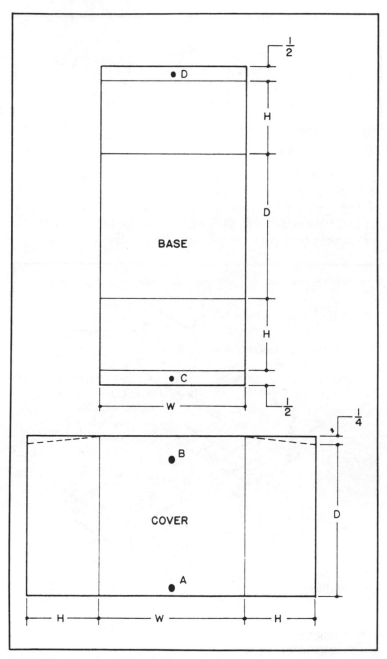

Fig. 6-20. Height (H), width (W) and depth (D) dimensions are plugged in on base. H, W and D of cover are simply made to fit the base.

Fig. 6-21. Easy component mounting with solder lugs.

To test its operation, go through the following procedure: press the switch and observe the tip start to smoke after it is heated; apply solder (40-60 rosin core) to the tip and allow it to melt. When satisfied with this, wipe off the tip with a rag, and you can use the gun for soldering all kinds of connections.

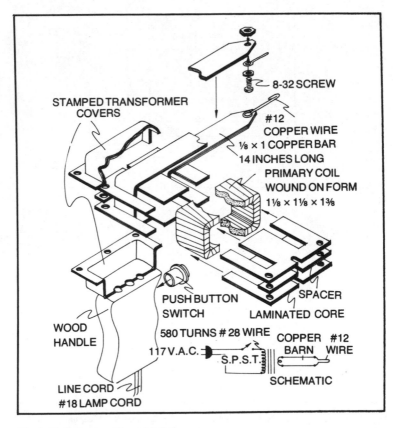

Fig. 6-22. Cutaway of the soldering gun.

A check list to assess the quality of workmanship follows:

☐ Does the tip melt the solder within eight seconds after the pushbutton is pushed?

☐ Is the coil filled with laminations and spacers (Fig. 6-23)?

☐ Are the covers on straight, and are they painted or plated?

☐ Does the pushbutton switch fit snugly?

☐ Does the handle have a smooth finish?

☐ Does the AC plug have an underwriter's knot?

☐ Is the copper bar secure? (No movement should occur.)

☐ Does the tip heat?

☐ Does the tip get hot enough to melt solder?

If the covers should get hot, check the copper bar to determine whether it is coming in contact with the cover.

This soldering gun has won several awards in competition, and we feel it offers a challenge to students, as well as affording the learning processes listed above.

EXPERIMENTER'S CHASSIS

Working with experimental circuits plays an important part in gaining understanding of the principles of electricity and electronics because when the principles to be learned are placed in operation observable measurements can be taken. A good chassis for building experimental circuits is vital. There is much to be gained from designing and building such a chassis.

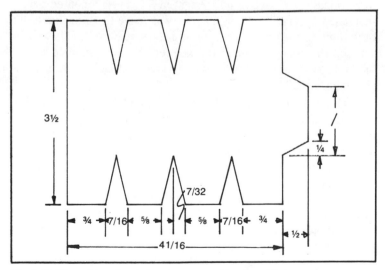

Fig. 6-23. The coil form insulation pattern used for the soldering gun.

Building a chassis such as the one described here accelerates learning about the various hardware and other materials that play an important part in electronics work. It can be completed at modest expense.

In preparing to build this chassis, serveral important design factors were considered. First, it was decided that the unit should be able to handle any type of tube or transistor available now or likely to become available. Another aim was to have a compact unit, without crowding, that would be able to handle several stages if necessary. Yet another was to provide connections that could be quickly and easily made and would be secure. For multiple stage work, there was to be an option of making certain leads (ground, filaments, etc.) conmon to mounted parts (transformers, relays etc.). Finally, the unit must be durable as it would be subjected to constant arranging and re-arranging of parts and components.

Examination of the drawings of Fig. 6-24 will show that the basic chassis will take various inserts (Fig. 6-25) to handle different tubes or transistors, and will permit adaptation to future designs. Quick, tight connections are made with Fahnestock clips. Also any number of basic units can be plugged together for multiple stage work. When units are plugged together, switches will give the option of making ground and/or filament leads common between units.

While each basic unit has a panel for mounting switches and potentiometers, space is not allocated for chasis-mounted parts. Instead, a special mounting adapter was made that fits above the basic unit to take transformers or relays required. This avoids wasting chassis space when not needed. Since each basic unit is only 4½" wide, a four-stage chassis would only be 18" wide, and a six-stage only 27" wide.

The parts list (Table 6-1) gives all the materials required for each basic unit. Since the drilling for the plugs and jacks in the side pieces is critical it is best to cut and drill side pieces together for the ultimate number required and save the extras. First, cut the hardboard top. Next, cut and drill at least one transformer-relay adapter plate.

Mount the switches, with the double-pole switch in the center hole. Mount four banana plugs (with solder lugs on the inside) on the right-hand side piece, and four banana jacks (with lugs on the inside) on the left-hand side piece. Cut and drill the front panel and fasten to the front support with two ½" #8 roundhead wood screws. The top three holes are for rotary switches and poten-

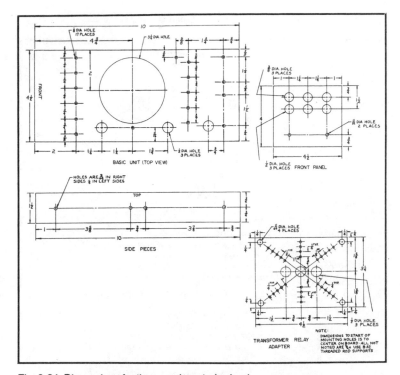

Fig. 6-24. Dimensions for the experimenter's chasis.

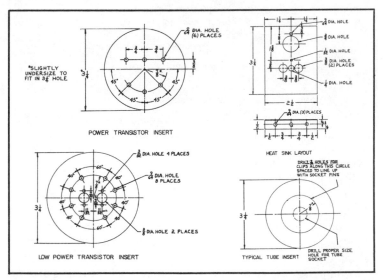

Fig. 6-25. Dimensions of tube and transistor inserts and the heat sink layout.

Table 6-1. Parts List for the Experimenter's Chassis.

Amount	Size and Description Basic Unit
1	1/8" × 4½" × 10" hardboard (top)
2	1/8" × 1½" × 10" hardboard (sides)
1	1/8" × 4" × 4½" hardboard (panel)
2	½" × 1½" × 4½" white pine (ends)
17	#10 Fahnestock clips
4	G-C 33-034 banana plugs
4	G-C 33-192 banana jacks
13	Soldering lugs
2	S.P.S.T. toggle switch
1	D.P.S.T. toggle switch
14	3/8" × 6-32 machine screw and nuts

For Each Tube Insert

Amount	Size and Description
1	1/8" thick 3¼" d. disk hardboard
1	Tube socket as desired, with mounting screws and nuts. No. 10 Fahnestock clips, ½" × 6-32 screws and nuts for each tube pin.

For Each Low Power Transistor Insert

Amount	Size and Description
1	1/8" × ¾" d disk hardboard
2	Transistor sockets, Elco 3304
8	#10 Fahnestock clips
4	¼" × 2-56 machine screws and nuts
8	¼" × 6-32 machine screws and nuts

For Each Transformer-Relay Adapter

Amount	Size and Description
1	1/8" × 3¼" × 4½" hardboard
4	8-32 brass thread rod, 3½" long
8	8-32 brass nuts

For Each Power Transistor Insert

Amount	Size and Description
1	2½" × 3¼" piece 18-20 gage aluminum
1	1/8" thick 3¼" d. disc hardboard
3	#10 Fahnestock clips
3	Soldering lugs
3	¼" × 6-32 machine screws and nuts

tiometers. The bottom three holes are for toggle switches, pushbottons and pilot lights.

All inserts are slighly less than 3¼" in diameter, so the inserts fit snugly in the insert hole in the top, and all have clip mounting holes on a 7/8" radius. They can be made of hardboard. Bakelite or other pastic, or any other insulating material that is no more than 1/8" thick.

For power transistors, where some sort of heat sink is usually desirable, a special insert is used. The holes in the aluminum plate will accommodate a number of different power transistors, and space is available for other configurations and transistors.

In using the chassis, many items can be connected between clips (including tube socket clips), and short lengths of wires can be used to make connections between related clips.

DIP SOLDERING PRINTED CIRCUIT BOARDS

The soldering of printed circuit boards can be a long and tedious task if each individual joint is soldered by the point method. Industry has found that a much more efficient technique is to solder all the joints on the circuit board at one time.

The soldering of printed circuit boards can be done in a few seconds by dipping the circuit board into molten solder. It is one of the methods most frequently used by industry to solder components to the printed circuit board. Moreover, its relative simplicity affords the practical assurance that anyone can work with it safely and effectively.

The first step after the circuit board has been etched and drilled is to mount the individual components. If a part is to be mounted a certain distance from the circuit board, use small scraps of plastic between the component and the circuit board. After the wires have been passed through the circuit board, they can be bent to an angle of approximately 60 degrees (Fig. 6-26). This prevents the component wires from being pushed out of the holes when the board is dipped into the flux or molten solder.

Fig. 6-26. Components mounted on the printed circuit board prior to dip soldering with the leads bent to an angle of 60 degrees.

Some Precautions Are Necessary

It is not necessary to keep the long ends of the different component wires from touching each other for they will be cut off very close to the printed circuit board in a later operation. However, do be sure that none of the component wires touch the copper clad at any place other than where they pass through the circuit board. If the wires touch the copper clad, they will be soldered to the circuit board at every place that the component wire touches the copper clad. This would require a considerable amount of time spent in melting solder to loosen wires and to eliminate excess solder; it could result in the elimination of the circuit board.

After a final inspection assures the proper placement of parts, the printed circuit board is ready for cleaning. The cleaning operation, as indicated in Fig. 6-27, is accomplished by dipping the bottom of the circuit board into a liquid flux. Be sure that the flux does not cover the top of the circuit board. It should touch the copper-clad side of the board and the component wires which are on the underside of the board. This cleaning must be done immediately before the soldering operation—not several hours or days before—otherwise the copper-clad will tarnish.

A Lamp Speeds Drying

An infrared heat lamp may be used to speed up the drying of the flux on the printed circuit board if proper care is taken. The printed circuit board, for instance, should not be placed too close to the heat lamp nor be subjected to the heat for too long a time. Damage may result to those components designed not to be exposed to excessive heat.

Once the flux if dry, the circuit board is ready to be soldered. If the board is not dry and it is lowered into the molten solder, an explosion may result because of the presence of moisture.

The solder must be heated and maintained at approximately 410° F. If the temperature of the solder drops too low, it will begin to solidify; too high a temperature may result in damage to components, lifting of the copper-clad, or scorching of the phenolic plastic board.

The Board Takes a Dip

The dipping of the circuit board into the molten solder is a relatively simple operation. The board is placed in a jug to facilitate handling of the circuit board. The dross is removed from the top of the solder before the soldering operation is started.

Fig. 6-27. Applying liquid flux to the printed circuit board before soldering.

Instead of immersing the circuit board into the solder at a right angle, lower the board at an angle of 5 degrees. Figure 6-28 shows one of the edges of the circuit board being lowered appropriately into the molten solder; the remainder of the printed circuit board is then lowered into the solder.

As soon as the complete circuit board has made contact with the solder, remove the first edge of the board from the molten

Fig. 6-28. Dip soldering the printed circuit board with a rolling action.

255

solder with a continual rolling action until the entire board is removed from the solder. Immediately after removing the circuit board from the molten solder, remove the excess solder from the bottom of the board to eliminate any shorts between two or more of the conductors. Throw off the excess solder with a sharp flick of the wrist. Because of the molten solder, remember that safety precautions must be observed at all times.

Allow the solder to harden and then examine the board for possible shorts caused by solder between two or more conductors. Remove these with a low-heat soldering pencil. If any solder is removed near transistors or diodes, use a heat sink.

Easy-to-Construct Equipment

The equipment for dip soldering is simple to construct. The author used a melting pot made from 3/16" plate steel welded in the shape of a box. A cast iron box can be used in place of a metal pot. The size of the largest circuit board to be dip soldered should determine the size of the solder pot.

Use an electric or gas stove burner to maintain the heat of the solder and devise a way of determining the temperature of the solder. If the solder is heated to a high temperature, damage to the circuit board or components will result. This can be overcome by preheating the solder on a gas burner, thereby reducing the time necessary to heat the solder to its melting point usually required with electric burners. After the solder is preheated, it is placed on the thermostatically controlled electric hot plate which controls the temperature of the solder.

The heart of the hot plate can be an oven temperature control which maintains a constant temperture of the solder. The heat element—at one time inside the oven of the stove—is coiled underneath the hot plate element, which is connected to the oven temperature control. This thermostat controls the flow of electricity to the burner.

Since the oven element is not immersed in the solder, the temperature control will have to be calibrated using a pyrometer. Maintain a temperature of 410° F. for a period of time and permit only a slight variation of the required temperature. If the burner maintains a constant temperature, this reading will be indicated on the control. If the temperature reading is too high or too low, the oven control will have to be adjusted to compensate for the difference. If several different temperatures are required, then this same operation must be repeated until the desired temperatures are marked on the oven control.

To prohibit someone from changing the temperature setting, place a cover box over the oven temperature control. A shield can also be placed around the solder pot to prevent someone from accidentally making direct contact with the pot. The shield and cover box are pictured in Fig. 6-29.

Determine the size of the hot plate box before construction by considering the burner, the oven temperature control and the on-off switch.

Again, if you take care to regulate the solder's temperature with the oven control to reduce the possibility of damaging components or connections, you will facilitate the soldering of circuit boards via the soldering of all connections at one time. In this way you can reduce shop soldering problems and acquaint yourself with one of the soldering methods most commonly used in industry today.

NO FUSS, LOW MUSS CIRCUIT BOARD ETCHING

Designing and etching your own circuit boards with good results can be done without exotic and expensive equipment and involves little effort or time. Here's how to do it.

Cover a piece of styrene foam with paper. Cut the leads of the components of length and following the circuit diagram, stick them through the paper into the foam (Fig. 6-30). Route things to avoid

Fig. 6-29. Solder pot being removed from the electric stove. Note the safety shield around the solder pot and the cover box on the left front of the stove.

Fig. 6-30. Stick the component leads through the paper into the foam.

crossed copper lines; if the situation is unavoidable, you will later have to jump the crossed line with insulated wire.

Stand up capacitors and resistors to save space. For good air circulation, leave adequate space around resistors, and stand up those which will be dissipating excessive heat. In general, use common sense in your layout; e.g., keep the output of an amplifier away from the input to avoid feedback. You should experience less trouble here than in point-to-point wiring on a chassis.

Now take a felt pen and draw lines to connect the circuits. Draw small circles at point where leads are stuck into the foam. These are your eventual drill points. When you remove the components, you will have a *mirror image* of the exact circuit pattern.

To get the actual etching pattern (Fig. 6-31), turn the paper over and trace it through a light source. If you are going to use the pattern often, make a ditto master.

Cut a piece of circuit board to accommodate your design, plus a little extra for mounting. Lightly sand the copper with fine grade paper to clean off the dirt, grease and oxides. Don't touch the surface with your fingers. Place a piece of carbon paper on the copper with the template on top. Paper clip this sandwich together. Trace the pattern with a ball point pen (red will show what you have traced).

Remove the paper and check that all lines are visible on the copper. Now trace these lines with a resist pen. This is a tube of an asphalt compound which you can purchase from most radio/electronics outlets for about $1.25. Let the resist dry for about 10 minutes. We have learned that it's better to draw in the solder points first, and then trace the connecting lines.

We use an etching solution of diluted nitric acid in a ratio of eight parts water to one part acid. *Handle this stuff with care!* Pour the acid into the water—not vice versa: Pouring the water in first will cause splattering from an exothermic reaction. Don't breathe the fumes and don't get it on your skin or clothing. Work in a well-ventilated area wearing a full face mask. If you get any solution on skin or clothing, use plenty of water to get rid of it—and do it immediately.

Why bother with nitric acid? The acid has the advantage of allowing you to see how the etching is progressing at all

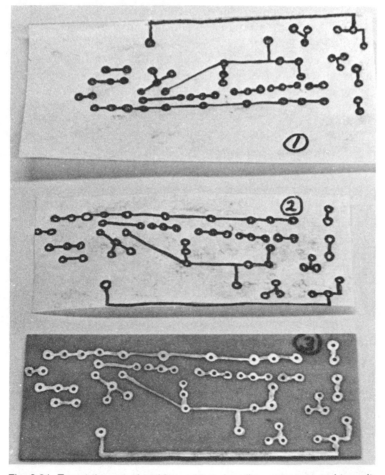

Fig. 6-31. To get the actual etching pattern, turn the paper over and trace it through a light source.

times—and it's much faster than ferric chloride, which quickly becomes clouded.

Put some solution in a glass baking dish—just enough to cover the circuit board. Place the board copper side up in the solution and slowly rock the dish to slosh the solution around. With agitation, the board should be completely etched within from 15 to 30 minutes. Remove the board with tongs or tweezers and wash it under cold water. While the board is under the tap, scrape off the resist with a piece of plastic (the resist is quite brittle when cold). Final cleanup can be done with kerosene.

Drill the mounting holes with a 3/64" bit. Lightly sand the copper again just before soldering. Your board should resemble the one in Fig. 6-31. Use a soldering pencil of around 35 watts or less with 60/40 solder.

Be prepared to dent your budget slightly for materials used in this technique—that includes the cost of a resist pen, kerosene, a baking dish, nitric acid and a pair of tongs. You will be able to produce dozens of boards before supplies run out—all of excellent quality, too.

DIODES IN COMPONENT PROTECTION—THREE WAYS

Occasionally meters and electronic circuits are fused to protect them against accidental current overload. With sensitive meters and circuits, fusing is not very practical because of the high and variable resistance a low-current fuse adds to the circuit. For sensitive circuitry, a semiconductor diode can be used to provide component protection. The use of a diode as a safety device can save many dollars in repair bills.

Since there are numerous applications of the semiconductor diode as a protection devise only three representative examples will be discussed.

Diode Protection for Current-Meters

Current-meter protection takes advantage of the fact that a P-N diffused diode requires a few tenths of a volt of forward bias before it is conductive. The circuit is shown in Fig. 6-32. Resistance R is put in series with the meter resistance R_m so that, when the current through the meter is two or three times the full-scale deflection, the potential across $R + R_m$ will be sufficient for the diode to conduct and bypass some of the current around the meter. Assume that a 50 micro-amp-meter is to be protected. This meter would typically have a resistance of 2000 ohms. With a

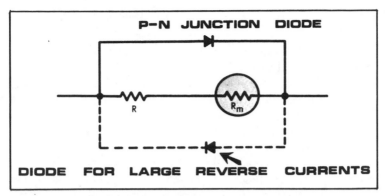

Fig. 6-32. Diode protecting an ammeter.

double overload (i=100 μA), the voltage across the meter movement would begin to conduct and shunt the current around the meter. A 1000 Ω series resistor would be required if the diode begins to conduct around 0.3V instead of 0.2V. Another diode could be added to protect against accidentally imposed large reverse currents.

Diode Protection for a Vacuum-Tube Voltmeter

A vacuum-tube voltmeter can be protected from the damaging effects of an overload by shunting a diode across the meter terminals, as shown in Fig. 6-33. The meter in a standard vacuum-tube voltmeter is sometimes regarded as burnout proof; however, overload conditions can cause damage to the meter since the current can be two to five times that of the rated full-scale value. Typical damage from an overload surge can result in a bent point or an open movement coil.

The characteristic curve of a typical low-cost diode has its "knee" at approximately 0.5V. The meter in a vacuum-tube voltmeter usually has a full-scale deflection of 0.2V. Thus, a diode connected directly across the meter terminals will have negligible effect on the meter accuracy. The shunt diode will provide good limiting action when the voltage applied to the meter appreciably exceeds the full-scale value.

In a vacuum-tube voltmeter circuit having a 200-μA meter movement (which required 195 mV for full-scale deflection), the meter current will be limited to 250 μA even though the input voltage was increased to 30 times the full-scale reading. Special diodes such as zeners diodes, could be used for protection devices; however, the cost of a common type diode is usually less expensive.

Diode Protection for a Transistor Emitter Follower Circuit

Many transistor projects utilize emitter follower circuits. One drawback of the emitter follower circuit is that if the output leads are shorted, the transistor in the circuit may be damaged. When such damage occurs, it is easy to lose interest in a project especially if you have to purchase replacement transistors and rebuild the project.

Damage to the transistors in a conventional emitter follower circuit, because of a short of output leads, can be prevented with the addition of two diodes, D_3 and D_4 as shown in Fig. 6-34. In normal operation e_2 follows e_1, and the presence of diodes D_3 and D_4 does not affect the operation of the circuit. The 510 Ω input resistor limits input impedance of the stage when short occurs at output. Under these conditions, diodes D_3 and D_4 clamp e_1 to the shorted output. Therefore, excessive current cannot flow through either transistor regardless of the input voltage.

SILK SCREENING A TEST INSTRUMENT PRINTED CIRCUIT

This project is an extinction-type neon voltmeter and continuity checker, using the printed circuit to eliminate wires (Fig. 6-35). This project, furthermore, was originally developed and produced with mass-production techniques. A parts list is shown in Table 6-2.

The voltmeter measures AC or DC, from 60V to 600V. It indicates whether the voltage is AC or DC without manipulations or switching. If both sides of the neon bulb glow, it is AC. If only one side of the bulb lights up, it is DC. The continuity checker is a visible type, employing a light bulb to indicate a short or a continuous circuit, adding to the functional value of this multipurpose instrument (Fig. 6-36).

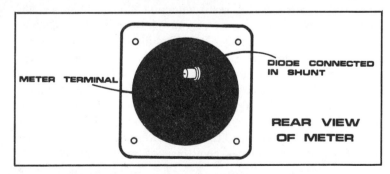

Fig. 6-33. Diode protecting a VTVM from overloads.

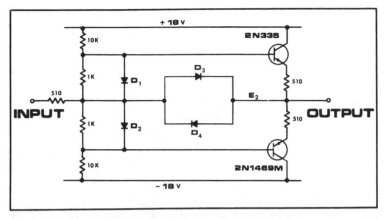

Fig. 6-34. Diode preventing damage to a transistor emitter follower.

It can be used to check voltages in radios, television sets and other electrical devices. Its application, you will have already surmised, is as broad or as is its maker's imagination.

The steps for making this project follow:

☐ Draw the circuit of the voltmeter-continuity checker (Fig. 6-37).

☐ Photograph the drawing.

☐ Develop the negative and make a positive print.

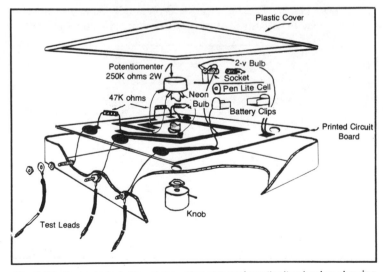

Fig. 6-35. An exploded view of the voltmeter and continuity checker showing how the parts fit over the circuit.

Table 6-2. Parts List for the Voltmeter and Continuity Checker.

2	Resistors: 47-K Ohms, ½-w, ±10%
1	Potentiometer: 250-K Ohms, 2-w (linear taper)
1	Neon bulb: Ne-2 or Ne-51
1	Lamp holder: screw-base socket
1	Lamp, flashlight: screw base, 2v
1	Penlite-cell: 1.5v, dry cell
2	Penlite-cell holders: made to size
3	Test leads: length to suit, flexible wire
3	Soldering leads: brass or steel, 3/16" hole
3	Machine screws: No. 6-32, brass or steel
3	Phone tips: for use with test leads
1	Knob: wooden dowel, 1" dia. × ¾"
1	Box: plastic sandwich box
1	Printed circuit: 4" × 4" Copperclad

☐ Prepare the silk-screen frame and place the picture of the circuit on the silk.

☐ Print the circuit on the copper-clad using acid-resistant lacquer with the silk screen process.

☐ Make the etching following this procedure: (*a*) Use a ferric chloride solution, 40 percent Baume 180° F. (*b*) Agitate the solution during the etching process. The etching can be done either by spraying the solution or dipping the copper-clad in the solution. (*c*) As soon as the copper disappears, wash off the solution under running water. (*d*) Remove the acid-resistant lacquer from the copper-clad with denatured alcohol, turpentine or lacquer thinner.

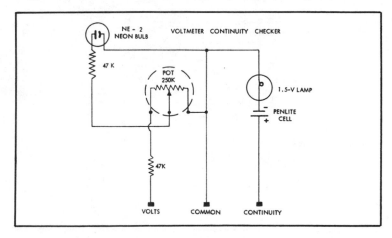

Fig. 6-36. Schematic of the voltmeter and continuity checker described in the text.

Fig. 6-37. The printed circuit ready for silk-screen processing on PC board.

The photographic silk screen process produces a sharper and more even-appearing line than the direct silk screen process. In a more complicated printed circuit this could be critical. In this project, though, the spacing between the lines is relatively coarse so that nothing is lost by using the conventional method although photographic silk screen printing was used here.

☐ Drill ⅜″ holes for the potentiometer in the plastic box, 1/16″ holes for putting parts and wires on the copper-clad, and three holes in the front of the box for three #6-32 machine screws.

☐ Solder the parts to circuit board.

☐ Make the knobs and the pointer from 1″ dowel rod and drill holes in the center of the dowel to fit the potentiometer shaft and press fit. Fix a nail to the knob after it has been painted.

☐ Assembly precautions and testing procedures that should be adhered to follow: (*a*) Do not overheat the copper (it will come loose from the board if it is overheated). (*b*) Double-check the solder joints and reheat the cold joints. (*c*) Intentionally short circuit the common and continuity side to see if the light burns. (*d*) Do not tighten the potentiometer nut too tightly or it will crack the plastic box.

☐ Calibrate the voltmeter portion of the tester. The variac (60V to 500VAC) is connected to the power supply; a regular voltmeter is connected to the variac; and the neon voltmeter, in turn, is connected to the regular voltmeter, in parallel. Turn the potentiometer completely clockwise and adjust the voltage from the source to 60V. Turn the knob clockwise until the light goes out and mark this point 60V. Continue this procedure until the range has been calibrated. Increase the voltage source 10V before repeating so that the meter will read in terms of 10V intervals. Calibrate it for both AC and DC.

☐ Make the test leads whatever lengths best suit the user's needs.

You now have an extinction-type voltmeter and continuity checker using a printed circuit—the kind that industry and the scientists use today.

TWO TYPES OF ELECTROPLATING

The answer to many finishing problems is the process of electoplating. A simply built machine, powered by a battery charger, is all that is needed. The cost of chemicals used will be more than offset by paint waste, for example, that is a yearly occurrence. The speed with which you get factory results will surprise all: five or six minutes, on larger jobs, to 10 seconds on a small screwdriver project. The rust-free, chip-resistant metal coating will please everyone.

The core of the machine is the complex of three 1½-gallon fish bowls purchased at a local dime store at about $2 each. Each tank contains a solution necessary for a nickel-plated or copper-plated finish. Current is carried across aluminum or brass buss bars. Meters pinpoint electrical flow. The control of the electrical flow is accomplished with a length of nickel-chromate resistance wire from a discarded heater or hot plate. A 50W, 25Ω, wire-wound power rheostat will make the job neater. A simple stand built from pine to support the tanks is advisable because of the caustic action of the chemicals on varnished surfaces.

Now we're ready to go to work. Here's how the machine works. Metal is transferred from the anode, which is either copper sheet or nickel bar, to the cathode, which is the product that is to be finished, via free ions in the plating solution. Amount of current flow will detemine the thickness and quality of the plate. The amount of voltage and amperage necessary is controlled by the surface area of the project being plated. To plate copper, for example, 20 to 30 amps per square foot of project area is called for (Fig. 6-38). Do not let the large number scare you. Electronics projects will not have the surface area to overtax the current supply.

The end product is the first consideration. A careful step-by-step procedure is necessary, but easy, for the bright controlled finish you desire. The first step is to establish a smooth surface for the metal to adhere to. Draw filing is suggested where possible. Treatment with coarse-to-fine emery cloth will remove the scratches made by the file. Final polishing with machine compound and a buffer will provide a luster that you will insist is good enough.

Before using the machine, a cleaning procedure will be needed or all your polishing time will be wasted. A soap-powder solution, in conjunction with a grit soap, such as Lava, will remove oil and grease. Rinse carefully with clear water as you suspend the project on a copper wire hook. Even the oil from your skin, transferred by a finger print, can cause trouble at this stage.

The first of the three tanks contains a diluted solution of 28% uncut sulfuric acid and water. Suspend the project on the cathode bar. The anode in this tank is a piece of scrap copper. Turn up the rheostat and let the current flow. This starts the action.

Bubbles will rise vigorously about the project. Each of these bubbles is lifting particles of dirt and oxidation. It might take several minutes to do a complete job. All ferrous metal to be plated directly must be etched with this method.

Remove the project from the solution and rinse again in clear water. Moving from tank to tank without rinsing can be bad due to the contamination of solutions.

Flash plating in the nickel solution is next for ferrous metals. Place the project in the nickel solution until a smooth silvery

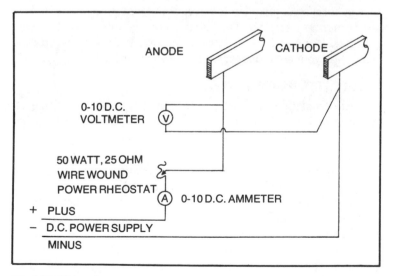

Fig. 6-38. Wiring diagram of an electroplating machine.

coating is deposited. Agitate the solution around the project and brush off bubbles that cling to the job. Experimentation is necessary to determine the correct time period. Be careful to keep the hanging hook in contact with the buss bar while lifting the project to observe for the desired silver coating. This is called *flash coating* and is necessary to make a good bond for the copper that is to come.

Rinse and move on to the copper bath. Agitate around the project and turn at regular intervals. The time period is regulated by the thickness of the deposit desired. It is not desirable to try to deposit too much metal in one period. A dark spongy appearance will result if the project is left too long in any one bath. It is better to build up the plate by layers. Small scratches can be filled by this process. Final buffing after either the last deposit of copper or nickle will bring a luster that your students cannot ignore. Plating of nonmetalic objects can be done by first painting the job with a mixture of bronze power and varnish.

Materials are easily found at through a local supply house. Use sheet copper for the anode in the copper solution. A visit to a plating job shop will provide at low cost the pure nickel anode. In a commercial operation, the nickel anodes used must be removed from the solution before they are completely dissolved. The butt lengths are 12″ to 14″ in length and are just right for our purpose. See Table 6-3.

Sulfuric acid, mixed in the proper proportion, can be found at the corner gas station. This is the same solution used in car batteries. Boric acid is sold at the drug store. White corn syrup is at home. Mix your solutions carefully and you're in business.

SILVER ELECTROPLATING

Electroplating is a process in which a metallic coating is deposited electrochemically. With the inexpensive and easily obtained materials described here, a unit on electroplating can be introduced with little difficulty. In the process, metal particles (ions) are transferred from a soluble anode through a plating bath onto the item being plated which acts as the cathode.

Metals which are often used for plating include gold, silver, German silver, chromium and rhodium. Silver is a very desirable metal to use in art metalwork, but its cost is high. Articles made of copper, brass and other base metals can be plated with silver to increase their value and beauty, and the cost is considerably less than it would be if the pieces were made from sterling silver. It is

Table 6-3. Materials Necessary for Copper and Nickel Electroplating.

Copper

28 oz. Copper Sulfate
4 fl. oz. Sulfuric Acid
   (28 percent battery acid)
½ tsp. White corn syrup
1 gal. Water

Nickel

32 oz. Nickel sulfate
6 oz. Nickel chloride
4 oz. Boric acid
   (dissolve in a little hot water)
1 gal. Water

highly desirable to plate surfaces which often come into contact with food or beverages.

Electroplating Equipment

Very little equipment is needed for electroplating, and it is relatively inexpensive. Three pieces of equipment are essential: two plastic or Pyrex glass tanks of approximately two gallons capacity each and a low-voltage, direct-current generator or a rectifier which is used as the electroplater. The electroplater should be one which allows the electroplating voltage to be adjusted between 0 and 6 volts. Dry cells may be used instead of the electroplater. See Fig. 6-39.

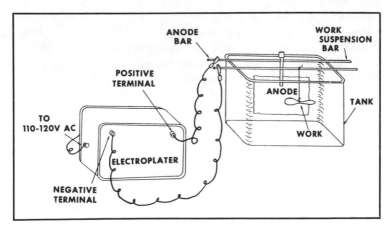

Fig. 6-39. Electroplater for tank plating.

Preparing the Base Metal for Plating

Two important operations must be performed before an object can be electroplated. The first of these is the buffing and polishing of the object to the desired degree of surface smoothness, since the plated surface will be only as smooth as the surface to which the plate adheres. This point cannot be overemphasized. To obtain a plated surface with a high polish, it is necessary for the base metal surface to have a high polish *before* it is placed in the plating bath. Pits, scratches or other irregularities on the base metal will be visible on the plated surface, because the deposit of silver is relatively thin. Attempts to cover defects by plating are usually expensive, unsatisfactory, and accomplished only with a heavy silver coating and considerable buffing.

The second operation is a thorough cleaning of the piece to be plated. This should be done after buffing and polishing and just prior to the electroplating. The purpose of cleaning is to remove all surface oils, fingerprints, oxides and grease, some of which may not be visible. The surface must be chemically clean. The presence of foreign matter will result in poor adhesion which will cause the silver deposit to crack and peel. If the cleaning is done superficially, the silver plate may not be deposited at all. The principal cause of most difficulties encountered in electroplating can be traced, directly or indirectly, to inadequate preparation of the article prior to plating.

The Water-Break Test

The water-break test is a simple means of determining whether or not the surface of the base metal is free of foreign matter. The article is dipped into clean, cold water and withdrawn. If the water runs freely from the surface leaving an unbroken film of water, the surface is clean. If, however, the water clings in small drops, the presence of oil or grease is indicated and the cleaning operation should be continued until the piece can pass the test. Under no condition may any surface to be electroplated be touched by the hands from the time it is found to be chemically clean until it has been plated. Clean rubber gloves should be used for handling.

To ensure proper cleaning, the following steps should be taken. After buffing and polishing, the piece should be washed thoroughly in hot, soapy water to which a small amount of ammonia is added if the piece is unusually greasy. If intricate details, such as deep engraving, are present, a soft brush should be used on those areas. The washing should be followed by a hot-water rinse to remove the soap. At this point all visible traces of buffing and polishing compounds should have been removed.

Buffing and polishing compounds usually have a grease base and, for that reason, pieces which have been cleaned in hot, soapy water commonly fail to pass the water-break test. To further cleanse them either of two processes is recommended. One is the application of a degreasing compound (available from any dealer in electroplating supplies) with a soft cloth or brush, rubbing it over the entire surface and rinsing under hot, running water.

The other method—a simple and efficient one—is the electro-cleaning process. The cleaner, a chemical which is also available from dealers in electroplating supplies, is dissolved in water. The solution is then heated to approximately 200°F or just below the boiling point of water. Before beginning the actual degreasing operation, a copper wire loop is fastened to the piece to serve as a connector to the wire leading from the electroplater. The loop is also useful for moving the piece after degreasing without touching it. In the electro-cleaning process, the loop is used to connect the piece to the negative terminal of the electroplater, and the piece is immersed in the cleaning solution. A stainless-steel anode, which is connected to the positive terminal of the electroplater, is also placed in the solution. The electroplater is set to deliver 6 volts over a period of three to five minutes. The piece is then removed from the cleaning solution and immediately rinsed in hot water. If the article still cannot pass the water-break test, the process should be repeated until a successful test is made. When the water-break test is finally passed, the piece should be plated immediately.

Many plating salts and solutions are extremely poisonous. Anyone working with them should obey the following safety rules:

☐ Carry on electroplating in a well-ventilated room.

☐ Do not allow plating salts or solutions to come in contact with open wounds or skin abrasions.

☐ Wash hands thoroughly in running water after handling plating salts or other materials.

☐ Do not eat food, chew gum, or smoke while electroplating.

☐ Keep your hands away from your face while electroplating.

☐ Clean up immediately any spillage of electroplating solution or salts.

☐ Clean up thoroughly when electroplating is finished.

☐ Store all electroplating salts and solutions under lock and key.

☐ Store solutions in dark, tightly closed jars or bottles, properly labeled and marked "POISON."

☐ Wear rubber gloves when handling electroplating salts and solutions, and when removing pieces from the electroplating bath.

Silver is readily deposited on copper and its alloys when they are immersed in a silver cyanide plating solution. Silver will be deposited even though the electrical circuit is not complete. Unfortunately, such a deposit of silver has poor adhesion and, if additional silver is deposited electrochemically, the coat will probably peel or flake off. To overcome this difficulty, the base metal is usually plated with a light coating of silver (called a *silver strike*) in a solution similar to a normal silver electroplating solution except that the strike solution is higher in cyanide content and relatively low in silver content. The piece is then transferred to the regular solution where a heavy coat of silver is deposited on the surface. A formula[1] that makes a satisfactorily silver strike solution is:

Silver cyanide	0.5 oz.
Sodium cyanide	8.0 oz.
Water	1 gal.

The silver strike solution is placed in a plating tank, and a stainless steel anode is connected to the positive terminal of the electroplater. A copper wire from the negative terminal of the electroplater is also placed in the solution. The electroplater is turned on and operated at 6 volts. The piece to be plated is connected to the copper wire coming from the negative terminal and immersed in the silver strike solution. It is essential that the circuit be complete and the electroplater in operation before the piece is immersed so that the thin coat of silver will be deposited electrochemically. The silver strike operation should continue for a few seconds until the entire surface of the base metal has received a light deposit of silver. The piece is then removed and immediately immersed in the silver-plating solution with the current on.

Silver electroplating salts can be purchased from any dealer in electroplating equipment and supplies. The salts should be dissolved in warm water and allowed to cool to room temperature before use.

---

[1] William Dixon, Incorporated, *Sal Hyde Plating Manual*, Newark, N.J.: William Dixon, Incorporated. P. 11.

The article to be plated is connected to the negative terminal of the electroplater, and the fine (99.95+ percent pure) silver anode is connected to the positive terminal. Fine silver must be used for the anode, because sterling silver will not plate satisfactorily. The fine silver anode must have a surface area equal to or greater than the surface area of the piece to be plated.

The voltage used during the electroplating process should approximate 1 volt. In any case it must be maintained within the range of ¾ to 1½ volts. The process should be continued until the desired thickness of silver deposit is achieved. This usually requires from 30 minutes to one hour. The piece is then removed from the plating solution, rinsed in cold water and dried.

The normal color of freshly plated or pickled silver is a matte white. Polishing will result in the usual shiny color we know as silver. The piece should be rubbed with a paste of bicarbonate of soda (baking soda) and water, washed and dried. It is recommended that it be hand polished with a good quality silver polish applied with a soft cloth.

Chapter 7
Gadgets

Just name the gadget, and if it's electronic, you'll probably find it in this chapter. Dozens of projects will teach you electronics while providing you with a handy device.

BRIGHTEN THAT STROBE—CHEAP!

Strobe flashers—AC powered—are popular with many hobbyists. Although a good commercial reflector for the Zenon tube costs $3 to $5, you can make your own from heavy aluminum foil from a local variety store. A 12″ × 25″ sheet costs less than $1. The foil comes in a variety of color combinations with different colors on opposite sides of the foil. The silvered reflector surface for our strobes is often preferred.

Procedure

Lay out a full-sized pattern of the reflector on lightweight posterboard, as shown in Fig. 7-1. The size of the reflector may be increased or decreased by changing the outside diameter of the pattern. Use a fine-point laundry pen to draw around this template—pencils and pens will not register well on the surface of the foil. Be careful not to bend or crease the foil. After cutting, form the foil into a cone so that notch "A" lines up with point "B." The small end of the cone should form an elongated rectangle for the Zenon flash tube. Be sure the tab is on the back of the reflector and secure it with cellophane tape.

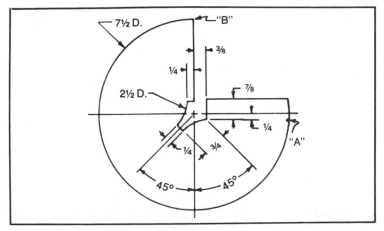

Fig. 7-1. Reflector detail of the strobe units.

Box size will depend upon individual taste. We cut ½″ particle board for the sides to allow for a 6″ × 6″ inside measurement. One edge of each side is rabbetted to take a ⅛″ standard hardboard back, fastened in place with screws. The PC board strobe circuit is secured to the inside of the back with insulated stand-offs and machine screws.

Cut a piece of 16-ply posterboard to fit the front of the box. Make a round cut-out, slightly smaller than the od of the reflector opening in the posterboard, and fasten the open frame to the front of the box with glue. Place the reflector against the inside of the posterboard front and secure it with dabs of silicone adhesive. The Zenon flash tube will protrude through the slot at the back of the reflector.

The completed box is sanded, sealed and painted or covered with contact paper. If contact paper is used, the surfaces should be well sealed and smooth to ensure good adherence. Cover the front first, allowing at least ½″ to extend around the corners of the box. Cut the front opening paper so that ½″ or longer tabs can be folded back inside the posterboard opening. The side pieces of contact paper are placed slightly back from the front edges of the box and the back edges should fold around into the rabbetted edges of the box. Be sure to finish the exterior before installing the reflector.

The Strobe Circuit

Figure 7-2 shows the full-sized PC board layout and component placement. *R3*, the variable speed potentiometer, is mounted on the hardboard back of the box (Fig. 7-3). The larger the pot, the slower the strobe's flashes can be regulated. A line switch can be installed to turn the unit on and off. When mounting the circuit board to the back, stand-offs should be chosen to allow the flash tube to protrude through the reflector opening. Leads should not be allowed to touch the foil. Insulate them with sleeving. See Table 7-1.

Safety

There is some medical evidence that exposure to strobe rates from 6-10 flashes per second could cause an epileptic attack, even in a person *without* previous history of epilepsy. Therefore, *extreme caution* must be used with strobes—both by the builder and by observers.

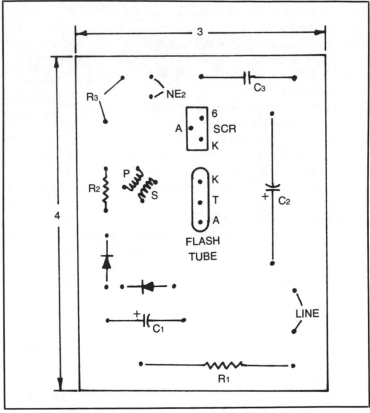

Fig. 7-2. PC pattern at 78 percent and component placement of the strobe unit.

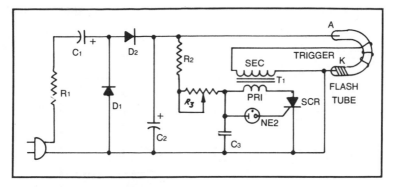

Fig. 7-3. Schematic of the strobe unit

Caution must also be exercised with the high voltage involved in most strobe circuits. Be sure all capacitors are fully discharged before making any circuit modifications.

A "JUNK" TV BUGGER AND ITS USE

The junk approach works best with simple projects which are easy to troubleshoot, allow a good deal of variance in component values and quality, and permit some freedom in printed circuit layout. The TV Bugger project meets all these requirements. It is a simple circuit (Fig. 7-4).

When *S1* is closed, the TV Bugger transmits a low power signal which can scramble reception of selected television stations. Since the power output is quite low, the range of well-constructed projects is about 30'. Its low power output is intended to avoid breaking FCC regulations.

Table 7-1. Parts List for the Strobe Unit.

R1—150 Ω/10 W
R2—82 KΩ/½ W
R3—1-5 MΩ pot
C1—4.7 μF/250 V
C2—22 μF/450 V
C3—.1-.47 μF/200 V
D1, D2—1 A/400 PIV diodes
SCR—Silicon controlled rectifier,
    4 A/200 PIV
NE 2—Neon lamp
Flash tube—FT 106
T1—Trigger transf., 4 KV PC-type
    leads

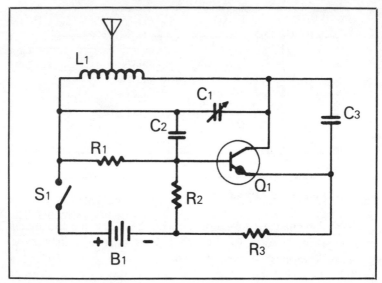

Fig. 7-4. Schematic of the TV bugger.

Constructing the Circuit

The component values shown are not critical in most instances, and they can be substituted as needed. If an NPN transistor is not available, the circuit could be altered to use a PNP. You can successfully construct the circuit inside the case that housed the original junk transistor radio. The parts list of Table 7-2 assumes that this will be done.

Sometimes a gimmick—a capacitor made of two insulated wires tightly twisted together—will be required in series or parallel with $C1$ or $C3$. The on-off switch on the transistor radio's volume control will work well for $S1$, while the tuning capacitor is excellent for $C1$. The battery connector is easily salvaged, but if it is missing one can be made from the top of an old battery.

The antenna can be the telescoping one built into the radio or a 15″ -30″ wire. Experimentation will help here. The printed circuit board and the battery are the only parts which cannot be obtained from junk.

The gimmick may be needed on some projects to tune in more channels or to compensate for varying amounts of unintended capacitance or inductance due to circuit layout. Tuning capacitors from AM-FM radios work best, but regular AM ones will bug several channels. Tuning is very sensitive and must be done patiently or the channel will be skipped.

278

Coil *L1* may be easily wound by making nine turns around a 3/16″ dowel rod or drill with a stiff, 16-20 ga. wire, and soldering on a center tap. The coil should be expanded to a length of about ½″. Most NPN transistors found in the audio amplifier section of a radio will work in this circuit. The project may be constructed on a very small printed circuit board scrap.

A SIMPLE, INEXPENSIVE IC DECISION MAKER

This simple inexpensive circuit uses a 7400 quad two input NAND gate TTL IC (about 30 cents from most surplus mail order houses), and two indicator driver transistors as active devices. The result is shown in Fig. 7-5.

Two of the NAND gates are connected to form a clock generator with C1 and C2 as frequency determining elements. The clock frequency may be altered by changing these two capacitors. The other two gates are connected to produce a flip-flop which is cycled by the clock when S2 is depressed. The Q and Q outposts of the flip-flop (IC pins 3 and 6) are connected to indicator driver transistors Q1 and Q2. The transistors specified are 2N2222, but many other NPN types could be used. The state of the flip-flop is shown by the condition of LED 1 and LED 2.

The circuit can be constructed on either perforated board or an etched circuit board. An IC socket is recommended to protect the IC from high soldering temperatures and to make replacement and testing easier.

Table 7-2. Parts List for the TV Bugger.

Part no.	Description
R1	10 kΩ*
R2	4.7kΩ*
R3	1kΩ*
C1	Variable tuning capacitor from transistor radio
C2	.001 μF
C3	10 pF (7-15 pF will usually work)
S1	Switch on transistor radio volume control
Q1	NPN transistor from audio circuit of radio, medium power
B1	9 V transistor battery
L1	9 turns of stiff wire, ½″ length, 3/16″ dia, with CT
	Small printed circuit board to fit components
	Case and hardware, as required

*The values of R1 and R2 may be altered for differing transistors and many other components are not critical. A gimmick may be needed with C1 or C3.

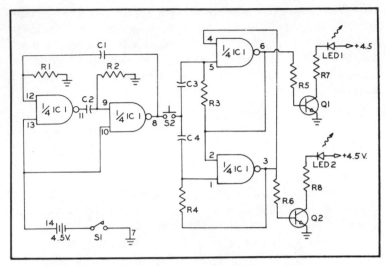

Fig. 7-5. Schematic of the simple, inexpensive IC decision maker.

A 1-½" × 3" × 4" plastic box houses the circuit board and three AA size penlight cells. The aluminum lid is marked with dry transfer letters and sprayed with clear lacquer. See Table 7-3. If the clock fails to oscillate, a 10 k Ω resistor connected across C1 and C2 should solve the problem.

HIGH-FREQUENCY TRANSFORMER

The materials needed for the project are as follows: A ¾" × 9" × 14" piece of wood; 350' of #30 B & S magnet wire; a cardboard

Table 7-3. Parts List for the IC Decision Maker.

Part no.	Description
R1,2,5,6	1000 Ω, ½-W, 10%
R3,4	270 Ω, ½-W, 10%
R7,8	47 Ω, ½-W, 10%
C1,2	10 mfd each 10 v
C3,4	.01 mfd each 10 v
Q1,2	2N2222 or equivalent NPN transistor
LED1,2	10 mv or equivalent
IC1	7400 quad two input gate TTL IC
S1	SPST slide or pushbutton
S2	NO pushbutton
Misc.	battery (4.5v), case, IC socket, sockets for LEDs, battery clips, hardware, wire and solder

tube, approximately 2" in diameter × 18" long; a 7" high container, such as a ½ gallon ice cream container; corrugated pasteboard (flat on one side, corrugated on the other); a 25' roll of aluminum foil; a vibrator spark coil; three pounds of paraffin wax; two 8-penny nails; a quart juice can; and a train transformer or other 12-V power supply. See Fig. 7-6.

In making the transformer, the following suggestions may be helpful:

Wind the secondary coil of the transformer on the pasteboard tube, making sure the tube is clean and free from foil coverings of any kind. The winding consists of the magnet wire spaced with carpet thread. Dead end the windings by threading the wire through several small holes punched near each end of the tube. Wood discs fitted in the open ends will help in mounting the coil for winding. It will also support the coil and discharge electrode. The coil should be given several coats of shellac and dried thoroughly in an oven at about 250°F.

The primary coil of the transformer is a 1" strip of aluminum foil doubled lengthwise to make a flat tape ½" wide. (It will be necessary to make up about 20' of this tape.) The tape is wound around a 7" form (I used the ½ gallon ice-cream container) with corrugated pasteboard as a spacer between turns. About eight to 10 turns are generally needed to obtain resonance. I started out with 10 turns and removed winding until the maximum spark was obtained from the secondary coil. The outfit shown here has an eight-turn primary.

The condenser is made from two pieces of aluminum foil, 4" × 36", and two pieces of corrugated pasteboard, 6" × 42". The foil is centered and temporarily taped to the smooth side of the

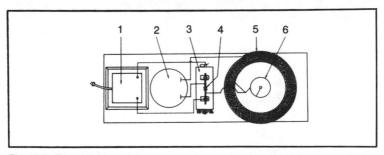

Fig. 7-6. The scheme of the layout of the high-frequency transformer. No. 1—10V to 12V power transformer; No. 2—capacitor; No. 3—vibrating spark coil: No. 4—spark gap; No. 5—Oudin coil primary winding; No. 6—Oudin coil secondary winding.

pasteboard. The two elements are then laid one on top of the other (corrugated sides down) and rolled into a compact bundle. This unit is slipped into the juice can and impregnated with hot wax. It should remain in an oven hot enough to keep the wax melted for at least 30 minutes.

The spark gap is made between the heads of two 8-penny box nails, held in the spring clips on top of the spark coil. In the event that the spark coil being used does not have spring clips, they can probably be removed from old dry-cell batteries or they can be purchased for a few cents each at a radio supply store.

The apparatus is now ready to be connected as shown in the diagram. After setting the spark gap 1/16" wide, a temporary connection to the primary coil is made that can be moved to cut in or out turns until resonance is obtained (maximum length of spark at secondary coil). When the right number of turns is established, the excess winding is removed, and the connection made permanent. If all apparatus is properly constructed, connected, and adjusted, a high-frequency spark of about 30,000 volts should be obtained (approximately 1½" spark).

AN ELECTRONIC FLASH

Are you looking for a gimmick or simple project that's almost fool-proof, has a high interest level and is versatile in its application. Well then read on!

Initially, the project appears as a gimmick with little functional value, yet it is highly intriguing to the beginning hobbyist in electronics.

Mention has been made as to this project's versatility. The following represents areas of possible use: study of component parts and their mathematical values within the circuit, a study of RC constants and how they are affected by changes in value, the use of numerous pieces of test equipment, and even application into sawtooth and sweep generators for horizontal detection systems.

Oscillator Circuit Process

Figure 7-7 represents the basic oscillator circuit. By deduction, one can readily see the simplicity of the entire network. Fundamentally, the circuit is composed of a low wattage—high value resistor, a capacitor, and a neon tube (Table 7-4). At point A, or the input, a DC voltage is applied to the RC branch of the circuit. The value of the input voltage must be equal to or greater than the

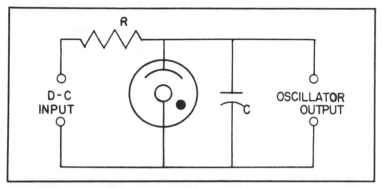

Fig. 7-7. Relaxation oscillator of the flash.

ionizational potential of the neon tube, which is shunted across the capacitor. As the capacitor begins to charge toward the value of the input source, its value is also reaching the ionization potential of the neon tube. When the capacitor voltage reaches the input value, the capacitor acts as a source for the shunted tube, the neon gas ionizes, and the tube conducts. During conduction, the resistance of the neon tube drops to a very low value and the capacitor discharges rapidly through the tube. When the capacitor voltage drops to a value lower than that which is required to maintain ionization, the tube is extinguished. The tube now acts as an open circuit and the entire process is repeated.

During both the build-up and the ionization periods, the resistor within the circuit serves a twofold function. First, it drops a portion of the input voltage so that the entire source is not applied directly to the tube, and secondly, while charging and ionization are occuring, the resistor acts as a current protection device.

Due to the fact that the resistance of R is much greater than the resistance of the conducting neon tube, the time constant of the circuit during charge is much greater than the time constant during discharge. Figure 7-8 represents the voltage time characteristics for the charging and discharging rates of the neon tube. Note that there is a gradually rising and rapidly falling voltage produced, and that after the initial charging of the capacitor the voltage does not fall below the deionizing level.

Table 7-4. Parts List for the Electronic Flash.	C—.1 μF, 400 VOLT CAPACITOR R—22 MEGOHM RESISTOR TUBE—NE-2 NEON BULB INPUT—90 VOLTS D-C

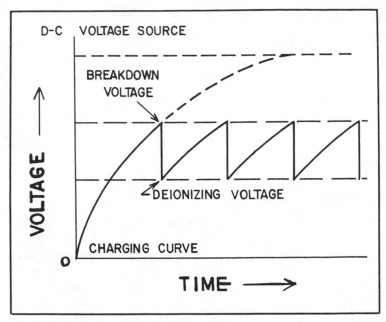

Fig. 7-8. Voltage time characteristic.

Thus far, consideration has been given only to fixed values of R and C. If the series resistor is varied, the frequency of the output is changed. A variation in frequency can also be produced by making the capacitor variable. When the variation of either resistance or capacitance is such as to increase the time constant, the output frequency will be reduced. Under these conditions, a longer time is required for the capacitor to charge to the breakdown voltage. Each cycle of voltage uses a longer time interval; therefore, its frequency is reduced. If the time constant of the RC network is lowered, the output frequency is then raised. The capacitor now charges more rapidly, and each cycle is completed in a shorter period of time.

Although variations in the output frequency are achieved by changes in value, a more satisfactory method can be employed. By increasing or decreasing the magnitude of the DC voltage applied to the circuit, the linearity of the voltage rise is improved.

When the DC voltage is increased, the capacitor charges to a higher voltage. With the same time constant as before, the same percentage of the full charge is reached in the same amount of time. Under these new conditions of higher applied voltage, the breakdown voltage of the neon tube is reached more quickly, and a higher frequency is generated. If the supply voltage is lowered, the opposite is true.

Although our discussion has centered primarily around an oscillator device, emphasis can be shifted to the neon tube and its use in other circuit functions. For example, neon tubes are used widely for the protection of circuits. If there is a possibility of circuit damage by the application of high voltages for a short period of time, a neon tube can be connected across it. The tube conducts during high-voltage periods, acting as a low resistance which prevents the high voltage from damaging the circuit. When the high voltage is removed, the tube stops conducting and the circuit performs normally.

In addition, neon tubes are also used as regulation devices. By substituting a load resistor for the capacitor in Fig. 7-7, a regulator circuit can be devised. The resistance, R, is high enough to limit the tube current when the load current through RL is low. As the load current increases, the voltage drop across R increases and reduces the tube voltage. A small reduction of tube voltage within the operating range results in a large decrease in tube current, consequently reducing the voltage drop across R. Since small variations of load current cause compensating variations in tube current, the voltage across the neon tube remains essentially constant.

Finally, the neon tube oscillator, in its simplest form, can be used to produce a sweep voltage for cathode-ray circuits. By adding a blocking capacitor to the output of Fig. 7-7, the DC component of the varying voltage is removed before applying to the horizontal amplifier. The dependability of ionization is not high, since small variations in operating temperature or gas pressure change the amount of breakdown voltage needed. Also, the time required for deionization is long, thus reducing the sweep efficiency at high frequencies.

TWO SATELLITE-MONITORING WIDGETS

With the growing congestion in astro-traffic, it seems possible everyone needs a device which will provide "erroneous-spectre-scan-monitoring" (whatever that is) of the heavens. This satellite-monitoring widget may not do just quite that job; but with a small dosage of over-verbalizing, you can make the gullible believe you have one.

Let us consider the monitor in relation to its separate functions. Refer to Fig. 7-9 for the schematic diagram of the individual RC widget circuits. To determine the interval between flashes, the time can easily be calculated t = RC. The answer is in seconds. For example, a 10-meg resistor and a .1 μF capacitor would produce a one-second flash interval. Consequently, you can select from the random parts in the shop and —by multiplying the value of resistance by the value of capacitance—arrive at almost any time constant value you might want.

The NE-2 will fire at approximately 90-V DC so that a student could power the circuit with a 90-v DC battery and make the unit completely selfcontained. Or he can build the half-wave power supply of Fig. 7-10.

There is nothing critical about any of the values—in fact, you can, if parts are scarce, leave out the 500 to 1000Ω resistor, the second filter capacitor, and the volume control used to regulate the level of DC voltage.

Both Figs. 7-9 and 7-10 can be built into a 12″ × 2½″ × 2¼″ mini box with anywhere from four to 12 individual widgets to produce a unique conversation piece.

TWO TELEPHONE INTERCOM SYSTEMS

Looking for a novel electricity project? Try this inexpensive telephone-intercom system. It can be used from room-to-room or floor-to-floor in the house, house-to-house, or house-to-garage. Do not attempt to hook up with your local telephone company; that's illegal.

For a little over $10, a two-phone, four-wire intercom can be installed. Cost varies with the length of wire required between stations. Extras, like plugs and sockets on the wires, might be desirable to make quick disconnects or to allow several locations for the phones. These can be purchased from a local electronics supply house for about 75c, or more, per phone.

Surplus telephones can be purchased, less ringer and induction coil. The ringer will not work on 6-V DC, so neither the ringer nor coil are needed. A door buzzer and flush buttons from a local

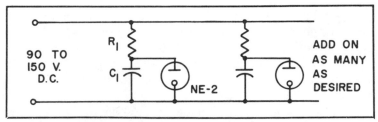

Fig. 7-9. Satellite-monitoring widget schematic.

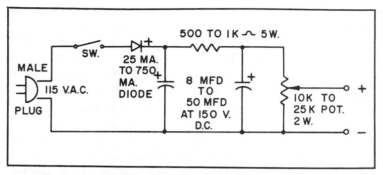

Fig. 7-10. Another satellite-monitoring schematic.

hardware store can be substituted. Special signaling buzzers, buttons, and transformers are available from suppliers. The phones should be tested and in good working order. A slight modification is necessary for good signaling, however.

The wire from the phone base to the wall outlet, usually either three-conductor or four-conductor wire, should be replaced with either four-conductor, or five-conductor, No. 22, stranded, vinyl-covered intercom wire. The number of conductors depends on the circuit used (refer to the circuit diagrams). This wire change is necessary because the original-equipment wire has too much resistance for the 6-V ringing equipment.

Four-Wire, All-Call System

A simple all-call system, using the three phone wires plus the added ringing wire, can be installed easily. The white wire (often yellow inside the phone base) is used for the common ground, the green wire (sometimes black inside the phone base) is the common positive, and the red is the voice wire. The fourth wire (usually black inside the vinyl cable) is used for the ringing circuit. The buzzers and push-button switches are in parallel, thus both buzzers ring when either button is pressed (see Fig. 7-11).

The advantage of this system is the lower cost of wire and connectors. For a long distance between stations, this would count up. It has one disadvantage: The buzzers are harder to adjust to sound properly, and occasionally, one will fail while the other one works. This is true when the current source becomes low. For long distances, extra batteries may be needed, hooked in parallel with the main one. They can be placed near each phone in the circuit.

With more than two phones in the circuit, relays, because they have a lower current consumption, can be inserted in the system where the buzzer normally would be. Each buzzer, then, could have its own power supply (Fig. 7-12).

Five-Wire, Return-Call System

This system features the return-call idea. Pressing the button on one phone rings the buzzer on the other, and vice versa. The phone wires are connected in the same way, but two ringing wires are used instead of one. Ringing is more reliable and buzzer adjustments are not critical. This is the recommended circuit.

Where more than two phones are desired, in this system, extra ringing wires and switching arrangements might be needed. Wire and connectors may be purchased.

Five-conductor wire will cost about 3½c per foot for 100′ spools. SW-1 refers to the receiver switch in the telephone base. SW-2 refers to the flush-type, door-bell switch which must be added to the phones. The Kellogg type K-1,000 phones (Fig. 7-13) require that some jumper wires be installed to complete the circuit within the phone base. This is because they were removed when the ringer and other parts were removed. A wire soldered onto a nail, or a similar metal probe, will suffice. Jumpers are placed from contact to contact as shown in Fig. 7-13.

THE LIGHT-BEAM RECEIVER

The use of the two-stage amplifier makes it possible to construct a modulated-light receiver. Although the circuit for the receiver (Fig. 7-14) is simple, the components need to be discussed. The photocell is a cadmium sulfide unit available from

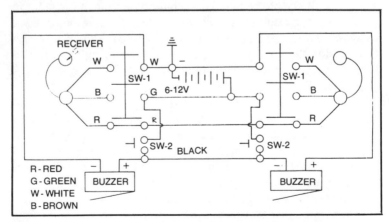

Fig. 7-11. Both buzzers ring when either button is pressed.

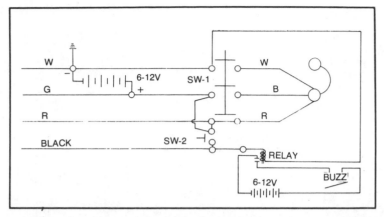

Fig. 7-12. Each buzzer could have its own power supply.

several sources (such as Radio Shack or Calectro). The resistance should be between 20 Ω and 1 MΩ. This type of photocell is preferred to the self-generating cell because of its higher sensitivity.

The voltage of the self-contained current source (B_1 in the diagram) should be between 3 and 6 V, with 3 V being the optimum for the lower resistance cells and 6 V used with those of higher resistance. The current through the transformer primary is controlled by the resistance of the photocell. When dark, the photocell displays a high resistance and there is little current through the transformer. When the photocell is exposed to light, more current will flow through the transformer due to the lower resistance of the cell. The changing illumination on the cell thus

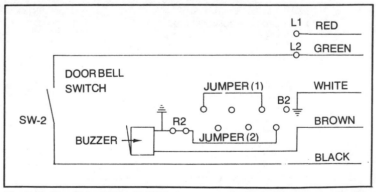

Fig. 7-13. The Kellogg type K-1000 phones require that some jumper wires be installed to complete the circuit within the phone base.

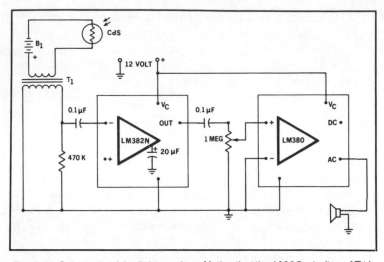

Fig. 7-14. Schematic of the light receiver. Notice that the 1000Ω winding of T1 is connected to the amplifier input. If difficulty is encountered during initial testing, reverse the leads.

produces a changing current through the current and an AC voltage across the secondary of the transformer. See Table 7-5.

Table 7-5. Parts List for the Light-Beam Receiver.

Qty	Part
1	470 KΩ, ½ W
2	0.1 µF 40 V capacitor
1	20 µF 15 V electrolytic capacitor
1	1 MΩ potentiometer
1	cadmium sulfide photo-resistive cell, (Radio Shack #276-129)
1	output transformer, 1,000 Ω C.T. to 8Ω, (Radio Shack #273-1380)
1	supplementary battery pack (3 to 6) size AAA or AA pen-light cells. See text.
1	integrated circuit, LM382N
1	integrated circuit, LM380
1	miniature speaker

*Part source supplied by authors implies no endorsement by publisher.

The transformer is the same as that used in the intercom (1000 to 8 Ω), with the 8 Ω winding connected in series with the photocell and the current source. The 1000 Ω winding is connected to the input of the preamplifier. Due to the high amplification, avoid placing the transformer in the vicinity of ac fields (such as those generated by a power transformer), since the transformer will detect these fields and produce an annoying background sound in the output.

The photocell itself will need to be protected from all stray illumination. It can be mounted in a box with a hole cut at the end to admit the beam of modulated light, or it may be mounted at the end of a long tube which acts as a light shield. Perhaps the best procedure is to mount the photocell in a box and arrange a simple lens over the hole in the box to concentrate the light on the surface of the photocell. This will be illustrated in a later issue when we discuss the completed receiver.

Effectiveness of the completed circuit can be tested in several ways. If a flashlight beam is used to illuminate the photocell, any interference with the beam will produce a sound in the output. Waving any object through the beam (such as a hand) will produce a thumping sound in the output. The same beam of light can be "sawed" by using a pocket comb. When the teeth of the comb are passed through the beam, a typical sawing noise is heard.

This is the same principle as that used in the projection of a sound motion picture. The sound track on the film is a variable density band that effectively modulates the light falling on a photocell in the projector. The result is the sound that was printed on the film.

If the flashlight used to illuminate the photocell is struck while the beam is illuminating the photocell, a ringing noise is heard, caused by the vibration of the filament of the lamp in the flashlight. The same effect can be obtained by introducing a vibrating tuning fork into the light beam from the flashlight. A tuning fork can usually be borrowed from the science department.

A match struck in front of the photocell so that its light falls on the surface of the cell, will produce a sound such as one would expect from a fire—loud with cracking and popping noises. Lamps in the room will produce a 60 or 120 Hz hum when their light falls on the surface of the photocell.

This is the principle that will be employed in the construction of the modulated light transmitter. This will be a light source whose illumination is caused to vary at an audio rate. The source is a flashlight lamp which is connected to the output of a LM380 amplifier.

THE LIGHT-BEAM TRANSMITTER

This light-beam transmitter can communicate with the receiver just described. The receiver and transmitter will operate over a distance of 10 feet in a darkened room, but when lenses are added to each unit, the distance is increased to over 30 feet.

The cases are made of sheet metal, fastened by spot welding. The two cover plates on each box are fastened to the case with self-tapping sheet metal screws. The use of the metal enclosures provides the necessary shielding for the circuits.

The snout of the photocell on the front of the receiver is a 1-¼" long, ½" id plexiglass tube secured to the case with epoxy cement. The snout's diameter was chosen to match the diameter of the photo-cell and therefore serves as the light shield for the photocell and its mounting support (Table 7-6).

The transmitter's light source is the reflector and cover glass assembly of a flashlight, with the lamp socket modified to accept the special GE112 lamp. The outer ring of the assembly is fastened to the case with epoxy cement, arranged in such a manner that the bulb can be changed is an overload causes it to fail. The GE112 lamp was specially chosen for this application. The modulation of

Table 7-6. Parts List for the Light-Beam Transmitter.

Qty.	Name
1	LM380 amplifier module
1	Miniature microphone jack
1	1 MΩ potentiometer
1	18Ω 2 W resistor
1	47 μF 25 V electrolytic capacitor
1	0.1 μF 50 V capacitor
1	Miniature toggle or slide switch SPST
1	Battery holder, 8 AA Penlite cells (12 V)
1	GE No. 112 prefocused miniature lamp
1	Flashlight lens and socket assembly or other suitable 1 amp holder
	Suitable metal enclosure

the light-beam is accomplished by causing the brightness of the lamp to vary at the same rate as the modulation signal. A heavy current lamp is slow in changing its light output with changing current conditions, while a lamp that draws little current is relatively fast in changing the light output with a change in current. The GE112 is a low current lamp and therefore ideally suited to this application, since it can change its light level rapidly enough to respond to a high frequency signal.

The schematic diagram (Fig. 7-15) shows the basic LM380 2 W amplifier with a 1 MΩ volume control for a high impedance input. A crystal or dynamic microphone will provide the necessary input to modulate the light beam. The output of the transmitter consists of the GE112 lamp circuit. The average current of the lamp is adjusted by the 18Ω resistor to provide approximately half brilliance to the lamp. This current for the operation of the lamp is obtained from the DC output of the amplifier.

The specifications of the LM380 indicate that the output voltage at the DC output terminal is approximately half that of the input voltage. The output current is a function of this voltage and the resistance of the load and changes with the output voltage. As the signal is amplified by the LM380, the output of the amplifier at the dc output terminal will vary between the limits set by the amplifier. The DC output terminal therefore has a voltage that has an average DC value and also an instantaneous value; since this value is always changing, it could be called the ac component of the output. The capacitor in parallel with the 18 Ω resistor passes this ac component to the lamp, thus increasing the efficiency of the total circuit.

The GE112 lamp has a small lens built into the end of the lamp, but an additional lens can be used to improve the focus of the light output on the photo-cell. This could present a challenge to students in electricity classes to determine what combination of lenses can achieve the longest transmission distance.

This article concludes a series of articles dealing with the LM380 2 W amplifier, which will, however, be used in some of the future articles. We'll turn next to other integrated circuit modules, beginning with those designed for voltage regulator projects.

CODE PRACTICE OSCILLATOR

This circuit has been designed to oscillate. The diagram (Fig. 7-16) shows a code practice oscillator that will provide loudspeaker volume. It is a form of the basic. Armstrong oscillator and uses the

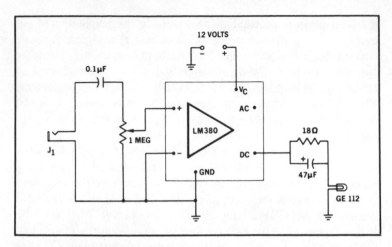

Fig. 7-15. This schematic diagram does not show the switch or the battery pack but both are included in the parts list.

8Ω winding of the transformer as the feedback path. The secondary winding has a capacitor *C3* connected in parallel to serve as a parallel tuned circuit. A small part of the resultant signal is connected through capacitor *C4* to the gain control *R1*. The key in

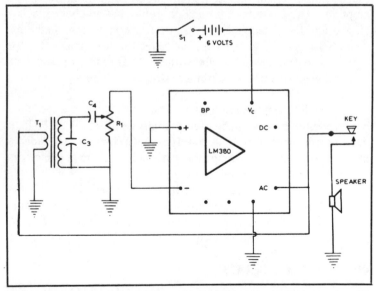

Fig. 7-16. Code practice oscillator schematic. C3 is a 0.l µ F, 50 V capacitor; C4 is a 260 pF, 50V capacitor; Rl is a 10K Ω potentiometer; S1 is a SPST switch; and T1 is a 1000 Ω to 8 Ω output transformer.

series with the loudspeaker produces the dots and dashes of Morse code.

Headphones can be substituted for the loudspeaker if desired, but the sound may need to be reduced for comfortable listening. The gain control *R1* also affects the pitch of the sound generated in the oscillator. It should be set to the lowest position that produces a suitable sound. If set too high, it will cause the amplifier IC to overheat with a resultant thermal cut-out. This does not harm the IC, but it does produce distortion and erratic operation. Should the oscillator fail to oscillate, reverse the leads to the 8 Ω winding of the transformer.

FOUR ELECTROMAGNETIC PROJECTS

With minimum construction and little cost you can demonstrate four electromagnetic principles in these projects: magnetization, solenoid action, telegraph operation and relay switching. The project is built for the experiments and then dismantled so that parts may be re-used.

Basic Construction

Use furring strips to build the wooden support bracket. No ripping of stock is required. Cut two pieces of furring 3″ long. Sanding or finishing the bracket is not necessary.

The coil used in all four projects is wound around a plastic soda straw. Cut the straw 2″ long, and tape at each end to keep the coil from slipping off. The coil must be hand-wound, 1-½″ long, using 10′ of #20-22 ga. double Formvar coated magnet wire. This makes a coil of about 2-½ layers.

Make a mandrel from a ¼″ dowel and shape it to fit into the straw. The mandrel provides a handle to aid coil winding and also prevents crushing the soda straw as the coil is wound. Tape is not needed to hold the coil in position since the wire is stiff enough to keep from unwinding. Leave about 4″ of wire on each end of the coil for the electrical connections. The enamel coating should be scraped off the coil ends and the ends fastened to Fahnestock clips stapled to the wooden support brackets.

Make a switch (SPST) from a 3″ long strip of furring, two wood screws, and ½″ by 2½″ piece of tin can metal. Commercial knife switches or pushbutton switches may be used instead of the hand-made switch. If the hand-made switch is used, caution should be exercised because the tinplate strip will get hot under long repeated use. However, the projects are not designed for continual use, but merely for demonstrating the appropriate electrical principles.

Project 1—Magnetization of Metal Objects

Connect a simple circuit with the coil, switch, and dry cell. Depress the switch, insert a straightened paper clip into the coil, and withdraw slowly. Turn the power off and test the paper clip for magnetic properties. Try magnetizing various metals such as pins, nails, brass clips, etc., to demonstrate which kinds of metals hold their magnetism the longest.

Project 2—Solenoid Action

Connect a wire hanger to a vertical bracket. Drill a hole in the bracket and insert the support wire. A coat hanger can be cut and shaped for this purpose. Support a small nail from the wire so that half of the nail is suspended in the coil. Use a rubber band to provide spring action for the solenoid action. Connect the circuits, activate it, and observe the solenoid action.

Project 3—Telegraph Operation

Solder a 6″ piece of copper wire under the head of an 8d common or box nail. Nail the coil to the wooden base and secure the free end of the wire to another Fahnestock clip. Cut a strip of tin can metal approximately ⅜″ by 2-½″ and nail to the top bracket so that it rests about ⅛″ above the nail head. Connect the circuit and demonstrate the telegraph action. If the telegraph does not operate properly, check the air gap and the stiffness of the striker plate.

Project 4—Relay Switching

Cut an L-shaped piece of tin metal and attach to the telegraph set-up so that the two tin pieces touch and make an electrical contact. Connect the circuit and test the relay switching action. One lamp should be lit with the switch open; the other lamp should light when the switch is closed. Note that the tin can strips will tend to magnetize after repeated use and stick together. This is particularly true if the cut edges are ragged.

Variations

These four experiments can be improved and expanded. For instance: a compass can be used in Experiment 1 to show the polarity and direction of the magnetic field; the solenoid can be improved to simulate a railroad gate crossing; the telegraph can be circuited to another telegraph so that you can send and the other receive Morse code; different metals can be used in the relay to see which kind of metals work best.

ELECTROSTATICS

Three-year-old children know that an inflated balloon will stick on a wall if the balloon is first rubbed in someone's hair. Why not apply what a child knows to demonstrate the laws of electrostatics?

All of the classic pith ball experiments can be performed with aluminum coated rubber balloons, but the balloons' greater visibility and the necessity for the experimenter's greater physical involvement in performing the demonstrations amplifies the dramatic impact of the experiments.

Ordinary rubber balloons, spray coated with aluminum paint, can be charged with a rubber rod rubbed against a piece of fur, or with a glass rod rubbed against a piece of silk, but the larger size of the balloons means that more rubbing will be required.

However, the balloons can be charged instantly with a small Van de Graaff generator with an output of about 100,000 volts.

The electrical charge is harmless because of the low energy storage capacity of the generator, but will produce deflections of a foot or more. There is no danger from sparks, except possibly to a person wearing a pacemaker. The actual current flow in any given spark from the Van de Graaff generator is surprisingly high, but the discharge time is of the order of a microsecond or less.

If one of the balloons is touched with a finger, a small electrical discharge can be felt, and the balloon, losing its charge, is no longer repelled by the generator and immediately falls back to it. Even before touching, a small spark jumps to the balloon, and the balloon is again repelled, giving dramatic proof that the phenomenon is electrical in nature. Each balloon is repelled by the other balloons also. What is not so obvious is that each spark striking a balloon burns a tiny hole in the rubber, causing an almost imperceptible leak. By the next day the balloon will be shriveled. It is best to blow up and paint the balloons on the day they are to be used, saving a supply to be blown up the following day.

Transferring the Charge

A charge transfer device can be built by rolling up a sheet of thin plastic, and attaching the balloon with a piece of thread down through the tube, taped to the far end. A strip of plastic tape will keep the plastic sheet from unrolling. It is important that the plastic be kept clean to avoid losing the charge on the balloon. Usually it is best to shut off the generator just before performing this experiment, so that the voltage will remain relatively constant.

The captive balloon is then brought close to the generator, but not quite so close that a spark will jump between them. At this point, when the charge distribution is drastically changed by the close proximity of the balloon to the generator, the opposite side of the captive balloon is touched with the finger. If done correctly, the former repulsion between them will be immediately replaced by an attraction. The finger is then removed before withdrawing the balloon. The balloon will then retain a net charge opposite to that of the generator. If two balloons are suspended by thread so that they touch each other in the uncharged state, they can be charged by touching them with the captive balloon. Successive applications with the captive balloon will make them repel each other. Then, if the captive balloon is touched to the generator, it can be used as a test device to check the polarity of charge on the free hanging balloons and on the balloons hanging over and repelled by the generator. This balloon will then repel the balloons next to the generator, but will attract the balloons hanging at a distance from the generator.

The inverse square law of electrostatic repulsion can be checked by sliding a small metal nut down the threads attached to a pair of charged aluminum coated balloons. Limit the travel of the nut by suspending it from another piece of thread. Deflection angles can be measured by the show of the threads on the wall made by an iodine movie lamp some distance away, or from a photograph.

Further experimentation is possible by filling the coated balloons with helium or a mixture of helium and air. Hydrogen should never be used because of the danger of explosion.

Preparing the Balloons

Use an aluminum paint that contains a relatively fast drying solvent, such as Derusto One Coat Spray 873-D25 aluminum finish. Painting the balloons in the draft from an electric fan aids in rapid evaporation of the solvent.

A UNIT FOR SOUND DETECTION AND RANGING

Radar is a fascinating word to anyone interested in electronics. It bears the connotation of the all-seeing eye, able to detect and locate objects beyond the limits of human sight. Except for the mode of transmission and detection (sound), the electronic circuitry is very similar to radar.

The timing constants were developed on the basis of the velocity of sound travel as 1140 fps (corrected for 25°C.). The maximum distance was selected as 25', or a total travel of 50'. This gives a pulse duration of .04 seconds. Adding .01 seconds for system-recovery time, gives a .05 second total time per pulse.

The specifications of the equipment are as follows. Scope Presentation: Type A (distance) and P.P.I. (Plan Position Indicator). Range: 25 feet. Rate of rotation: 6 rpm. Pulse rate: 20 per second. Frequency of pulses: 10,000 cycles per second. Power requirements: 120-V AC at 2.5 amps. High voltage: 8,000-V DC.

The equipment was constructed in four units, but can actually be combined into two: power supply and indicators and parabola.

Many types of sound transducers were sampled for the most efficient means of power transfer. An ordinary 4" loudspeaker finally provided the most acceptable operation.

A crystal microphone provides the sound pickup. The microphone must be shock mounted so that the noise from the improvised gear train does not overload the receiver. The gear train used was salvaged from a wind-up clock.

The P.P.I. scope is provided with its rotating trace by a TV yoke, belt-driven by a selsyn motor, with the yoke connections made by slip rings through contact springs we salvaged from a relay. See Table 7-7.

Circuit Explanation

The triggering pulse which synchronizes the entire operation of the units is obtained from an NE-2 relaxation oscillator circuit comprised of R-1, R-2, R-3 and C. It is discriminated by R-4 and C-2, inverted, and amplified by V-1. This sharp peaked pulse then triggers V-3 and V-4 which act as a multivibrator. See Figs. 7-17 through 7-20.

The keyed multivibrator is an electrical device for producing a square wave every time there is an input. Because of this, it is also called a "one-shot" multivibrator. Its input grid in V-3 is biased so that it is in a cut-off position.

The second tube, V-4, is biased to full conduction and at the instant of the pulse input, V-3 conducts and V-4 cuts off. When C-12 finally discharges, V-3 again cuts off and V-4 conducts. This way a square wave is produced.

Table 7-7. Parts List for the Sound Detection and Ranging Unit.

Bill of Materials
Condensers

(All values are in mfd 600 v unless it is noted otherwise.)

C1	.05	C21	.001
C2	.001	C22	.02
C3	.001	C23	.0005
C4	.001	C24	.02
C5	.05	C25	.0005
C6	.05	C26	.001
C7	.05	C27	.25
C8	.01	C28	.002
C9	.25	C29	.001
C10	.22	C30	.01
C11	.001	C31	.02
C12	.01	C32	.02 10kvdc
C13	.01	C33	.02 6kvdc
C14	.015	C34	.5
C15	.00039	C35	40 × 40 × 40
C16	.002		450vdc
C17	.002	C36	8 450vdc
C18	.001	C37	80 450vdc
C19	.0005	C38	2 400vdc
C20	.02	C39	8 150vdc

Resistors

(All values are 10 percent, ½-w carbon unless it is noted otherwise.)

R1	1 meg	R29	18 k 2 w
R2	2.5 meg pot.	R30	100 k
	pulse rate	R31	68k
R3	20 k 10 w	R32	250 k pot.
R4	8.2 k		cutoff bias adj.
R5	500 k	R33	39 k 2 w
R6	1 meg	R34	15 k 1 w
R7	18 k 2 w	R35	1.2 k
R8	2 meg	R36	1 meg
R9	39 k 2 w	R37	8 k 5 w
R10	470 k	R38	100 k
R11	3 meg	R39	1 meg.
R12	600 k pot.	R40	1 meg.
	cutoff bias adj.	R41	100 k
R13	50 k	R42	1 meg
R14	2.2 k	R43	100 k
R15	3.3 meg	R44	1 meg
R16	1 meg	R45	100 k
R17	33 ohms	R46	10k
R18	18 k 2 watts	R47	1 meg
R19	2 meg	R48	68 k 1 w
R20	10 meg	R49	100 k pot.
R21	100 k		contrast
		R50	1 meg pot.
R22	100 k pot.		brightness
	centering	R51	3.3 meg
R23	470 k	R52	50 meg 1 w
		R53	22 meg
R24	3 meg	R54	500 k pot
R25	600 k pot.		video clipping
	cutoff bias adj.	R55	200 k
R26	39 k 2 watts	R56	7.5 k 20 w
R27	2 meg	R57	100 k
R28	18 k 2 w	R58	270 k

Inductances

L1	10-50 mhy	L6	50 mhy vertical
L2	10-50 mhy		deflection yoke
L3	10-50 mhy		53 degrees max.
L4	10-50 mhy		deflection
L5	16 hy 200 ma		

Tubes

V1-2	12AX7	V16-17	12AU7
V3-4	12AU7	V18-19	6U8A
V5	6AV6	V20	5FP7
V6-7	12AU7	V21	1X2A
V8	6L6	V22	1X2A
V9-10	12AU7	V23	5U4GB
V11-12	12AU7	V24	6W4
V13	6L6	V25	OB2
V14-15	12AU7	V26	6AL5

Transformers

T1	5-w audio transformer 4,500 ohms to 3.2 ohms ac impedance	T3	scope transformer 117-vac primary to 1.25 vac 0.5 amp and 1,200 vac 4 ma
T2	cope transformer 117-vac primary to 1.25-vac 0.5 amp and 1,200 vac 4 ma	T4	power transformer 117 vac to 6.3 vac 10 amps, 5.0 vac 3.0 amp, and 800 vac ct. 200 ma

Miscellaneous

S1	SPST power on-off	FS	10 amp 250-v fuse
S2	SPST selsyns on-off	SPK	f" tweeter 3.2 ohms impedance
S3	SPST motor on-off	MIC	Microphone high impedance crystal replacement cartridge
NE-1	neon lamp		
NE-2	neon lamp		
SE-1	selsyn motor 117-vac 60 cps	MN	magnetic focus ring
SE-2	selsyn motor 117 vac 60 cps	M	parabola drive motor.

The pulse is fed through C-3 and the negative pulsed, .05 second square wave, obtained from the K.M.V., is coupled through C-5 to V-5 where the sawtooth wave is produced.

The square wave biases V-5 to cut-off, and C-6 slowly charges through R-5. This charge is almost linear and is eventually used to produce the linear line on the screen. The sawtooth, linear wave is

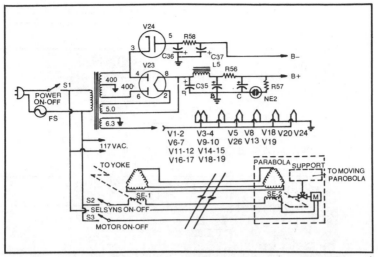

Fig. 7-17. Schematic of the power supply used in the sound detection and ranging unit.

is coupled through C-7 to V-6 and after being amplified by V-6, it passes through C-9 into V-7 where it is again amplified.

V-8, the cathode follower output, is biased to cut-off, but the positive pulse input causes conduction. The sawtooth, then, is converted into a linear current rise which sets up flux lines in the yoke. These flux lines in turn move the electron beam from the center to the edge of the screen. A portion of the current is impressed across R-17 which is fed into the cathode of V-6 where it acts as a deterrent to any difference between input of V-6 and output of V-8.

V-2 also inverts and amplifies the output of the discriminator circuit. The output of V-2 passes through C-11 where it triggers another K.M.V. composed of V-9 and V-10.

The output of this K.M.V. is positive and has a pulse duration of 400 microseconds. This pulse keys another M.V. with V-11 and V-12 as the oscillator, but a bit differently in that it provides a sine-wave output instead of a square wave.

During its 400 microsecond period, approximately 10 sine waves are produced and they are coupled through C-16 into the final output. The pulse of sound is amplified and passed into the output transformer which matches the impedance of the speaker-receiver used for pulsing.

The job of the receiver is to separate the noise in the room from the echoes returning from different objects and to amplify the

301

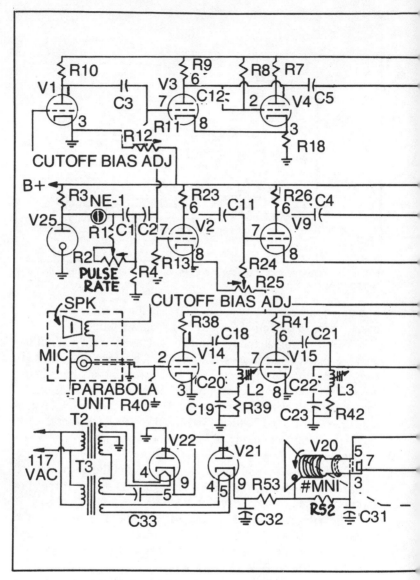

Fig. 7-18. Schematic of the sound detection and ranging unit.

weak echo enough so that it may be indicated on a scope. It basically consists of four tubes used as amplifiers. The inputs are tuned so that noise is rejected. V-17 acts as a cathode follower and V-18 acts as a detector. The pulses are clipped by the detector and are integrated by R-27 and C-75. The integrated signal is amplified

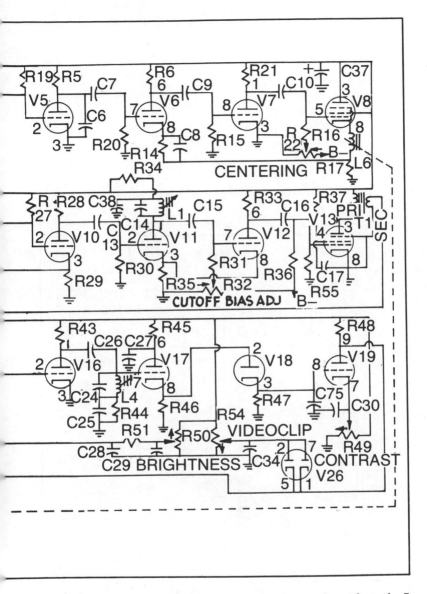

by V-19 and modulates the intensity of the electron beam from the 5 FP7 indicator scope.

The power for the units is obtained from the filtered outputs of a 5U4GB tube and a 6W4GT tube providing 400 v and 115-vdc respectively. The high voltage for the cathode-ray tube is supplied by a voltage doubler which is formed by two 1X2A tubes, T-2 and

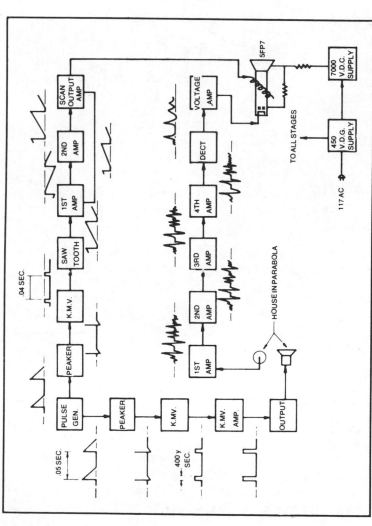

Fig. 7-19. Block diagram of the sound detection and ranging unit.

T-3, C-33, C-34 and C-32. The output is about quadruple of T-2 or T-3's output, approximately 7,000 v DC.

LASER DEMONSTRATOR

The laser demonstrator was designed and built to illustrate the principles of an actual laser. The model consists of a cavity, or housing, in which a light is flashed and then emitted through the cavity's optical system.

The power supply for the cavity is a separate unit controlled by a timing system. When this timing system is triggered, a flashtube, within the cavity, is flashed simulating the laser light beam.

The interior parts of the cavity may be observed through a resin window on the side. The observer is able to view a helical acrylic rod simulating the photoflash tube, a straight acrylic rod simulating the flashtube electrode, and a red resin rod simulating the ruby crystal. An actual flashtube with its electrode is mounted in the forward end of the cavity. While the model is functioning, the window is covered to permit the full flash of light to be concentrated through the aperture of the lens. See Figs. 7-21 and 7-22.

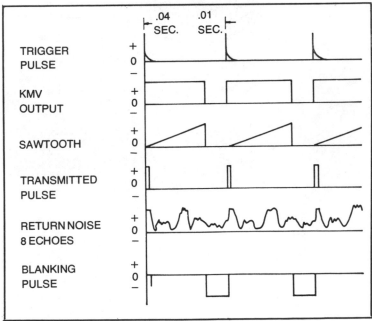

Fig. 7-20. Waveshapes and timing pulses of the sound detection and ranging unit.

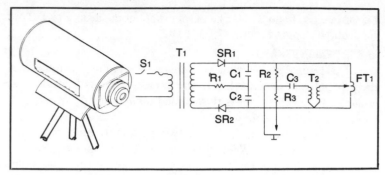

Fig. 7-21. Laser demonstrator and schematic.

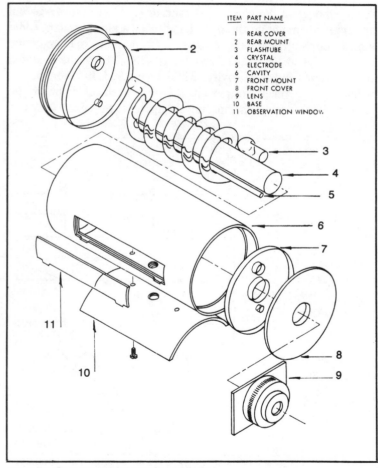

ITEM	PART NAME
1	REAR COVER
2	REAR MOUNT
3	FLASHTUBE
4	CRYSTAL
5	ELECTRODE
6	CAVITY
7	FRONT MOUNT
8	FRONT COVER
9	LENS
10	BASE
11	OBSERVATION WINDOW.

Fig. 7-22. Cutaway of the laser demonstrator.

Construction

The cylindrical cavity was constructed of aluminum (¼" × 4" × 8"). The front and rear covers were constructed of aluminum (¼" × 4"d). Two acrylic discs were used as mounting supports for the crystal rod, photoflash tube, and simulated helical flashtube and electrode. The lens was mounted on the front cover. The back cover acts as an anchoring base for the cavity components. The curved resin window (¼" × 1" × 5") was cast in the cavity window. The cavity base was shaped, tapped, and fastened securely to the cavity. Clear red automobile ignition wire was used to connect the power supply to the cavity.

The acrylic base of the power supply is 1¼" × 9" × 13". A straightforward, strobe, power pack is used. The low-voltage power supply is 470 V and the photo-tube-trigger voltage is 7,000 V. The timing mechanism cams are cut to pulse every seven seconds when automatic firing is selected. It also functions on single fire and manual. The power-supply panel is made of acrylic and is mounted at a rake angle from the base to the curved top. The control labels are engraved and laminated in clear and blue acrylic.

Suggested Demonstrations

Pour resin crystals, with phosphorus pigmentation, and then mount within focusing distance of the lens. When the laser fires, the crystal will light and continue to glow.

Pour resin crystals with phosphorus pigmentation. Then pipe the light from the center crystal to additional crystals mounted in spokes around the center crystal. When the laser fires, not only will the center crystal glow, but the entire wheel glows.

Place a mixture of equal portions of hydrogen and chlorine in a graduated cylinder, and place a ping pong ball on top of it. When the laser fires the light will ignite the gas causing a harmless explosion but placing the ball into orbit.

Arrange a series of optics on stands made from crystal resin. Focus and deflect the light beam in different configurations.

Place an emission indicator, connected to an oscilloscope, within focus of the cavity and determine the frequency and wave shapes being emitted.

Theory

Laser is an alphabetical shorthand for light amplification by stimulated emission of radiation. Light as radiated from the incandescent bulb quickly spreads in all directions, therefore requiring excessive energies to increase the illumination level at a large distance. With a flashlight beam, some improvement in focusing is accomplished, and more light is concentrated in a smaller area, increasing the illumination level, but not the amplification of the original source. The increased brightness level is accomplished by packing the light (photons) closer together so that most of the energy is concentrated into a type of beam. The light packets will not adhere to each other so a dispersion is experienced, but not as great as that of the incandescent light.

The light emitted from the crystal laser is a concentrated coherent light beam. The synthetic ruby crystal is an aluminum oxide crystal and contains .05 percent of chromium atoms, which are impurities that give the ruby its color and produce the laser effect. The ruby crystal is ¼" × 2" with its ends ground parallel to within two seconds of arc, and then the ends are mirrored. One end is 100 percent silvered, and the opposite end is silvered to approximately 96 percent which forms the window that allows the beam to escape from the crystal.

To raise the energy level of the chromium atoms to the excited level, for them to lase, the ruby crystal is placed in proximity to an exciting light source. This light source is triggered, showering the ruby with photons. These photons excite the orbiting electrons of the chromium atoms, thereby raising them to the higher energy levels. This excitation is caused by the absorption of photons by the electrons. Since they do not readily accept to much excitement, the electrons tend to rid themselves of this excess energy by emitting a photon as the electron returns to a lower energy level. The photon then travels toward the mirrored ends of the ruby. At this point, it is noted that the crystal did not destroy an atom, as in nuclear reactions, to produce energy. The photon reaches the mirrored surface and is reflected toward the opposite end. As it transits the crystal, more chromium atoms are excited, thereby starting the process of amplification. The photons shuttle back and forth between the mirrored ends until the energy level is greater than 50 percent. Then the light beam escapes through the crystal window. In the continuous ruby laser, this will be a continuous light beam whose wave length will be approximately 6943 Angstroms. In the case of the pulsed laser, it will be present in bursts. These wavelengths are detected as well as measured by phototubes.

The energy to power the laser is accomplished by two power sections. One section charges a large capacitor bank, which, when discharged, flashes the phototube that excites the ruby crystal.

The second power section is used to produce a triggering voltage that is then applied to the triggering anode of the phototube.

What the Laser Can Do

The laser has been used for an "optical catalyst"; drilling microscopic holes in wood, metal, carbon and diamonds; spot welding of plastics and metals; spectroscopy; standard for measurement; photography; medicine; optical ranging; and space communications and ranging.

PRINTED CIRCUIT MOTOR CONTROL

The control illustrated will produce up to 500 mA at operating speeds. The majority of miniature electric motors on the market perform at operating currents of from 420 to 500 mA. Some manufacturers, such as Pitman, Bonner and Wilson, do offer models operating at as much as 1.06 amps at full speed. If the control is to be used with the higher current motors, corresponding transistor changes must be made to compensate for the increased voltage and current handling requirements. Problems such as these provide the student with an excellent opportunity to become familiar with electronic supply catalogs and their numerous transistor listings. This is one of the best and most practical means of acquainting students with transistor characteristics, BV_{CB}, h_{FE}, Po, Ic, BV_{CES}, BY_{CBO}, Pc, and Tj. Try these yourself and see how many you can identify. These are all given in supply-catalog, transistor, column headings.

Refer to Fig. 7-23 and you will see that the control is a simple two-transistor, high-gain, DC amplifier. Q_1 is an NPN, high-gain transistor equivalent to types 2N168, 2N169, or 2N170. These transistors have a maximum collector current of 300 mA to 400 mA. Q_2, the output transistor, is a Motorola 2N554 or equivalent. The breakdown voltage on this transistor is 16 V, collector to emitter; 15 V, collector to the base. Caution students to consider carefully their d-c power source and its maximum output when selecting transistors for the control. The power source should be fully rectified for optimum performance of the control. The Q_2 bias voltage is obtained from a single 9-V transistor battery. Here again, the student must experiment and find for himself the best bias voltage. Remind the student that too much bias voltage is worse than not enough. Snap-on battery connectors, soldered to the printed circuit board, facilitate easy removal and replacement of the bias batteries. P_1 is a 10K linear taper potentiometer which may be replaced with any other value deemed necessary to obtain a specific output. (See Table 7-8).

A SIMPLE ELECTRONICS GAME

This electronics game may be built in any number of ways. The one illustrated here resulted in a table-top display at an open house staged by our industrial-arts department.

Figure 7-24 is a sketch of the game board. It can be made from cardboard, plywood, hardboard, plastic, or almost any other material.

The circuit is very simple, although at first glance it may seem complicated. It is simply a series of switches operating individually, a series of NE-2 neon lamps, and 100K resistors. Energy is provided from a 120 VAC line, readily available in any receptacle. The complete schematic diagram is shown in Fig. 7-25.

Either NE-2 or NE-51 may be used. The NE-2 has leads from its glass envelope for d-c connections. The NE-51 has a bayonet socket requirement. The 100K is a ½-W resistor which can be used to drop the voltage to assure longer life of the neon glow lamps. Small on-off toggle switches complete the circuit components. Use a lightweight parallel zip cord and a mole plug to complete the circuit to the 120-VAC source.

Start the game by turning on all of the lights. Then turn off any one light. By making moves as in checkers, eliminate one light at a time. In other words, jumping over a glowing lamp to a vacant spot allows you to shut that light off. The object is to finish the game with only one light remaining on the board.

CHARACTERISTIC CURVES PROJECT

This exercise in circuit construction and operation will yield a handy device for graphically depicting the characteristic curves of

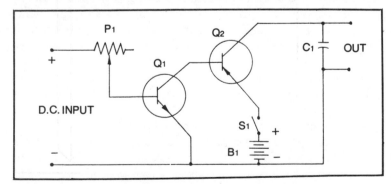

Fig. 7-23. Schematic of the printed circuit motor control.

Table 7-8. Parts List for the Laser Demonstrator.

Symbol	Quantity	Description
Q_1	1	Transistor, NPN-2n168
Q_2	1	Transistor, PNP-2n554
S_1	1	Slide switch
C_1	1	Capacitor, .02 mf
P_1	1	10K. Linear pot.
B_1	1	Transistor battery, 9 v
	1	Chassis, aluminum—2½ × 3¼ × 1⅝
	1	Copper laminate board, XXXP paper-base phenolic
	4	Circuit board connectors
	2′	Hook-up wire
		Miscel. hardware

electrical components when connected to an oscilloscope. Applicable to the testing of individual components as well as the conduct

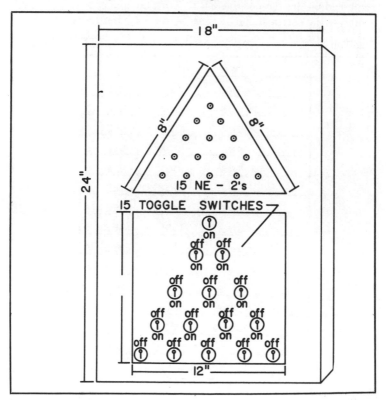

Fig. 7-24. The game board.

of in-circuit tests, the easy-to-operate unit can also be put to a variety of other uses . . . from the sorting of components for identification to a wide range of lab demonstrations.

As a glance at the accompanying schematic diagram (Fig. 7-26) will verify, the unit will fit into a "minibox" as small as 2″ × 3″ × 5″. The filament transformer is a 6.3 VAC with no center tap. The terminals marked V, H and G could be five-way binding posts to accommodate all types of leads.

To operate the unit, connect to 120 VAC line, 60-cycle source of power. The binding posts labeled V, H and G should be connected to the vertical and horizontal inputs of the oscilloscope. The binding post marked G is connected to the ground terminal on your oscilloscope. (Red binding posts should be used for V and H, and black for the ground terminal.) Almost any kind of wire is usable on these connections.

Turn on oscilloscope and set horizontal gain for a 1″ to 2″ line centered on the face of the CRT. Turn component tester on and calibrate the oscilloscope by pushing the push button switch on the tester. Adjust the vertical gain on the oscilloscope and the

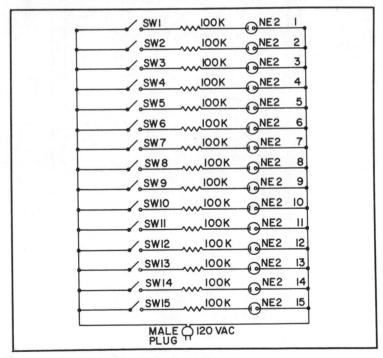

Fig. 7-25. Schematic of the electronic game.

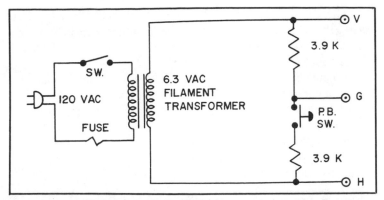

Fig. 7-26. Schematic of the characteristic curves project.

horizontal gain until the horizontal line becomes a line at a 45° angle. The unit and scope are now ready to be used to observe component curves.

Almost any electrical component will present a characteristic curve, and an analysis can be conducted to sort components and compare them to find good and bad ones. Arrayed along the bottom of these pages and in the boxed figures are some typical test patterns you can expect when the components being analyzed are connected to the ground and horizontal terminal on this test unit.

NIGHT LIGHT

This night light has proven itself a winner. In constructing it, you develop skills in plastics, metals and electricity. Moreover, there's the matter of budget-pleasing cost: 50 cents for the entire project!

The electrical wiring is so simple that even someone with no previous electrical background can do it. On the other hand, the project may be easily upgraded for the more advanced hobbyist. Work sheets can be developed to show power consumption, current drain, circuit impedance and cost of operation. Whatever the approach, the end product is impressive.

Construction Procedure

The back and two side pieces are made from opaque plastic and the front piece is made of translucent plastic (Fig. 7-27). Begin by cutting and filing the parts of the plastic box which will house the lamp. The protective paper should remain on the plastic until it is ready for heating and bending. The edges of the back and side pieces should be filed, sanded, scraped, and buffed. A jig will prove useful for bending the front piece.

313

The AC prongs are made from 20 ga. sheet brass. After making the ¾" fold, the edges should be filed and the ends rounded. A 3/32"-d hole should be drilled in the end which is not folded (Table 7-9).

To put the prongs into the back plastic, lay out the back as shown in detail #1. Drill a 1/16" hole and insert a jeweler's saw blade into the hole to make the 3/16" slot. After slotting the back piece, remove the paper and heat the plastic in an oven until it is pliable. Force the prongs through the hot plastic. When the plastic cools, the neon lamp and the resistor should be connected. Make sound mechanical connections before soldering.

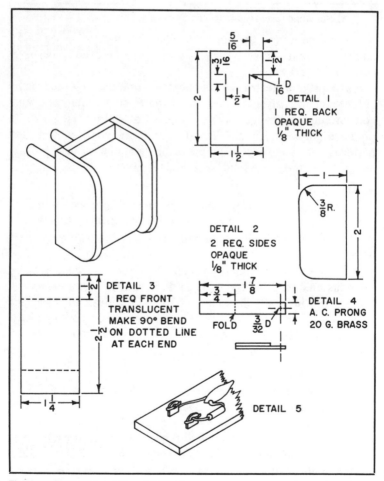

Fig. 7-27. Dimensions of the night light.

Table 7-9. Parts List for the Night Light.

Quantity	Description	Size
1 pc.	opaque plastic	$\frac{1}{8}'' \times 2'' \times 4''$
1 strip	brass	20 ga., $\frac{1}{4}'' \times 3\frac{3}{4}''$
1	neon lamp	NE-2
1	resistor	$\frac{1}{4}$w, 33KΩ
1 pc.	translucent plastic	$1\frac{1}{4}'' \times 2\frac{1}{2}''$

Carefully brush the ethylene dichloride in the seams. Do not permit any ethylene to run between the clamp jaws and the plastic sides.

However, if the ethylene should run, the side can be buffed. Allow 24 hours for the assembly to set. Then steelwool and soap the affected side or sides and wipe off the soap with a damp towel. Dry completely and buff.

TWO-SPEED UNIVERSAL MOTOR CONTROL

Projects in electronics can furnish learning experiences even if they are totally simple. This circuit has got to be, without doubt, the easiest two-speed motor control that can be built.

The accompanying circuit schematic (Fig. 7-28) for the two-speed control shows what occurs every half cycle. In the case of a $\frac{1}{4}''$ drill, the unit's speed will be cut approximately in half. This is very helpful when using larger drills with a high-speed $\frac{1}{4}''$ unit.

As can also be seen, the reason why this is used only with universal motors is the fact that this type of motor will run on AC or DC. This circuit puts out both, depending on the position of the toggle switch on the control.

The diode used (Fig. 7-29) was of the larger amperage, acorn-style. It is connected to one terminal of the switch.

The unit can be housed in a mini box or similar type of case. A mini box is suggested since this unit will find itself in use around the workshop, and aluminum mini boxes will take a beating and hold up for long use. Parallel zip cord is used for all the wiring of the motor control.

The male plug and female socket can be of the rubber type or of any other material which will stay together after much use and abuse. The switch can be a toggle-type on-off.

The diode must be a least two amp or larger. Many of these are now available at reasonable prices, some at less than a dollar each.

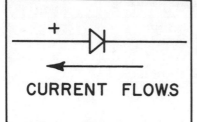

Fig. 7-29. What occurs every half-cycle when the two-speed motor control is employed.

CURRENT FLOWS

The front piece is shaped by using a strip heater. Make sure the bend is 90 degrees. To assemble the back and front pieces, use rubber bands at each end to hold them together. With a small art brush, apply ethylene dichloride to the seams to be cemented. After drying time has elapsed (20 min for handling; a full 24 hrs for working), check to see if the side pieces fit properly. Sand the edges of the two-piece assembly to insure a tight fit. To assemble the sides, use a 4" wood clamp. The parts are arranged inside the jaws of the clamp, which are tightened just enough to lightly hold the assembly and permit adjusting the individual parts of the assembly. When all parts are aligned, tighten the clamp just enough to hold the pieces in place. Be careful not to overtighten the clamp.

The unit works as follows: When a ¼" drill or any other apparatus using a universal motor is plugged into the control, its speed will be determined by the position that the toggle switch is in.

If the switch is closed on the control, then current will take the path of least resistance and flow through the closed switch. The portable drill will run at full speed.

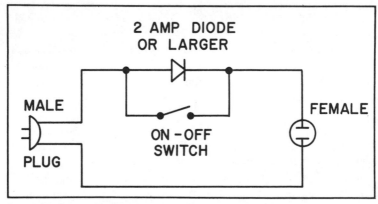

Fig. 7-28. Schematic of the two-speed motor control.

When the toggle switch on the control is open, then current must flow through the diode. The diode only conducts when its anode is positive with a potential applied to it.

ELECTRIC BUG KILLER

The basic parts of this bug killer include a three-color bulb and an electro-grid. The light attracts all flying insects toward the electro-grid. The grid consists of two wires which are melted into four ¼" acrylic rods (Fig. 7-30). Since the wires don't touch, current doesn't flow in the grid until an insect touches any two wires and forms a circuit.

A three-color bulb (green, white, and blue) was used since different colors attract different types of insects. Colored bulbs can easily be made from white bulbs and some acrylic paint. If you don't wish to paint the bulbs, the ordinary white bulb works nearly as well since it includes all colors of the spectrum. See Table 7-10.

Functional Design

It was made of two pine discs, cut with a circle cutter, and six ¼" dowel rods. The top and bottom discs were identical with the exception of four additional ¼" holes in the bottom one to hold the grid. The dowel rods were glued to the bottom disc while the top was fastened with two small screws. This allowed the top to be removed when the bulb had to be replaced.

The grid was constructed from four ¼" acrylic rods and copper wire. Tinned or uncoated copper wire is the best; however, if this is not available, the coating on magnet wire may be burned off with a propane torch to make a suitable wire.

The wire (18- to 22-ga.) is heated and melt-notched into place around the acrylic rods. A fixture was used to hold the acrylic rods in position while the wires were wound around them. The fixture was designed so that one end could be placed into a hand drill.

The first wire should be wound onto the rods so that the wires are no farther than ¼" apart. The second wire should be wound in between the coils of the first wire so that none of the wires touch. To operate properly, the grid should have at least nine wires per inch. The finished grid should be connected in parallel with the bulb. In order to decrease the current in the grid, it is necessary to connect a resistor of 30 to 45 ohms in series between each of the two grid wires and the bulb socket. For this much continuous power, it would be normal to use resistors rated at about 25 W; however, since power only flows in the grid when an insect touches it, we found it possible to use inexpensive 5-W resistors.

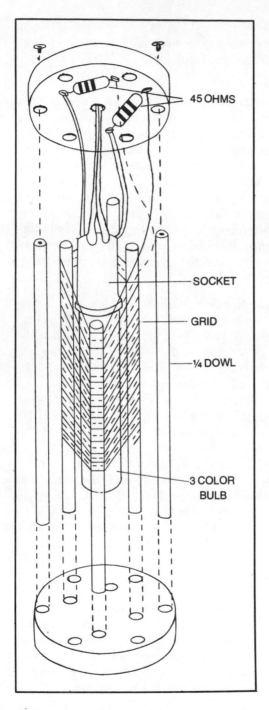

45 OHMS

SOCKET

GRID

¼ DOWL

3 COLOR
BULB

Fig. 7-30. The electric
bug killer.

2 top bottom	4"-d × ½" pine
6 posts	¼" dowel rods, 8¼" 1.
4 grid posts	¼" acrylic rods, 7½" 1.
2 tinned wire, 18-22 ga.	14' lengths
2 resistors	30-45 ohms, 5-w
1 socket	standard thread bulb
1 bulb	40-w, 5" 1.
1 cord	lam cord, 6'-8'
1 plug	bakelite AC plug
6 screws	#6, ¼"

VOICE-CONTROLLED TRAIN

Basically, the voice-controlled consists of a modified audio-freqeuncy (AF) amplifier whose output is rectified. The accompanying block diagram of Fig. 7-31 shows the essential components. This AF amplifier is modified to the extent that the output does not terminate in a loudspeaker, but in a circuit consisting of a semiconductor rectifier and a relay (see accompanying circuit schematic). Up to a point, this amplifier functions as a normal AF amplifier—i.e., the sound is picked up by the microphone and amplified by the 7199. The amplified signal is further amplified by the 6BQ5. This signal is then impressed across the primary of transformer T2 which is not the usual type of output transformer but rather an interstage type having a 1:3 or 1:5 ratio. For the best transfer of power, the primary impedance should be matched to the output tube.

The signal appearing on the secondary of transformer T2 is rectified by CR2. It is this rectified signal that actuates the relay.

Easy-to-Find Components

We dug into the junk box for the components used in the construction of this unit. The layout should follow accepted

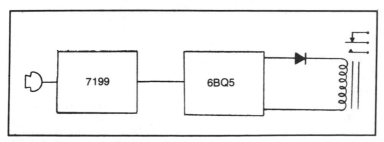

Fig. 7-31. Block diagram of the voice-controlled train.

practices for construction of an AF amplifier. Resistor R2 is used as a sensitivity control rather than in its traditional role as a volume control. Once the sensitivity level for a given microphone has been established, the shaft of control R2 can be locked in place. If you have relays suitable for use with semiconductors, it is easy to modify this unit for use with semiconductors instead of tubes. See Table 7-11.

The train used in conjunction with this unit should have a self-contained reversing switch. The "normally open" (N.O.) contacts of the relay (Fig. 7-32) are connected in series with the transformer used to power the train. After the sensitivity control, R2, has been adjusted so that the relay "pulls in" when a word is spoken into the microphone, the setup is ready to use.

We found it simpler and more attention-getting to use only an engine and coal car. To operate the train, turn on the power to both amplifier and train. Allow about 30 seconds for the amplifier to warm up, then speak one word—e.g., "Go"—into the microphone. Contingent on the position of the self-reversal switch on the train, it will move either forward or backward. Say "Stop," and the train will stop. Then say either "Forward" or "Back" (depending on which direction the train was running before you told it to stop).

The relay functions on rectified audio pulse—the words used are not important. This audio-actuated device may also be

Table 7-11. Parts List for the Voice-Controlled Train.

RESISTORS*	CAPACITORS
R1—500 K potentiometer	C1—10 mfd-25 w.v. electrolytic
R2—3.9 K	C2—.01 mfd-400 w.v. tubular
R3—100 K *all ½w	C3—.02 mfd-400 w.v. tubular
R4—470 K unless	C4—.01 mfd-400 w.v. tubular
R5—470K other-	C5— 10 mfd-25 w.v. electrolytic
R6—1.5 K wise	C6— 10 mfd-25 w.v. electrolytic
R7—100 K noted	C7—20 mfd-150 w.v. electrolytic
R8—470 K	C8—16 mfd-150 w.v. electrolytic
R9—100 ohms	
R10—150 ohms-2w	

TRANSFORMERS

T1—125 v. secondary at 50 ma; 6.3 v. secondary at 2 amps
        Stancor: PA-8421 or equivalent
T2—see test

SEMICONDUCTORS	MISCELLANEOUS
CR1—1N2069 or equivalent	F—1-amp fuse
CR2—1N2069 or equivalent	Potter & Brumfield Relay— Type LM1600

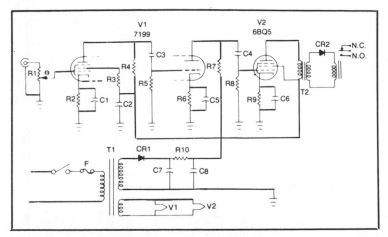

Fig. 7-32. Schematic of the voice-controlled train.

employed as an alarm by using ultrasonic frequencies (those above 15,000 cps) and interrupting the beam of sound. The transducers for ultrasonics may be of the type used by some television manufacturers for their remote-control devices.

SOUND-LEVEL METER AND ITS USE

A sound-level meter is just what its name implies; it records the level of sound surrounding it. A meter indicates the amount of noise or sound, and a sensitivity control sets a reference for comparison in two or more different locations.

The circuit is just a three-transistor audio amplifier connected in a grounded-emitter configuration. Note the input portion of the circuitry and the output.

Notice that a speaker is used as a microphone (Fig. 7-33). A speaker is usually less expensive and a little more rugged than a microphone. The output of the speaker is produced as the sound energy strikes the cone of the speaker and causes the voice coil of the speaker to move back and forth in the field of permanent magnet. This movement of the voice coil in a magnetic field produces a weak audio voltage in step with the varying sound. The voltage appears across the primary to the transformer, T1. The secondary voltage appears across R1, a sensitivity control which functions to control the amount of signal voltage to be fed to the audio amplifier through coupling capacitor C1. See Table 7-12.

Output circuitry is an output stage which feeds a full-wave rectifier bridge through a capacitor, C4. The bridge-type rectifier circuitry produces a d-c voltage to be read on the meter.

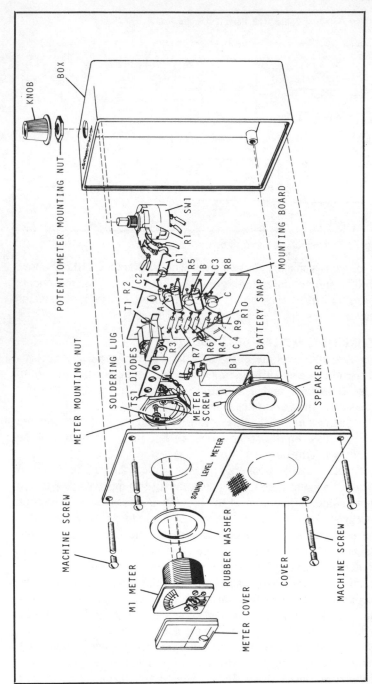

Fig. 7-33. Cutaway of the sound-level meter.

KNOB

BOX

POTENTIOMETER MOUNTING NUT

SW1

R1

C1

C2 T1 R2

R5 B C3 R8

MOUNTING BOARD

A C

R10 R9 C4 R4 R6 R7 R3

BATTERY SNAP

B1

METER MOUNTING NUT

SOLDERING LUG

TS1 DIODES

METER SCREW

SPEAKER

SOUND LEVEL METER

MACHINE SCREW

M1 METER

RUBBER WASHER

COVER

MACHINE SCREW

METER COVER

322

Table 7-12. Parts List for the Sound-Level Meter.

1	Meter, 5 ma d-c full-scale
1	Rubber washer (part of meter)
1	Meter cover (part of meter)
1	Meter-mounting nut (part of meter)
2	Meter screws (part of meter)
1	TS1, terminal strip, 2 mounting lugs
1	T1, transformer, Stancor TA-33 or equivalent
1	Battery snap (for 9-v transistor battery)
1	Battery, 9-v
1	Speaker, 2", 3.2-or 8-ohm (Utah SP 27A or equivalent)
1	Potentiometer, 10-K, ½-w
1	SW1 switch (on-off mounted as part of potentiometer)
1	Mounting board, plastic or insulating board, 1/16" × 2-½" × 2-½"
1	Knob (to fit shaft of potentiometer)
1	R1, potentiometer mounting nut
4	Machine screws, 2" × 3" × 5-⅞" (Bakelite, Daka-ware, Chicago)
1	Cover, ⅛" ×3 2-13/16" × 5-11/16" (Bakelite, Daka-ware, Chicago)
4	Diodes, 1N64 or equivalent
3	Transistors A,B,C—Sylvania 2N1265 or Texas Instruments 2N1372 or General Electric 2N1097
3	R2, R5, R8 resistors—150-K, ½-w, ± 10 percent
3	R3, R6, R9, resistors—100-ohm, ½-w, ± 10 percent
3	R4, R7, R10, resistors—2.2-K, ½-w, ± 10 percent
4	Capacitors, electrolytic, 10 mfd, 25v d-c
1	Grille cloth, 3" × 3"

Mount the electronic circuitry components onto the mounting board and solder. This is not a printed circuit board. However, a printed circuit board can be etched very easily for this circuit.

The sensitivity control is mounted into the box, long leads soldered to the potentiometer terminals, and the on-off switch mounted on the same shaft as the potentiometer. With long leads on the sensitivity control it will be easier to open the box to check wiring, troubleshoot, or change batteries.

Glue the speaker to the cover with duPont's Duco cement. The meter is fastened into place with a large meter-mounting nut.

The bridge circuit of diodes can be mounted on a terminal strip. Battery, B1, is held in place at the bottom of the box by a large "X" of electrician's plastic tape. Grille cloth is glued into place after the electronics have been checked to assure their proper operation.

This device can be used to check the level of applause at all types of meetings and events. It is an impartial judge when it comes to indicating the winner in a contest judged by audience applause.

Point the device toward the center of the crowd and then stand to one side so as not to interfere with the sound reaching the speaker. Also make an allowance for reading the meter at an angle.

Turn on the meter and adjust the sensitivity control so that the level of sound at its highest does not cause the needle to go beyond full-scale deflection.

Many uses may be found for this type of meter. It can be used to determine the level of sound or noise in a room—for instance, to see if television commercials are really louder than normal program material, or it can be calibrated with another sound-level meter to indicate the intensity of sound or noise. The meter cover is easily removed by simply pulling up on it. It can be marked to suit the purpose for which the meter is to be used.

CODE PRACTICE OSCILLATOR

A code practice oscillator is a necessity for anyone learning to send or receive code. A quiet code practice oscillator makes it a little more convenient to practice without disturbing others. This code practice oscillator is inexpensive since it uses only one transistor, a couple of capacitors, three resistors and a potentiometer. The headset is of the magnetic type. It is used as an inductor or part of the oscillator circuit. A 9V power supply lasts a long time. The key serves as an on-off switch.

The Circuit

The oscillator circuit uses R3 and R4 as a voltage divider network to establish the proper bias voltage for the base of the transistor. The voltage drop across R3 gives the base a negative potential sufficient to allow the transistor to conduct immediately, once the power is supplied by closing the key. C1 provides feedback to keep the oscillator functioning, C2 and the phones determine the frequency of the oscillator, and R1 adjusts the tone of the oscillator. See Figs. 7-34 and 7-35.

Construction Hints

The battery holder is designed to hold six 1.5-volt penlite cells. This provides more power than a nine-volt transistor battery, thus giving more hours of code practice. Two terminal strips provide a secure mounting for transistor, resistors, and capacitors. The battery holder can be held in place within the box by a large "X" of plastic electrician's tape.

Fahnstock clips can be replaced with a phone jack similar to the key jack if that type of plug is on the headphones. All parts are mounted on the cover for easy removal. This allows easy access to

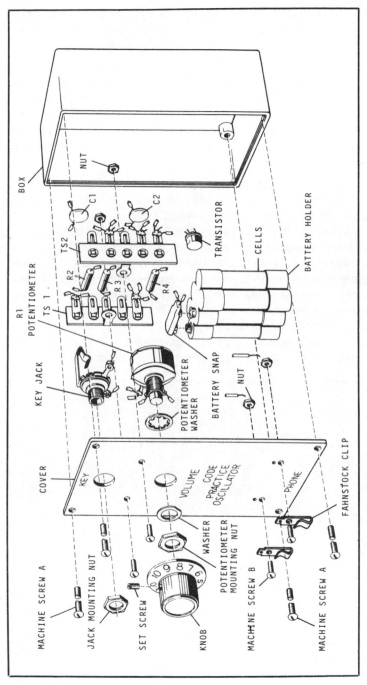

Fig. 7-34. Cutaway of the code practice oscillator.

325

the power supply when the cells in the battery holder need replacing. Space the terminal strips so that the resistors and the capacitors can be mounted easily. See Table 7-13.

Operation

Magnetic type headphones are necessary to operate this circuit. A key is needed to operate. Insert the key plug into the key jack and attach the headphones to the Fahnstock clips. Close the key and adjust the tone or pitch of the oscillator with the potentiometer knob to a pleasing tone. The oscillator is now ready to operate.

ANOTHER LIGHT BEAM TRANSMITTER

Everyone is familiar with the use of a flashlight to transmit code messages. This is the equivalent of a CW transmitter using on-off keying. While this system might be useful in an emergency it never seems to inspire much interest under normal conditions.

Fortunately, a modulated light beam system such as described here arouses the curiosity of nearly everyone who sees it. A light beam that can talk or play a teenager's favorite music is something to be given serious consideration. This is the equivalent of an amplitude modulated transmitter.

As you may have guessed, the transmitter is a flashlight, modified so the output of an audio amplifier (modulation) may be

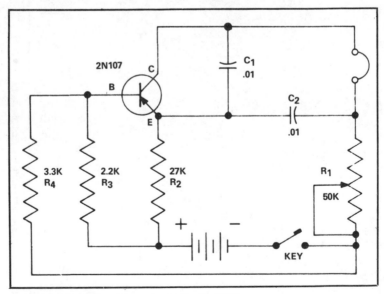

Fig. 7-35. Schematic of the code practice oscillator.

Table 7-13. Parts List for the Code Practice Oscillator.

4	Machine Screw A, 2″ # 6-32 NP
4	Machine Screw B, ⅜″ # 6-32 NP
4	Nut, 6-32 NP
1	Box ⅛″ × 5-⅞″ (Daka-Ware, Chicago)
1	Cover ⅛″ × 2-13/16″ × 5-11/16″ (Daka-Ware, Chicago)
1	Set Screw (part of knob)
1	Knob (to fit potentiometer shaft)
1	Washer (part of potentiometer)
1	Potentiometer Mounting Nut (part of potentiometer)
1	Potentiometer Washer (part of potentiometer)
1	R1, Potentiometer, 50-k, audio taper
1	Key Jack
1	Jack Mounting Nut (part of jack)
1	TS1, TS2, Terminal Strip, 5 soldering lugs
1	Transistor, 2N107 or equivalent
1	R2, Resistor, 27-K, ½-W, ± 10%
1	R3, Resistor, 2.2-K, ½-W, ± 10%
1	R4, resistor, 3.3-K, ½-W, ± 10%
1	C1, Capacitor, .01-MF, 100-VDC, Ceramic
1	C2, capacitor, .01-MF, 100-VDC, Ceramic
2	Fahnstock Clips
1	Key, Telegraph
1	Headset, 2-K, Magnetic
4	1.5-Volt dry cell
1	Battery Snap
1	Battery Holder

applied in series with the battery and bulb. This is basically the same system used on an AM transmitter. I have found that a convenient source of modulation is the phone jack on a transistor radio. Apply modulation (turn up the volume control) very gently to avoid burning out the flashlight bulb. Caution is especially in order if the flashlight cells are fresh and up to full voltage. If everything is working properly you should notice a slight flicker in the light caused by the modulation.

By now you are probably thinking the transmitter was easy enough to assemble so it must be the receiver that is the difficult item. Actually, the receiver is easier to obtain than the transmitter. All of us have been around light beam receivers for many years. I am referring, of course, to the phototube and amplifier used in a 16mm sound projector.

Operating the System

For preliminary testing, simply aim the modulated flashlight beam at the cathode of the phototube and sound will be heard from the speaker of the movie projector. Moderate amounts of daylight won't interfere with the system, but room lighting from an AC source will produce a loud hum, so turn off the room lights when using the transmitter. If considerable hum is still present with the lights out, you may have to add some electrical shielding for the phototube and amplifier to make up for that which was originally provided by the case of the movie projector.

LEYDEN JAR

To emulate the experiment of the Dutch physicist who discovered the basic principle of condenser action, use an ordinary glass jar of the metal, screw-type lid kind which housewives use for canning fruit and vegetables.

After cleaning the jar well and drying it thoroughly, cement a single thickness of tinfoil to the jar's lower three-quarters. To insure that the tinfoil is smooth around the jar's circumference, press it with the fingers or roll it with a paper-hanger's seam roller. All the air bubbles must be removed.

Fill the jar with crumpled tinfoil to about ⅔ of its height, being sure not to crumple it too tightly or make a firm mass out of it. Solder a small brass ball to the end of a 6″ length of ⅛″ rod. Force the rod through the center of a cork, the free end extending far enough to make contact with the tinfoil in the jar when the unit is assembled.

A piece of cardboard should be cut large enough so the cork fits snugly. The cardboard is cut to a disc shape and to a diameter that will fit inside the screw lid. This disc must be a good fit, with cardboard of a thickness that will withstand some strain without wrinkling.

Assemble as in Fig. 7-36, screwing the lid down so it compresses the cardboard snugly against the rim of the jar.

To show the action of the Leyden Jar, rub the ball at top with a piece of fur, silk, or lamb's wool. Touch a wire from the tinfoil on the outside of the jar to the ball, and the result will be a shower of sparks as the static electricity induced and stored in the jar is released. The faster and longer the ball is rubbed, the greater will be the intensity of the sparks.

ELECTRONIC APPLAUSE METER AND ITS USE

The electronic applause meter shown here (Fig. 7-37) is for use in auditoriums or other places where an audience is to decide the winner of a contest by the intensity of applause given to the contestants. On its illuminated large scale with linear divisions of 0

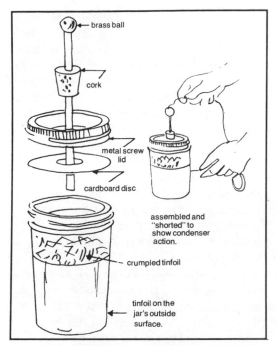

Fig. 7-36. Leyden jar.

brass ball

cork

metal screw lid

cardboard disc

assembled and "shorted" to show condenser action.

crumpled tinfoil

tinfoil on the jar's outside surface.

to 100, the pointer can be made to stop at the maximum applause reading. It can be reset to zero by a hand switch on a long remote-control cord, or reset manually at the applause meter.

Basically, the electronic applause meter consists of a microphone which is energized by the applause of an audience. The electrical impulse from the microphone is fed to an amplifier. The output of the amplifier is rectified by a full-wave bridge rectifier, and the rectifier supplies direct current to a solenoid, which pulls a

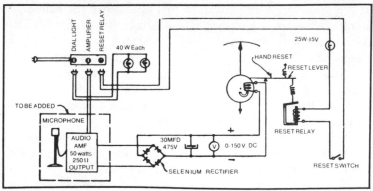

Fig. 7-37. Schematic of the electronic applause meter.

movable soft-iron core that is part of the pointer movement mechanism.

The intensity of the applause determines the power output of the audio amplifier and thereby the amount of electrical energy supplied to the meter solenoid. The greater the DC power supplied to the solenoid, the greater will be its magnetic field which, in turn, will cause a greater movement of the movable soft-iron core.

In order to have the pointer stop at maximum reading, the pointer is attached to a painted iron disk, 10″ in diameter and 1/16″ thick. A sharp chisel edge on the end of the reset lever (Fig. 7-38) lightly rests on the rim of the disk. The disk and the pointer both move as current is applied to the solenoid. When the current in the solenoid ceases, the disk will be held at this point by the sharp edge of the reset lever. The disk and pointer are returned to zero by a spiral spring if the reset lever is raised away from the rim of the disk, manually or by the remote relay.

The remote zero reset control consists of a relay attached to the reset lever. De-energizing the relay coil causes a spring to place the reset lever on the rim of the disk, thereby preventing the disk from being pulled back to zero by the force of the restoring spring. When the relay is energized by the reset hand switch, the relay lifts the reset lever away from the rim of the disk and causes the pointer to return to zero.

The frame is constructed of welded steel angle. The handle is made from 1″conduit welded to the rear brace angles. The front and bottom panels are made of ½″ plywood. The scale is painted on a sheet steel plate. The pointer is made of aluminum, with a weight on a 12″ tail for balancing.

The meter movement mechanism consists of a mild steel disk, 10″ in diameter, 1/16″ thick. To this is attached the pointer, a spiral return steel spring, and a semicircular soft-iron core. The disk assembly is supported by a ¼″ diameter steel shaft which moves on ball bearings. Brass balancing weights are attached to the disk. The restoring-force spring is a spiral spring taken from an old alarm clock. The return spring is attached to a brass lever ½″ wide, ⅛″ thick, and 8″ long. This is used for adjusting the return spring tension when calibrating the meter.

The zero rest position is fixed by a bolt which also supports the scale. The semicircular soft-iron core moves freely into the solenoid and the solenoid is held by brass supports.

The electrical circuit is the result of experimentation. The selenium rectifiers are of the 400-mA type. A DC voltmeter is used as a reference for calibrating the applause meter. A 25-W, 115-V lamp is connected in series with the relay coil for better relay operation.

The meter solenoid is a coil from a 10-W old-type dynamic cone speaker. It has a DC resistance of 600 ohms. The coil has an outside diameter of 3¼″, an inside diameter of 1¼″, and is 1½″ high. Two turns of .015 thick tin-plated iron is wrapped around the outside of the coil in order to concentrate the magnetic field. A dynamic-type microphone is used.

The electronic applause meter is calibrated by adjusting the tension on the return spring attached to the pointer movement disk. The volume control on the amplifier is adjusted so that an audio signal supplied to the microphone will produce a reading of 100 on the applause meter when 100 VDC is supplied from the rectifier to the solenoid. Shown on this page is a graph on the applause meter reading versus the DC voltage applied to the solenoid. This was found to be helpful for recalibrating the electronic applause meter.

When the electronic applause meter is first used before an audience, it will be necessary to obtain the maximum applause from the audience in order to adjust the volume control on the amplifier. At maximum applause the volume control is turned up so that the DV voltmeter reads 100 volts. At this point the applause meter should also indicate 100. Once the volume control setting is obtained, it should not be moved during the entire performance for accurate reading.

DIRECT-CURRENT TRANSFORMER FOR TWO PROJECTS

How to transform the 4½ volts from three common flashlight cells so as to light a lamp that requires a minimum of 90 direct-current volts for its operation is the basis for a very interesting experiment and a practical demonstration of the storage of electric power in an inductive circuit. The direct-current transformer, shown in Fig. 7-39 not only can be used to explain the electrical principles, but also has practical applications as a hook-up for practicing the Morse code and as a safe night light for a child's room.

Before describing the construction of this direct-current transformer, a brief resume of how electric power is stored in an inductive circuit may be of interest. A length of insulated copper wire coiled around an iron rod furnishes the makings of an electromagnet. If the ends of the wire are connected to an electric

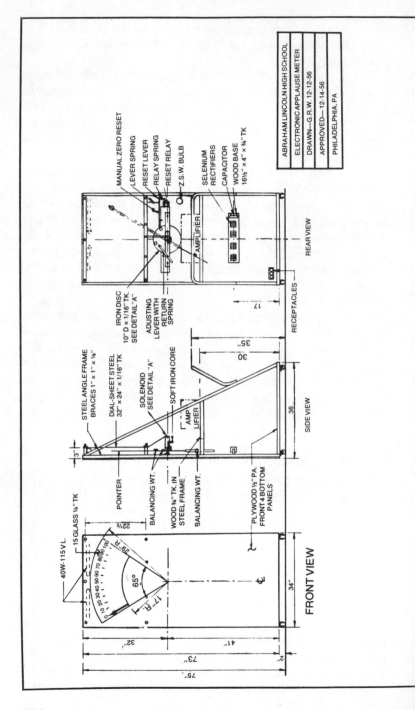

332

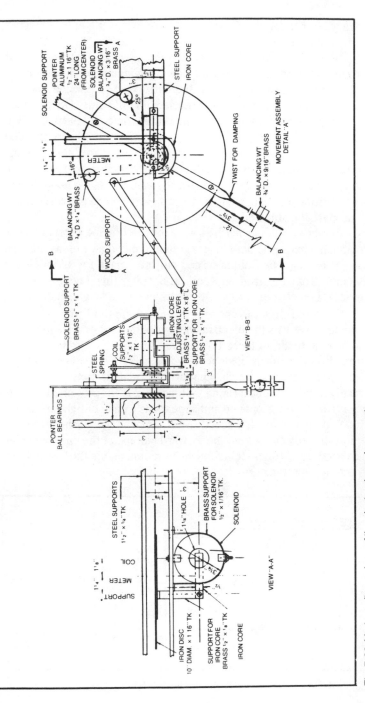

Fig. 7-38. Various dimensions of the electronic applause meter.

333

battery so that current flows through the wire, the electromagnet is in working order.

Now, if before coiling the length of wire, an instrument that measures the amount of current flowing is connected between the ends of the U-shaped loop and the battery, it is noticed that the current almost instantaneously reaches its full value. On the other hand, after the wire is coiled around the iron core, particularly if there are many turns of fine wire, a peculiar thing happens. The current rises more slowly to its full value and there is an appreciable time from the moment the circuit is closed to the moment the current reaches its steady value. It is as though there is an additional resistance to be forced away before the current can be properly established.

Actually, this is what happens, for the flowing current has to establish a magnetic field in and around the coil as well as the iron rod. This magnetic field is forced out into space for an indefinite distance from the iron and coil, and doing this takes a certain amount of electric energy while establishing the magnetic stress.

When the circuit is opened, this magnetic field collapses, and in doing so returns the electric energy to the copper wire in such a way that it tends to force the current to continue flowing in the same direction as it had been going. This energy cannot be destroyed and must be dissipated into heat or light or stored. Usually it will develop a rising voltage until it jumps—sparks—the gap as the circuit is opened and appears as a flash of light and heat. This voltage can easily exceed the circuit voltage. If the circuit is rapidly closed and broken, the forward voltage of the circuit can be increased many times. It is this phenomenon that is the basis of the direct-current transformer.

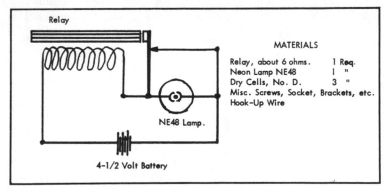

Fig. 7-39. Schematic of the direct current transformer.

The necessary parts for building this project can usually be found in any radio parts box and they are very easy to assemble. They include a small relay, a ¼-watt neon lamp, three small dry cells (the common size "D" flashlight batteries), and a few screws, bracket, wire, etc.

For the assembly use a 5.3-ohm, single-contact, single-throw, nominally closed relay mounted on a block of wood. To increase the frequency of vibration, a headless nail is driven into the block so the tension spring can be stretched to about twice its length and anchored on the nail. To further increase the frequency a small adjustable screw on a bracket is set so as to check the return motion at the instant it made contact. This arrangement will cause the armature to sing like a hornet and then the 90-volt neon lamp will glow with a bright red light. Thus we are actually lighting a 90-volt lamp with a 4½-volt battery.

To use this project as a hook-up for practicing the Morse code, the buzz of the relay should be set to a lower frequency so as to be audible. Then either the audible or the red visual signals can be sent and received. The neon lamp has instantaneous quenching, so the light flash is exceedingly sharp.

To use the hook-up for a night light in a child's room, the relay should be set to a high and almost inaudible pitch and then the transformer should be placed in a small box and packed with glass wool to smother the sound. Since the current drain is less than 70-ma, the three cells will have a long life, about a month at one hour per night. The youngster cannot be hurt if he plays with the glow light, for it is as safe as a three-cell flashlight.

FLUORESCENT LIGHT BLUEPRINTER

This blueprinter has proved to be a good incentive builder as well as a worthwhile addition to a drafting room. The box is made from ¾" × 8" pine, using ¼" plywood for the bottom. The top is ½" thick and it has a ¼" sponge rubber pad which fits on the inside. Double-strength window glass is used for the printing surface. The top is fastened firmly down with two window locks to prevent poor contact with the glass.

The printer uses five 15-watt black light fluorescent lamps and five 15-watt single ballasts connected in parallel with a single-pole switch to turn the lamps off and on. See Figs. 7-40 and 7-41.

Placement of the lamps is important. The layout of the mounting plate on this page shows how to get the proper distribution of light. The distance between lamps should approxi-

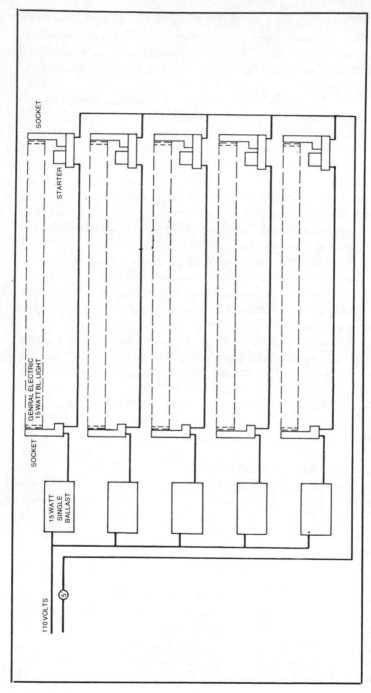

Fig. 7-40. Wiring diagram for the fluorescent light blueprinter.

336

mately equal the distance from the top of the lamp to the top of the glass.

With this printer, it takes about five minutes of exposure time to make blueprints using sunlight paper No. 213L. Some experimentation will be necessary to obtain desired results from different kinds of paper.

ELEPHANT NIGHT LIGHT

This project appeals to youngsters because of the circus motif and is useful for the small brother or sister who needs a light in the bedroom at night. The night lamp is an excellent educational project as it presents problems in wood, metal, leather, soldering, electricity, turning and geometric figures. See Figs. 7-42 and 7-43.

The legs and the head are made from ½" wood. A full-size pattern can be made by enlarging the blocks in the drawing to ½" squares. The tail is made from hardwood, as it is thin and must be rugged.

A number 10 can is used to make the base. A pair of dividers or a compass with the legs set 2" apart is used to score a line for cutting the base to the proper depth. A pair of tin snips and a file to smooth the edges are the only tools needed. A piece of tin 4" wide and 13½" long is needed for the body. Holes are made in this piece for the nipple of the light socket, screws to fasten the head, and a slot for the tail. The tusks are cut from pieces of tin and are left unpainted and held in place with brads.

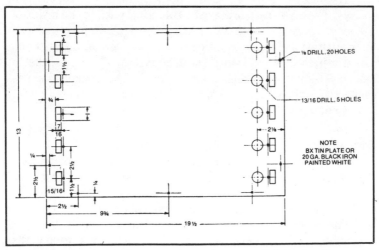

Fig. 7-41. Mounting plate dimensions for the fluorescent light blueprinter.

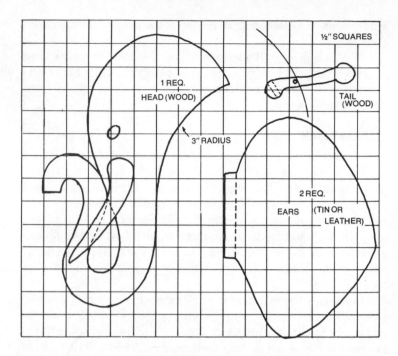

Fig. 7-42. Dimensions of the elephant night light.

The large flapping ears of the elephant can be quite realistic if made from mottled leather.

The head is fastened to the body by means of round-head screws. The tail is held in place by soldering a nail to the inside of the body. The nail must first go through the tail, as it pivots to snap the switch on and off. The tin body is positioned on the legs and nailed on both sides with small escutcheon pins. The light socket is mounted by use of a nipple and a finial. The socket chain is connected to the tail. The chain will have to be cut and tested for proper operation. The ears are added after painting.

Two coats of gray paint should be given the body. Details of the legs and eyes can be added later with black enamel.

Screw the legs on the base from underneath. A geometrical design can be marked out and painted on the sides of the base with bright circus colors. For weight, the base can be filled with Keene cement or plaster of Paris so the lamp will not tip easily.

The sub-base of wood can be turned from ¾" wood, 1" larger in diameter than the metal base. A groove should be turned in the wooden base so the edge of the can will fit into it. The two are held

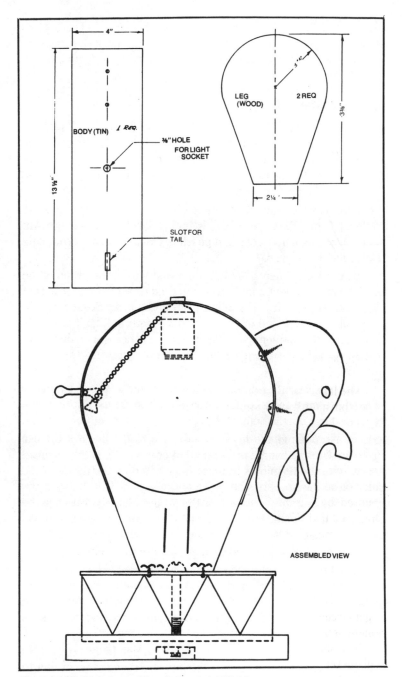

4"

BODY (TIN) *1 REQ.*

⅜" HOLE
FOR LIGHT
SOCKET

13½"

SLOT FOR
TAIL

LEG
(WOOD) 2 REQ

5"r

3⅜"

2¼"

ASSEMBLED VIEW

Fig. 7-43. Various views of the elephant night light.

together by means of a bolt and a counterbored nut. The bottom should be covered with felt. The inside of the metal part of the body should be shined with steel wool so it will make a good reflector.

ELECTRIC HOT PLATE

In this project, a coffee can is cut so as to be 2" high. A template is made of the bottom and transferred to the can. The vents are made by using a fruit-can opener to allow for the dissipation of the heat. The rectangular hole is made with a cold chisel and then filed smooth.

After transferring the template of the reflector to the top of the can and drilling, the edges of the ⅜" hole are rolled so as to prevent abrasion to the heater cord and possible damage. Brass grommets may be used if desired.

The measurements for the holes and wire are marked off on the asbestos element panel after it is filed to size. Twenty feet of No. 24 nichrome wire is wound tightly about a ⅛" Bessemer rod. The coil is then pulled out evenly to a length of 22". The coil is mounted on the element panel as shown in the drawing, using small pieces of nichrome wire to hold the coil firmly in its circular position.

The spacers are rolled completely around a ¼" rod by means of hand pressure and a combination pliers and are used to separate the reflector from the body of the can.

In assembling, the legs are fastened to the body of the can first. Be sure the handle is fastened to one leg. Then the terminal board, with the terminals in place is fastened on the inside of the can. Next, the spacers and bolts are attached. The nuts on the central post are now adjusted for the proper distance between the element panel and the terminal board. The connectors are fastened to the terminals on the element panel, inserted into the 5/64" holes on the terminal board, pulled taut, and fastened into place.

The reflector is slipped onto the heater cord. The cord is then fastened to the terminals and the reflector is bolted into place. The entire outside of the hot plate is given a black wrinkle finish. The plate is turned upside down, plugged into an outlet, and the finish is baked on.

The grill is placed on top of the can to complete the project. No grill pattern has been given, as this may be changed according to the taste of the individual. See Table 7-14.

Table 7-14. Parts List for the Electric Hot Plate.

No.	Description	Material
1.	Bottom and side—standard coffee can.	
2.	Element panel—¼" × 4⅝" d Transite.	
3.	Reflector—top of coffee can.	
4.	Terminal board—¼" × 1" × 3" Transite.	
5.	Spacers (3)—⅝" × 2½" excess coffee can.	
6.	Legs (3)—⅛" × 3½" band iron.	
7.	Handle—¾" d × 2" dowel.	
8.	Top leg fastenings (3)—6.32 × ¾" rh machine screw and nuts.	
9.	Bottom leg fastenings (3)—6.32 × ½" rh machine screw and nuts.	
10.	Reflector fastenings (3)—8.32 × 1" rh machine screw and 6 nuts.	
11.	Terminal board fastenings (2)—6.32 × ½" rh machine screw and nuts.	
12.	Terminals (2)—6.32 × ¾" rh machine screw and 6 nuts.	
13.	Heating element—20' of No. 24 nichrome wire.	
14.	Nichrome terminals (2)—6.32 × ½" rh machine screw and 4 nuts.	
15.	Connector—2' of No. 16 copper wire.	
16.	Center post ¼" × 2" rh stove bolt and 3 nuts.	
17.	Cord and plug—6' standard heater cord and plug.	
18.	Handle fastening—1" No. 6 rh wood screw.	
19.	Nichrome hold-down (18)—2" long No. 24 nichrome wire.	
20.	Grill-cast aluminum.	

THYRATRON MOTOR CONTROL PANEL

The no-load motor speed may be varied from 200 to 2300 rpm by armature voltage control, from 2300 to 4600 rpm by field voltage control with this control. DC voltmeters may be connected across the motor windings to show the correlation between rectified voltage variations and the speed of the armature. Since the control unit consists of two phase-shift variable voltage supplies, it may be used without the motor to demonstrate "stepless dimming" of incandescent lamps, or for any other purpose requiring variable DC voltage within its voltage and current range.

The circuit shown in Fig. 7-44 has been reduced to the fundamentals by omitting the current control and constant-speed control features found in commercial models. An oversize chassis with plug-and-jack connections was purposely used in this particular equipment to simplify possible later additions to the circuit. Simple breadboard mounting and permanent connections would be entirely satisfactory for the basic circuit.

The DC motor used is a 1/6 hp shunt machine rated at 115V, 1.3 amp. Motor voltage in this general range is desirable, as it permits a simple and economical auto-transformer winding on the main power transformer.

Both the field and armature rectifiers are of the bi-phase, full-wave type. This gives better motor operation than half-wave; also, two small thyratron tubes are likely to be cheaper than one large one. Armature tubes are type 393-A mercury-vapor triodes, rated at 1.5 amp. average current and 1250 inverse peak volts, with 2.5-V, 7-amp. filaments. Field tubes are type 2051 gas tetrodes, rated at .075 amp. average current and 700 inverse V, with 6.3-V, .6-amp. heaters. Both of these tube types are obsolete and can be obtained quite reasonably from surplus sources. However, any available triodes or tetrodes of similar current rating and not less than 500 inverse peak V will serve for a motor like the one described above.

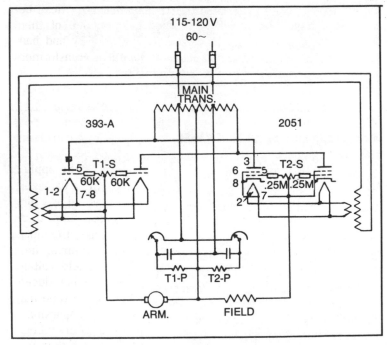

Fig. 7-44. Schematic of the thyratron motor control panel.

The main power transformer was made from a shell-type core with the center leg approximately 1½" square. This core was rewound with 680 turns of 20-gage wire, tapped at 205, 340, and 475 turns. When connected as shown in the circuit diagram, it provides approximately 150 volts on each side of the center tap, permitting a maximum d-c voltage somewhat above motor rating. This transformer also supplies center-tapped line voltage to the phase-shift circuits. Separate filament transformers were used by the author simply because suitable ones were at hand, but, with an auto-transformer winding, there is likely to be plenty of window area in the main core for adding filament secondaries. With a center-leg core area of about 2.25 sq. in., these windings would require, respectively, five turns of 10-gage wire, center-tapped, and 14 turns of 19-gage wire for 393-A and 2051 tubes.

Each phasing control consists of a 1.0 microfarad, 200-V capacitor and a potentiometer of 15,000 ohms. Separate control knobs for the low-speed and high-speed ranges are preferable for teaching purposes, and should be connected to increase speed with clockwise rotation in each range. Two low-powered class "B" driver transformers were used for reversing the phasing voltage to the thyraton grids. These transformers are not very critical, but each must have two high impedance windings, one of them center-tapped, with the turns ratio between primary and half-secondary not far from unity. Inter-stagte audio transformers might do if the winding reactance is high at 60 cycles.

Provision should be made at the output leads of the rectifier circuits for convenient connection of a voltmeter or oscilloscope for observation of the voltage variations and patterns. The oscilloscope patterns across the motor armature and lamp loads are extremely interesting and instructive.

The voltage at the thyratron anodes can safely be applied simultaneously with the filament voltage. Tubes will be damaged, however, unless the motor or lamp load is disconnected by a plug or switch until the tube filaments have been preheated. Since there is no automatic armature current control in the circuit, the motor should always be started with armature voltage at minimum, field voltage at maximum. During motor operation the field voltage should always be at maximum when armature voltage is reduced, and vice versa. Weakening both voltages at the same time may cause stalling or even reversal of the motor, with heavy sparking at the commutator and excessive peak current in the armature tubes.

One last word: don't be afraid to experiment with thyratron

circuits. They are much more simple and free from "bugs" than most vacuum-tube circuits. Just remember the preheating requirement and protect the tubes from excessive peak currents.

LOW-COST SMOKE-GAS DETECTOR

This smoke and/or gas detector costs less than $10 to build, can be constructed by any hobbyist with minimum supervision, and is a high-interest project of practical value. The unit is self-contained in a 6″ × 3″ × 2″ box and operated on normal residential current. The heart of the detector is a semiconductor sensor about the size of a thimble. The alarm is sensitive to fumes and vapors of gasoline, alcohol, carbon monoxide, products of combustion, natural gas, bvutane, propane, solvents and other substances in normal household use. Circuit layout is not critical and a printed circuit board can be easily designed.

Construction Tips

Transistors Q2 and Q3 (Fig. 7-45) are packaged with the collector lead in the center. Be sure to redesign the circuit board for component placement if another configuration is used.

We used a minature potentiometer (R2). If a larger control is desired, its body diameter should not exceed ⅞-in. The associated "C" hole then should be drilled to accommodate a ⅜-in. bushing, instead of the specified ¼-in., and the two positioning "A" holes should be omitted.

Hole templates are shown full scale. All "A" holes are ⅛-in., the "C" is ¼-in., (note exception above), and the "D" hole is 1/16-in. The "B" mounting hole will be ⅝-in. or ¾-in., depending on the diameter of the 7-pin miniature socket selected for the

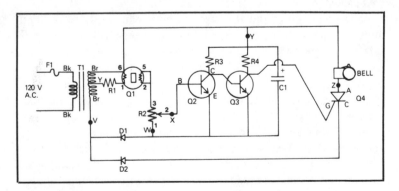

Fig. 7-45. Schematic and parts list of the low-cost smoke-gas detector.

sensor. If socket diameter is ⅝-in., drill the A1 holes; if it is ¾-in., drill the A2 holes.

How It Works

The transformer is not current limiting and must be fused. Half of the T1 secondary supplies energy to heat the sensor through R1, which is a voltage dropping resistor. The gas sensing semiconductor is designed to be heated by 1.5 V. The entire secondary of T1, D1, and C1 supplies collector voltage to the transistors. The 6.3 V are switched by Q4, causing the bell to ring.

As the concentration of gas or smoke increases, Q1 conducts, increasing the voltage across R2. A portion of this voltage increase is sensed by the base of Q2, causing it to conduct, or turn on. When Q2 turns on, Q3 turns off, making the gate of the SCR (Q4) positive and causing it to fire.

In the idle state (no smoke or gas), Q1 does not conduct enough current to turn on Q2. Therefore, Q3 is on and the anode of Q4 is close to zero potential and does not fire.

R1 is designed to run hot. The sensor *Q1* is designed to run warm.

Adjustment and Installation

Rotate the potentiometer fully counter-clockwise and let the unit warm up for five minutes. Slowly rotate R2 clockwise until the bell rings, them back off until bell stops. Test the alarm by blowing some smoke into the sensor. Normal household odors and chemicals will set off the alarm. Carefully adjust the control to eliminate most false alarms.

Mount the alarm on the wall close to the ceiling for maximum detection of smoke, close to the floor for maximum detection of gas, or about five feet up from the floor for a good compromise. Do not bend or cut the smoke sensor leads. Drill a suitable hanging hole in the bottom of the case at the end opposite the transformer.

A NEON BULB LIGHTS THIS CHRISTMAS CANDLE

This Christmas "candle" is a sure-fire project. The base requires two to six ounces of casting resin, hardener, coloring and felt for the bottom. The candle body consists of a 6-in. length of ½ or ⅝-in. plastic tubing. The lamp assembly requires a NE2 neon lamp, a .22 meg., ½-watt resistor; 6-ft. of 18-2 lamp cord; and a plug.

The halo can be made from ⅛-in. transparent plexiglass, and the decorations are cut from two sheets (one red and one green) of 4-½ × 4-in. vinyl.

A mold is required to cast the base, and in order to leave an opening for the candle stem, a removable plug made of dowel rod must be made (see illustration) for use when casting the base. Other requirements are mixing cups, a mixing rod, sandpaper, patterns for the leaf decorations, scissors, a soldering gun, and plastic film sealer. See Fig. 7-46.

Procedure

Place the mold on a flat surface and carefully install the fixture in the mold (make sure the plug is heavily waxed). Thoroughly mix the resin, hardener, and coloring in a cup. Pour the resin mixture in the mold and allow it to set overnight before removing from the mold.

Cut the tubing to the desired length, sand the ends smooth, mask the outside and paint the inside (using colored tubing would eliminate the painting operation). Using the plastic sealer, put a slot in one end for the halo.

Separate the lamp cord and cut one wire 2-½-in. shorter than the other. Connect the long wire directly to the bulb and connect the short wire to one lead of the resistor. Connect the other

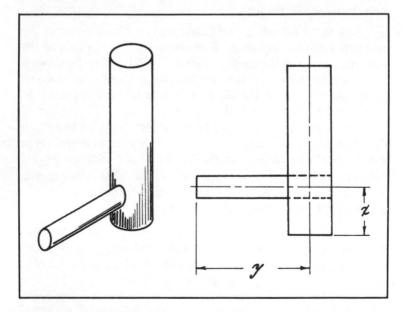

Fig. 7-46. This removable plug is used to leave an opening for the stem when casting the base of the Christmas candle. Demension X is the distance from the surface of the mold to the lip; Y is the distance from the center of the mold to the lip.

resistor lead to the other bulb lead. Solder the joints and wrap them with electrical tape—be sure to check students' work on this phase.

Cut the halo to shape, sand the outside edges, and cut a slot from edge to center for the bulb. Trace and cut out the leaf decorations, punch a ⅛-in. hole in the wide end, and attach with plastic lacing.

Assembly

Glue the halo in place on the stem, glue the stem to the base, and feed the wires through the stem until only the bulb sticks out. Make sure the wires are in a groove in the base, glue the felt on the bottom, and attach the plug to the free end of the wire. Finally, tie the leaf decorations in place.

GLOW IN THE DARK

Construction of this project is relatively simple, does not take more than a few days to complete, and uses easily obtainable materials: 1/16 by ¼ by 1-¼ in. long hard brass prongs, high intensity neon bulb with wire leads, 33 kilohm ¼ W resistor, and casting plastic. The mold is a 5 oz. paper cup. Make sure the cup is wax coated so the plastic will release after it has set. See Fig. 7-47.

Procedure

After cutting the prongs to length with a cold chisel, drill a 1/16 in. hole in one end of each piece and a ⅛ in. hole in the other end. Each hole should be 3/16 in. from the end (we made a jig for better accuracy). Insert a piece of ⅛ in. d wire through the larger holes, space the prongs 7/16 in. apart, clamp in a vise, and solder the resistor and bulb in place.

Cut the paper cup off so that, when the ⅛ in. wire is laid across the top of the cup, the parts are suspended about ⅛ in. off the cup bottom. Recheck prong spacing, center, and pour in the plastic. Be sure to leave ¾ in. of the prongs exposed to fit a wall outlet.

After the plastic has set, remove the project from the cup. File, sand and buff the plastic until a high polish is obtained.

AUTOMATIC TIMER

This automatic timer, designed for use with object-performance testing, is easily fabricated and can be adjusted for approximately three-minute intervals. The circuit will turn the relay "on" for approximately 2.5 seconds every three minutes, signaling the students to proceed to the next test station. Generally, a larger value capacitor will extend the time interval.

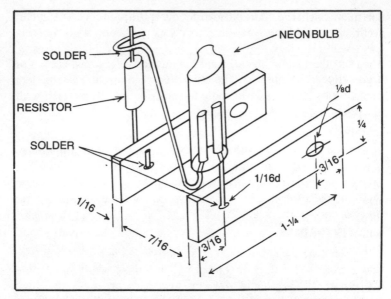

Fig. 7-47. Assembly drawing of the glow in the dark project.

In this circuit (Fig. 7-48) the relay keys an oscillator which is loud enough for all to hear. *R1* is a variable resistance of 2 megohm which must be adjusted by trial using a sweep second hand or stopwatch. Of course, several time intervals could be precalibrated on a dial so that the times could be easily set at a desired interval. When the circuit is first turned on, the first time cycle will be considerably longer than the set time. This is due to the initial charging of the capacitor. In subsequent cycles, the capacitor does

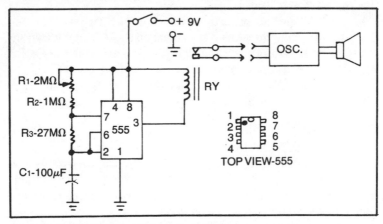

Fig. 7-48. Schematic of the 555 automatic IC timer.

not fully discharge. After the first cycle, the circuit will cycle accurately to within a fraction of a second. Although a 9 V battery was used in the prototype, voltages from 5 to 9 would also work. The readjustment of the variable resistance *(R1)* may be necessary if you change to a different voltage than the one first used. See Table 7-15.

THREE HIGH-PERFORMANCE AMPLIFIERS

The LM380N integrated circuit, a low-cost audio amplifier, makes an excellent phonograph, microphone, FM tuner, or intercom amplifier for a minimum investment in parts and time. Many distributers carry this 14-pin dual-in-line IC at a cost of approximately $1.50. It is manufactured by National Semiconductor Corp. By adding just a volume control, capacitor, 12V power source, and speaker, you will have an amplifier which rivals some $100 units. To appreciate its quality, test it on a good speaker system. Figure 7-49 shows the circuit configuration for a 2.5 W rms high fidelity monaural audio amplifier.

The decoupling capacitor C_2 may be eliminated if the wires running to the power source are fairly short. The tone control circuit, R_1 and C_1 may also be eliminated. This circuit requires a fairly high level input signal such as one from a ceramic phonograph cartridge. The power supply must be well filtered, and may be any voltage between 8 and 22 V DC. At high volume levels, this IC will draw over 100 mA, so forget about using a small 9 V transistor radio battery to power it. A 1 A 12 V transformer with full-wave rectification and a 1000 μF filter works well as a power supply. Beware of incorrect polarity.

Other Applications

A basic amplifier circuit was constructed on an 8 cm square printed circuit board shown in Fig. 7-50. A large copper foil area was needed to dissipate the heat generated by the IC.

Table 7-15. Parts List for the Automatic Timer.	1 - 555 Timer 1 - SPST Toggle Switch 1 - Potentiometer, 2 MΩ, ½W, 10% 1 - Resistor, 27 KΩ, ½W, 10% 1 - Capacitor, 100μF, Electrolytic, 6V 1 - Relay, hobby type (Radio Shack No. 275-004 or equiv.) Miscellaneous, copper clad board, solder, Utility Box, ¼ audio plug, etc.

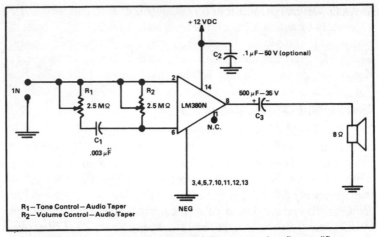

Fig. 7-49. Schematic of a 2.5W rms high fidelity monaural audio amplifier.

A stereo amplifier may be created by simply building two LM380N amplifier circuits. The LM380N is not designed to efficiently amplify signals from magnetic phonograph cartridges or tape heads. By adding an LM381 preamplifier IC, also manufactured by National, the amplifier will handle almost any input.

The LM380N also makes a sensitive intercom shown in Fig. 7-51. With this intercom circuit I exchanged messages over a 100 m distance. It is likely that you or your students will discover several other applications for the LM380N wherever an audio amplifier is needed.

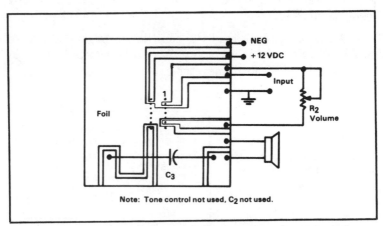

Fig. 7-50. Printed circuit board on which a basic amplifier was constructed.

2-WATT AMPLIFIER MODULE FOR TWO PROJECTS

This 2-watt amplifier module is designed for the LM380N integrated circuit manufactured by National Semiconductor Corp. This IC is a complete power amplifier (rated at 2.5 W at full output) which features a design gain of 50 (+34 db) with an output impedance of 4 Ω and an input impedance of 150 k Ω. Its functional diagram (Fig. 7-52) indicates the pin connections to the IC. The supply voltage (Vc) can be any convenient value between 9 and 18 volts with the ground terminal of the IC returned to the ground of the power supply. This negative return is connected to both pins 3 and 7. The output (pin 8) with no signal present has a voltage approximately one-half of the supply voltage. To block this dc voltage (and the 1.5 A available from the IC) the blocking capacitor (CI) is used to connect the loudspeaker to the amplifier. Pin 8 is the dc output of the amplifier module, the ac output is connected through the capacitor to pin 8.

The amplifier has two inputs, an inverting input (pin 6) and a non-inverting input (pin 2). Either can be used as needed in specific applications. The only precaution is that the leads to the (+) non-inverting input must be kept as far removed from the output leads as possible to minimize feedback and possible oscillation. Capacitor C2 is also included on the circuit board to prevent oscillation. Without this capacitor there is a possibility of high frequency oscillation which can cause an abnormal current through the IC. The grounded foil strip between the input and output sides of the IC on the circuit board also serves to minimize coupling and

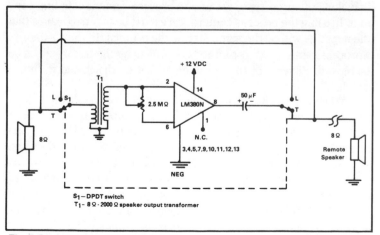

Fig. 7-51. A sensitive intercom circuit.

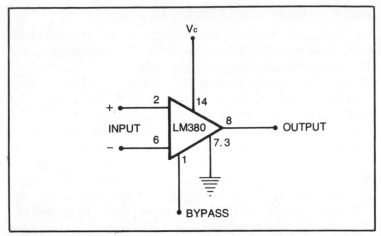

Fig. 7-52. Functional diagram of the LM380 IC.

feedback. In test circuits, this module has proved itself to be extremely reliable and almost non-destructable. See Table 7-16.

One reason for the ability of this IC to withstand abuse is built-in protection against a short-circuited output and also a thermal shut-down when the IC becomes overheated. To utilize the full output power of the IC, a heat sink must be connected directly to the IC.

Making the Circuit Board

Figure 7-53 is the full-size foil pattern for the amplifier module and can be used to produce a photographic negative (see page 354). In making the prototype, we used a process camera with line copy film. In using the process camera, care must be taken to insure that the negative will be the same size as the original. In the absence of a process camera, other methods of producing the negative can be employed. Several of these methods will be discussed in future articles.

When a satisfactory negative has been produced, the circuit board must be prepared. The size for the amplifier module circuit board is 2.5″ × 2.5″ (6.35 × 6.35 cm). This is cut from stock and carefully cleaned, the foil being cleaned "bright" with either fine steel wool or a scouring powder (such as Comet). The board is then sensitized in subdued light or in a darkroom using the yellow or red safe-light. The sensitizing consists of spraying two *even* coats of Etch Resist Sensitizer. The first coat is sprayed in vertical lines and the second coat sprayed in horizontal lines. The spray can should be held about 12″ from the surface of the board. Inspection

Table 7-16. Parts List for the 2-Watt Amplifier Module Projects.

COMPONENT LIST AND SOURCES*

A. Amplifier Module:

C1-0.1 μF, 50 V disc ceramic	Radio Shack #272-135
C2-500 μF, 25 V electrolytic (PC type)	Radio Shack #272-1030
	Calectro #A1-032
IC Socket—14 DIP, low profile, solder type.	Radio Shack #276-1999
LM380N-2.5 W IC amplifier.	Radio Shack #276-1725
	Calectro #J4-1203
P.C. Board—Phenolic copper clad, single side.	Radio Shack #276-1587 (6" x 9")
Sensitizer	Calectro #J4-622 (3 oz. can)
	General Cement #22-231 (16 oz. can)
Developer	Calectro #J4-630 (pint can)
Etchant-Ferric Chloride	Radio Shack #276-1535
Heat Sink	Allied Electronics #957-2668
Cement (any thermal bonding cement such as:	Allied Electronics #957-0377

B. Intercom System:

R1-1 M Ω potentiometer	Radio Shack #271-211
S1-S.P.S.T. switch	Radio Shack #275-612
S2-D.P.D.T. switch	Radio Shack #275-614
T1-Transformer 1200:8Ω 150 mW	Calectro #D1-724
	Radio Shack #273-1380
B1-9 V battery (two sets of 3AA cells held in two cell holders, Calectro #D3-058)	
Loudspeaker (4 to 8 Ω impedance)	2.5" replacement speakers (square type are easier to mount)

* Radio Shack and Calectro are listed as sources of electronic components because they are readily available throughout the country. These are not the only sources, and certainly not the least expensive sources. Several other distributors of IC's, capacitors, etc., that we have used are:

(1) Digi-Key Corporation
P.O. Box 677
Thief River Falls, MN 56701

(2) James Electronics
1021 Howard Avenue
San Carlos, CA 94070

(3) Quest Electronics
P.O. Box 4430
Santa Clara, CA 95054

(4) Allied Electronics
401 E. 8th Street
Fort Worth, TX 76102

A letter to these companies on school stationary will enable you to obtain a copy of their current catalog. If several students are building the amplifier module, you will find that bulk purchases of the components will frequently result in a reduced cost to the students.

of the sensitized board should show an even purple color which becomes more light sensitive as the board dries. Completed boards should be stored in a light-tight box until exposed.

When the circuit board is completely dry (20 to 30 minutes), it is ready to be exposed through the negative artwork. The graphic arts platemaker or even direct sunlight may be used as the light source. Make certain that the negative "reads correct" when placed on the sensitized board. With our platemaker (nuArc) the exposure time was 4 minutes. When exposed to direct sunlight, the exposure was 6 minutes. Some experimentation may be necessary in the exposure time to obtain optimum results.

After exposure, the circuit board is developed using a solvent type developer in subdued light. During the first 10 to 30 seconds of development, a positive image will appear on the circuit board. After this, continue to develop the circuit board until the unwanted resist is dissolved. When the development is completed, wash the circuit board in running water. Be careful—the image is very soft and can easily be smeared or scratched. Do not touch a dry circuit board with the fingers since a finger print will interfere with the etching process.

Dry the board to harden the resist pattern and when dry, place the board in commercial ferric chloride etchant for 15 to 30 minutes to remove the excess copper foil. To complete the circuit board,

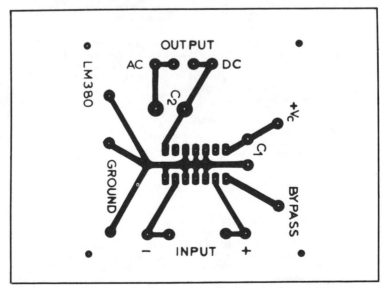

Fig. 7-53. Use this full-size foil pattern to produce the amplifier circuit board.

clean the remaining foil pattern with fine steel wool or scouring powder to remove the resist which remains.

Examine the circuit board to make certain that the foil pattern is satisfactory and if so, the board is ready for drilling. Component holes are drilled with a #59 twist drill (in a drill press—not a hand drill) and corner holes, if desired, are drilled with a ⅛" drill.

Since students frequently have difficulty in soldering semiconductors to circuit boards due to the application of excessive heat, we recommend that a 14 DIP socket be mounted to the circuit board and the IC plugged into that socket. In soldering the socket to the circuit board, a small (⅛" dia tip or smaller) soldering iron is used. Avoid creating solder bridges between the pins of the IC socket.

When all parts have been mounted and soldered to the circuit board, the excess lead lengths on the components are clipped as close to the solder joint as possible.

The Intercom System

With the addition of a transformer, a potentiometer, two loudspeakers, and two switches, the modulator amplifier can function as an intercom system. The circuit for this system is shown in Fig. 7-54. While the system can be housed in any convenient enclosure, we used standard boxes manufactured by the H.H. Smith Company with the markings engraved on the front panel. Many other marking systems are available and can be used when engraving equipment is not available.

We found that the intercom would function satisfactorily using a 9 V power source. However, the current requirement precluded the use of the familiar "transistor battery." Therefore two cell holders for three type AA (pencell) batteries connected in series were fastened to the bottom of the case to serve as a power supply. We also found that because of the excellent low frequency response of the amplifier module, the inexpensive replacement loudspeaker was the best choice to produce a clear signal. While these speakers are limited in their low frequency response, they are adequate in this application. The remote speaker is simply a box with a loudspeaker and jack for connections. The intercom and its remote speaker can be separated by up to 200 feet of two conductor cable.

The only construction technique in the intercom that may cause difficulty is in the connections to the talk-listen switch. This detail is shown in Fig. 7-55. With the two boxes connected, the system should function. If not, an ohmmeter and voltmeter can be

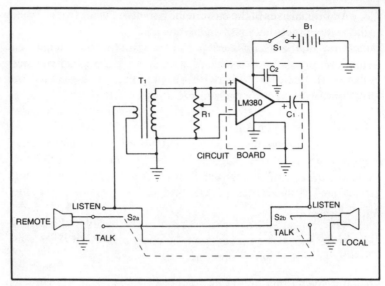

Fig. 7-54. Schematic of the intercom circuit that uses an LM380 IC.

used to find the trouble. In a system that does not function as it should the integrated circuit is seldom at fault—the trouble is usually in the connections.

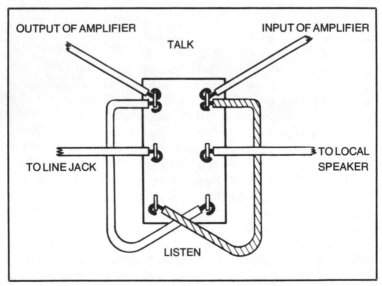

Fig. 7-55. Talk-listen switch construction detail of the intercom.

Additional uses of the basic amplifier module include the voice relay, code practice amplifier, and use of the intercom as a utility amplifier. Instructors might want to recommend to students that while they are making the basic circuit board they make two since subsequent articles will provide instructions for using two amplifiers in a stereo system.

TOUCH BUZZER

This circuit is fun, interesting, versatile, educational, and inexpensive to construct. It can easily be incorporated as a demonstrator and project.

Circuit Operation

Close switch *S1* prior to closing *S2* to prevent transients from activating the SCR. After *S2* is closed, open *S1*. Touch the base of transistor *Q1*, and static electricity will initiate a chain reaction which activates the buzzer. Transistor *Q1* amplifies the current created by touching its base and is direct coupled to the base of *Q2*. Transistor *Q2* amplifies the current and is direct coupled to the base of *Q3*, which in turn amplifies the current and is direct coupled to the gate of the SCR. The collector current of *Q3* triggers the SCR into conduction, sounding the buzzer. After removing the "touch," the buzzer continues to sound because of the characteristics of SCRs. Resistor *R1* prevents the CEMF (counter electromotive force) of the buzzer from turning off the SCR. (Reverse polarity stops SCR conduction.) Not all buzzers need *R1* to do this; some require different methods. Resistor *R2* limits the current passing through the buzzer, and *R3* limits the current through *Q1*, *Q2*, and *Q3*. To stop the buzzer, open switch *S2*. See Fig. 7-56 and Table 7-17.

Notice that the emitters of *Q1* and *Q3* are connected to a positive voltage source for forward biasing, and *Q2* is connected to a negative voltage source for forward biasing. Substitute various transistors (using transistor sockets) to observe the effect of transistor gain on the sensitivity of the circuit. Similarly, different SCRs may be used to observe changes in gate currents for triggering.

To make a proximity detector, attach a length of wire to the base of *Q1*. The length of wire needed depends on the sensitivity of the circuit, but should exceed 3 ft. To decrease the sensitivity of the circuit, remove *Q1* and touch the base of *Q2*.

A SIMPLE TWO-POLE MOTOR

This motor has proven popular with beginners in electricity because it is easy to make, operates on either AC or DC, and can be used to drive small toys. In spite of its simplicity, it involves practically all of the parts of a commercial motor and, therefore, serves as an excellent means for teaching the fundamentals of motor operation and construction.

As it really runs, you derive an immense amount of pleasure from seeing how the motor performs on both kinds of current, the effect of shifting the commutator on the shaft, and the reversing of the direction of the armature by means of a d.p.d.t. switch or by simply interchanging either the field or armature leads.

The following materials are needed:

One piece of ⅛″ × ½″ strap iron, 7½″ long.
Two pieces of ⅛″ × ½″ strp iron, 2½″ long.
One piece of ¼″ d crs, 2½″ long.
One piece of ⅛″ d crs, 4″ long.
Two pieces of spring brass ¼″ × 2½″, #28 gage.
One piece of 3/16″ id brass tubing, 22 gage, 1″ long.
One poplar or other soft-wood base, ¾″ × 3″ × 4″.
Eight ½″ #4 rh wood screws.

When making the field magnet, apply two pieces of friction tape 2¼″ long to the edges of the center section. Allowing 4″ leads and starting at the bottom, carefully and tightly wind on two layers of #22 B&S gage magnet wire.

The armature magnet is made by carefully winding two layers of #22 B&S gage magnet wire on the armature core, beginning and ending in the center. Allow 3″ leads and be sure to wind in the same

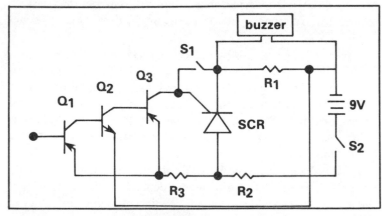

Fig. 7-56. Schematic of the touch buzzer.

Qty.	Description
2	2N2904 PNP
1	sN2218 NPN
1	SCR 1.5 A
1	3 Vdc buzzer
2	SPST switch
1	9 V battery
1	10 K, ½ W (R3)
1	2.2 K, ½ W (R2)
1	1 K, ½ W (R1)

Table 7-17. Parts List for the Touch Buzzer.

direction when changing over from one side of the shaft to the other.

The commutator is made by sawing the 3/16" brass tubing in half lengthwise. A 1" strip of paper is glued on the shaft for insulation. The tubing is then cemented to the paper. The armature leads are soldered to the commutator segments, one lead to each segment. When assembling, solder a washer on the commutator end of the armature shaft to take out end play.

Connections are made as follows for series: (1) Connect one field magnet lead to one of the brush terminals. (2) Connect the other field magnet lead to one of the Fahnestock clips. (3) Connect a wire from the terminal of the second brush to the remaining Fahnestock clip.

For parallel: (1) Connect both field magnet leads to the brush terminals. (2) Connect one wire from each of the brush terminal screws to the Fahnestock clips.

ELECTRIC PRACTICE KEYBOARD

With this electric keyboard, when a key is struck, the corresponding letter will light up on a panel behind the copyboard. With blank keys under the fingers and lights flashing on and off each time a key is depressed, the student's eyes are held near the copyboard level, thus eliminating the eyetravel from copy to key which resulted from the "hunt-and-peck" method.

To a great extent, another problem is also overcome with this method because someone learning the keyboard has a satisfaction, or reward, for striking the key. This reward is needed, for, unlike the typewriter, which leaves a copy of the character struck or the linotype which delivers a matrix on command, the conventional practice keyboard gives no satisfaction at all.

The Construction

To build this machine, I disassemble and rebuild a regular practice keyboard. Then add the indicator panel.

The Panel

The panel characters are set in type in a predetermined pattern and a production proof is made. This proof is photographed, developed and the film is painted on the reverse side with a light-colored, silk-screen ink to make the letters visible. The film is then mounted on heavy chipboard, with holes prepunched for the bulbs, and covered with a piece of ⅛" clear plastic.

To limit the cost, only one bulb is provided for each letter, number, space, and punctuation mark. This means the upper-case and lower-case letters have the same bulb and will light when either key is depressed, but the two keys are separated by the width of the board so the student will know which he has struck (Fig, 7-57).

The 110-v ac was stepped down to 6 VAC, and all bulbs were soldered to a piece of sheetmetal, making a common ground and eliminating the cost of bulb sockets.

The Key Section

Since each key must act as a switch, the keyboard was shortened by ¼" to allow brass pins, mounted in plastic, to be placed above each of the 15 sets of keybars (Fig. 7-58). With all keybars having a common buss and all panel lights having a common ground—the buss must be insulated from the ground—the circuit is completed each time a key is depressed and makes contact with a brass pin.

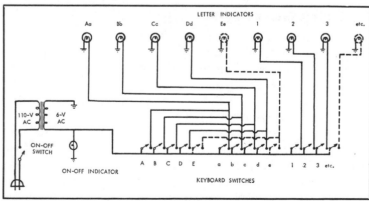

Fig. 7-57. Wiring diagram of the electric practice keyboard.

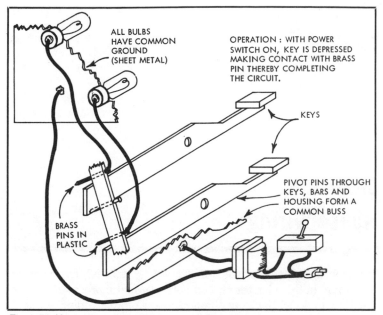

ALL BULBS HAVE COMMON GROUND (SHEET METAL)

OPERATION : WITH POWER SWITCH ON, KEY IS DEPRESSED MAKING CONTACT WITH BRASS PIN THEREBY COMPLETING THE CIRCUIT.

KEYS

PIVOT PINS THROUGH KEYS, BARS AND HOUSING FORM A COMMON BUSS

BRASS PINS IN PLASTIC

Fig. 7-58. Key section of the electric practice keyboard

Finishing

To finish the keyboard and make it as attractive as possible, all metal is polished (including each keybar), giving the unit a two-tone color because the original keyboard is made of copper. As a final step, a copy holder is bolted in place, and the characters are taken from each key and replaced with blank paper of the appropriate color.

Chapter 8
House Wiring

It's probably all around you—house wiring, that is! The projects you'll find in this chapter should help you feel more at ease with anything to do with wiring of a house—or other building, for that matter.

TEST THIS WIRING PANEL

This type of panel also provides "slide-in-and-out" arrangements for a variety of other electrical circuits. Bread-boarded electrical appliances such as irons, lamps, and small motors can be installed for trouble shooting. Layouts of electronic circuits such as power supplies, five-tube superhet circuits, and others can be built on panels for easy installation and demonstration by the instructor.

For power install individual 10-V transformers. The wiring components consist of miniature 10-V lamp bases, with each terminal screw connected to a Fahnestock clip. The switches are ordinary push-button switches with terminal screws connected to Fahnestock clips. By using lights you eliminate the necessity of bell maintenance and buzzers that are found in most signal circuits. Clips eliminate connecting wires to terminals of the components.

USE THE BUSWAY RECEPTACLE FOR SHOP POWER NEEDS

If you're planning to install a busway to improve power supply flexibility in your shop, take an extra step and add the busway receptacle to the electrical specifications of the remodeling project. You'll eliminate a few more problems, reduce your liability potential, save some money, and permit immediate use of all your equipment.

The National Electrical Code, Article 240-15, requires the use of overcurrent devices at the point where the conductor to be protected receives its power supply. However, there are exceptions to the article which apply to feeder taps of certain lengths. Exceptions 5 and 6 state the lengths of feeder tap and the proper tap conductor sizes and current ratings which must be used with each.

An overcurrent protection on the busway feeder for the lighting. The electrical system from which this derives is a four wire, 240-V, three phase system. As in most bus plugs which are attached to the duct of the busway, the voltage range is from 240 to 600 V, 30 to 200 amp, and two or three phase. The circuit breaker panelboard for the busway receptacle can be purchased from any electrical supply firm.

PORTABLE WIRING PANEL

Build a skeletal arrangement of a house wall, mounted combinations of the usual types of electrical boxes, and connect them with ½″ TW pipe. Next select 4 × 8 sheets of ⅜″ plywood, and lay out the openings for these boxes which are cut out with a sabre saw. Then place the plywood sheets over the framework, screw it on firmly, and connect a fuse box on to a pipe stub left hanging through the plywood sheet.

GROUND WIRE TESTER AND ITS USE

Simple construction, relative low cost, and the fact that it combines metalworking and electricity make this ground wire tester an ideal project. The completed tester gives you to use at home where electrical outlets may not be grounded or incorrectly grounded.

Fabrication of the sheet metal box provides experience in layout, cutting, drilling, and bending. Both hand and power tools can be used. Wiring the tester calls for soldering, cap wiring, circuitry and assembly. See Figs. 8-1 and 8-2.

Wiring Procedure

Screw the terminal strips to the front wall of the box, using the #6-32 machine screws and nuts. Connect one 330K resistor to lugs 2 and 4. Connect the other to lugs 2 and 6. Do not solder at this time.

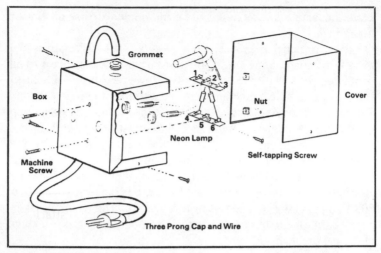

Fig. 8-1. Ground wire tester assembly.

Connect one NE-2 neon bulb (see Table 8-1) to lugs 1 and 4. Connect the other bulb to lugs 3 and 6. Do not solder the connections at lugs 1 and 3. Solder the two connections at lugs 4 and 6.

Insert grommets in each of the two 13/32-in. holes in the front of the box and push a bulb through each grommet. Insert a grommet in the 13/32-in. hole in the end of the box.

Strip the insulation from each end of the heavy duty cord. Knot the cord at one end, leaving enough cord for connections. Push the

Fig. 8-2. Schematic of the ground wire tester.

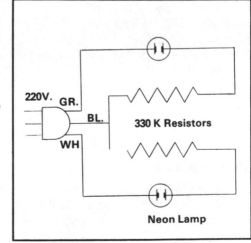

Table 8-1. Parts list for the Ground Wire Tester.

Req'd	Part	Specification
1	Box and Cover	9 × 9-in., 30 ga. tin plate
2	Resistor	330K ½w ± 10%
1	Parallel Ground Cap	
1	Plug Adapter	Optional
3	3-Lug Terminal Strip	
1	Heavy Duty Cord	18 ga., 3-wire, flexible, 5-ft.
3	Rubber Grommet	9/16 in.
2	Neon Lamp	N.E.-2
4	Self-Tapping Screw	Pan Hd., ¼-in., #4
2	Machine Screw	Rd. Hd., ½-in., #6-32
2	Nut	Sq. Hd., #6-32

other end of the cord upward through the grommet and attach the parallel ground cap.

Strip the three leads at the knotted end of the cord. Connect the green lead to lug 1, the black (ground) lead to lug 2, and the white lead to lug 3. Solder all three connections at each of lugs 1, 2 and 3.

To prevent danger from shock, and to assure proper operation, be sure all leads are connected to the correct lugs and that solder joints are good. Close the box with self-tapping screws.

To test the unit, plug the parallel ground cap into an outlet with a known ground. Both neon bulbs should glow when the outlet is grounded. Plug the cap into an adapter on which the ground wire is not connected. Only one lamp should glow.

ELIMINATING POWER-LINE INTERFERENCE TWO WAYS

Power-line interference originates in electrical devices and appliances; it can also originate when a tree limb brushes against a power line. The trouble begins whenever an electric arc is generated. This arc produces a series of radio waves which are radiated in one or more of the following ways: direct radiation from the arc; radiation from the power line near the appliance producing the arc; and feedback through the power line to the radio or TV set.

RF radiation traveling from its source to a radio or TV set is the cause that can seriously affect the reception of a particular receiver.

Have you ever been confronted with a question that asked for a remedy that would eliminate the dots and dashes that run across the picture tube, annoying harsh noises, and clicks from a TV receiver? These were caused by power-line interference. But, how did you answer the question? Did you take the easy way out and place the blame on some local radio amateur or on atmospheric conditions? If you did, you answer was not correct, because, for the most part, these agencies had nothing to do with these disturbances. The real culprit could be the drill motor in your shop, or the electric food mixer in the home — economics lab, or the sewing machine that is owned by the little old lady who runs the dressmaking establishment up the street.

There are many types of electrical equipment that are used in the shop and the home that are generating these annoying disturbances. If you suspect that your shop possesses such equipment, do not cosign it to the junk pile.

There are two basic methods for eliminating power-line interference. The first method, which is by far the more effective, involves eliminating the interference at its source. But, as in the case of the tree limb that is brushing against a power lime, there are times when this cannot be accomplished. For this reason a second method, that of eliminating the interference at the receiver, is presented.

Eliminating Interference at Source

Electrical appliances, such as brush-type fractional motors, switch contacts, heating units, spot- and arc-welding equipment, etc., are commonly the source of interference. The most effective solution for these disturbances is to install on the appliance .1 μF to .05 μF, mica, ceramic, or molded plastic tubular capacitors that will bypass the RF energy to ground before it can enter the power lines. Such a connection of capacitors is called a filter.

Present manufacturers of appliances usually equip their products with capacitors for the purpose of filtering out any RF energy that may be generated. If you are in doubt as to whether or not your particular appliance is so equipped, plug your appliance into an outlet that is also supplying power to an operating radio or TV set and turn on the appliance. If there are no signs of interference, you may rest assured that the appliance is properly filtered. If interference is present, get busy and install those capacitors as the interference that is present in your set may also be present in every set in the neighborhood.

Eliminating Interference at Receiver

The second method of eliminating power-line interference involves filtering the power line as it enters the radio or television set. This method is employed when the electrical interference is a neighborhood problem and cannot be directly remedied by you. It involves constructing a power-line filter and mounting it on the chassis of the radio or TV set.

For maximum effectiveness, this filter must be mounted on the chassis of the device being filtered. Under no circumstances should the filter be mounted at the ac wall outlet. The power cord of the set acts as an antenna that can pick up the radiation given off by the actual power line concealed in the wall. This, obviously, prevents the filter from operating as intended. The filter is installed on the chassis to keep the filtered ac leads as short as possible.

The construction of a satisfactory line filter that can be installed within a radio or TV set is not difficult and because of construction ease is an ideal, practical project.

Chapter 9
Integrated Circuits

These two projects should start you on your way toward learning more about integrated circuits. Today, of course, most projects empoly these little wonders.

GET STARTED WITH THE OP AMP

The operational amplifier is an ideal solid-state device for getting started with integrated circuits. Here is a basic experiment and a variation to help you get started.

The op amp is the basic building block of virtually every major linear-amplifier system from regulated power supplies and broadcast station control consoles to automotive electronics. It is inexpensive, commonly available and simple to use. Any electronics student from ninth grade up, and many middle school students, can work with it with no difficulties. Safe and rugged, it can be used in a variety of projects or experiments. It isn't necessary to know how the op amp works to use it, although a few words of caution in hookup can save time in troubleshooting.

The Black Box

The triangle in Fig 9-1 is the symbol for the ideal amplifier. In reality, no amplifier is ideal, but the integrated circuit op amp can be made to approach the ideal by using negative feedback techniques. For this reason, when an electronics engineer indicates the triangle in a circuit design, he often leaves off the power supply connections and the feedback networks for the sake of simplicity.

The black box approach assumes that the amplifier has certain ideal characteristics:

☐ Infinite input impedance so it will not load down a previous circuit by drawing input current.

☐ Zero output impedance, allowing it to deliver plenty of current.

☐ Infinite open-loop voltage gain without negative feedback.

☐ Infinite band width, from DC to infinity.

☐ Zero offset voltage to introduce no output voltage errors.

In reality none of these characteristics exist in most low-cost op amps, especially with regard to frequency response. However, when used in most conventional circuits, most recent op amps perform acceptably near the ideal. This is especially true with low-frequency signals.

A good first circuit to begin the study of linear integrated circuits is shown in Fig. 9-1. The op amp is the 741, manufactured by virtually all major solid-state electronics companies. It has the advantage of having a built-in frequency compensating circuit. This keeps the op amp from oscillating but it also limits its higher frequency response. It is thus an ideal device for beginners in IC experiments. When two 741 op amps are required, a 747 may be substituted; it is actually two 741s in the same package. The pin layout illustrated is a 14-pin dual inline package (DIP) IC. A metal case and a plastic mini-DIP package are also available, but since

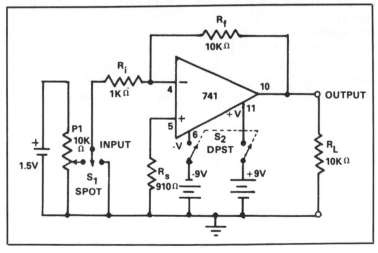

Fig. 9-1. The basic inverting amplifier using the 741 op amp.

they have only eight pins, the numbering is different. If used, check data sheets for pin configurations.

The plus and minus signs at the op amp inputs indicate that the op amp may give either an inverted of a non-inverted output. The circuit is set for an inverted output with a voltage gain of 10. By changing the circuit to that of Fig 9-2, the amplifier can be made to amplify without inverting the signal. This option is possible due to the op amp's use of the differential amplifier at the input.

The Power Supply

Each of the circuits shown uses a dual-polarity power supply. This allows the output of the amplifier to be at ground, or zero volts, potential. It also allows the output voltage to swing both in a positive and negative direction. The simplest way to acquire a dual-polarity power supply is to place two fresh 9V transistor radio batteries in series, with the junction of the two batteries grounded. (Be certain the same ground used for the power supply is used for the input ground circuit.)

A dual-polarity tracking regulator IC power supply designed specifically for use with operational amplifiers is preferred, but not necessary. Its output voltage would be the more standard ±15V.

The Real Op Amp Circuit

Some basic information on the function of the parts in the circuit in Fig 1 will be helpful before you assemble and experiment with it.

The DC and AC voltage gain is determined by the actual ratio of R_f R_i. That is, $A_v = -R_f F_i$. The value of R_i determines the input resistance of the inverting op amp circuit. The value of R_s is selected to give the best thermal stabilization to the circuit. Its value should be approximately equal to the resistance value of R_f and R_i in parallel.

Since the op amp is powered from a dual-polarity supply that is grounded at the center, the output voltage may swing in either a positive or negative direction. For the inverting amplifier, if the input voltage at some instant is positive, the output voltage will be negative at a value of $v_o = A_v(-v_{in})$. Of course, if the input is at zero—that is, shorted to ground—the output voltage should also be at zero volts. If the input voltage is too large for the amplifier to amplify without distortion, the output voltage will swing only to approximately the value of the power-supply voltage.

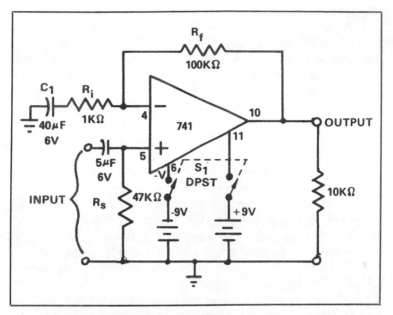

Fig. 9-2. The noninverting AC integrated circuit amplifier.

Since $jR_i = 1,000$ ohm, and $R_f = 10,000$ ohm, the voltage gain is $A_i = R_f/R_i = -10,000/1,000 = -10$. This will be true only if the actual measured values of the resistances are the same as the indicated values! The input resistance will be 1000 ohms.

If the input voltage is set at $+0.5V$, the output voltage of the operating amplifier will go to $-5V$. If the input voltage is changed to $-0.5V$, the output will be $+5V$.

The Op Amp Basic Experiment

The pin numbers in the diagram are shown in a top view of the IC package. *However, if the IC socket is wired from the bottom, if you are using a socket, be careful which pins you are connecting.* Most package systems have sockets already wired to connectors that are viewed from the top. Pin 1 of the IC DIP itself is generally indicated by a dot or other marking on the IC. Ceramic packages having a notch at one end always have pin 1 to the left of the notch as viewed from the top.

Assemble the inverting amplifier circuit. Measure the value of each resistor before inserting into the circuit. Be certain that the grounded end of R_s is connected to the same ground as the grounded center of the dual-polarity power supply.

Voltage Gain Measurement

After the circuit is assembled, connect the dual-polarity power supply and turn on the power. If two transistor radio batteries are used, this should give ±9V. If a standard IC power supply is used, the voltages will be closer to ±15V Do not exceed ±15V or damage may result to some devices. Be sure to use the double-pole, single-throw switch so that both power supplies are connected at once.

☐ Connect the input of R_i to ground. (This will set the input voltage to zero volts.)

☐ Set the electronic voltmeter on the −15V range and connect across the load resistance from output to ground.

☐ Measure the output voltage. (If the circuit is functioning correctly, it should read zero.) Reduce the voltmeter range one setting at a time until some measurement is available. Note:the *real op amp* has an input offset voltage that will be amplified, so some output reading will normally be found. The actual output voltage will rarely be a real zero.

☐ Remove the voltmeter and reset the meter to the +1.5V scale. Remove the input ground or switch to the battery-potentiometer voltage source as shown in Fig. 9-1.

☐ Adjust the output of potentiometer P_1, now connected to R_i is +0.5V. Disconnect the voltmeter.

☐ Reset the voltmeter to −15V scale and reconnect it across the load resistance from output to ground. Measure and record the output voltage.

☐ Repeat the last three steps for an input voltage of 0.8V. Meausre the output voltage, record and compute the gain A_v for both measurements. (They should be the same, although if 9V batteries are used, the second reading may approach the limit of the circuit and be less than the first A_v.)

☐ If a signal generator and oscilloscope are available, connect the generator to the input through a 10 μF capacitor, and the scope across the output. Set the generator on 1 kHz and slowly increase the generator output until the scope pattern of the amplified signal begins to distort. Reduce the input somewhat and accurately measure the output and input voltage signals; then compute the voltage gain. How did the gain compare to the dc gain?

☐ Shut off the power to the op amp and disconnect the equipment.

☐ From the measured values of R_f and R_i, compute the value of the predicated circuit gain. Compare with the measured gain values.

The AC Amplifier

The noninverting circuit of Fig. 9-2 is generally used to build an AC amplifier using an op amp. It allows one to use a much higher input resistance to better load the signal source than the DC amplifier of Fig. 9-1. The voltage gain is set by R_f and R_i, as in the inverting amplifier, but the gain is slightly larger. Note that C_1 provides the AC ground return. Without going into the math derivation, it is sufficient to say that the gain may be computed from the expression:

$$A_v = 1 + R_f R_i$$

Notice that the new resistance R_2 is used between the input and ground. This allows one to properly load the AC signal source such as a phonograph pickup cartridge. Notice also that the gain of the circuit in Fig. 9-2 is 101 since $R_f = 100,000$ ohm.

This circuit should be checked out for AC voltage gain just as any solid-state amplifier might be. It is designed to operate at low audio frequencies, not as a phono amplifier.

Conclusions

Once these two basic IC op amp circuits have been experimented with, you are ready to get into more sophisticated op amp applications. Remember, though, that, while the 741 is compensated so that it is stable, it does not operate well as an amplifier for frequencies beyond the mid-audio range. The wideband op amps such as the LM 101 or 709, however, must be compensated externally or they will oscillate badly. This will show up only on the oscilloscope. Therefore, students should be first introduced to the op amp with a device such as the 741 so they can determine how it works and what it can do without the frustration of intermittent or non-operation due to oscillation.

Many problems in IC experimenting originate in the power supply. It should be well filtered and not provide a feedback path which in turn will cause oscillation. Be sure the power supply is connected through a decoupling network if it is suspected as the problem source.

A 1-Hz IC BLINKER

The circuit here is a simple 1 Hz switch which when "on" lights a red LED and when off turns the LED off. The use of a variable resistor allows some adjustment of the blink rate. Since we wish to have a blink rate of about 1 s, it should be on ½ s and off ½ s.

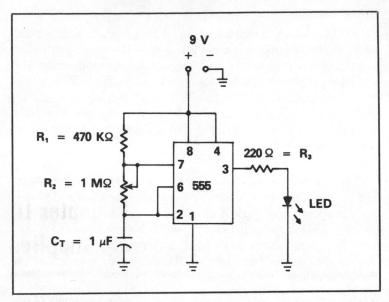

Fig. 9-3. Layout of the 555 IC.

The final circuit is shown in Fig. 9-3. R1 = 470 KΩ since it is a standard value and is close to the design value of 500 KΩ. The 220 resistor, R3 in series with the LED, is used to limit the current through the LED and prevent it from burning.

Chapter 10
Power Supplies

Almost every piece of electronic gear uses a power supply of one sort or another. The chances are that you'll find just the one you're looking for on the following pages. If not, you should learn enough from just reading about these projects to design the power supply you need.

A LOW-COST POWER SUPPLY FOR CAR TAPE PLAYERS

As the popularity of in-car tape players has increased, so has the demand to construct a project that will allow them to operate their 8-track or cassette players in their homes. This power supply is exactly the same as a commercial unit that sells for nearly $20. This unit can be constructed for less than $10, even if all new parts are used. Most of our students use surplus parts and make their own cases, thereby lowering the cost to $6 or less! The cost can also be lowered by using four individual diodes instead of a bridge rectifier (Fig. 10-1).

The unit consists of a transformer with a 12V secondary and a current rating of at least 1A. A rectifier like the one used can be purchased at most radio parts stores. The filter circuit is built up of a number of electrolytic capacitors, which are wired in parallel to give a capacitance of around 6000 μF. The voltage ratings of these capacitors should be around 25 WVDC. The case can be purchased, or fabricated from sheet aluminum and painted or covered with contact paper. See Table 10-1.

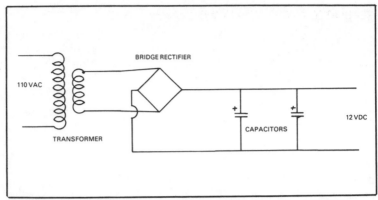

Fig. 10-1. The schematic of the low-cost power supply for car tape players.

Construction

To begin construction, determine the size of the case by laying out the components as they will be arranged in the finished project.

The size will vary with the size of the individual components. Next, machine the switch, light, output terminals and other attachments to the case. All holes should be drilled at this time. Then bend the case to shape and either paint or cover with contact paper. Mount the components in the case (pop rivets are excellent for this purpose), and wire the circuit according to the schematic diagram. If a fuse is desired, place it in series with the transformer secondary. Decals on the front add to professional appearance.

The students will find that the supply puts out around 18 volts with no load; when the tape player is turned on, however, the voltage will drop to around 13-14 volts, which is normal for car tape players. No voltage regulation was provided because the cost would have been greater, and the tape player is already regulated for constant speed.

Table 10-1. Parts List for the Low-Cost Tape Player Power Supply.

110/12v transformer, 1-amp sec. min.
Bridge rectifier or 4 diodes
Electrolytic capacitors to total 6,000 mfd
    at 25 WVDC

Case	Line cord and plug
Switch	Output terminals
Pilot light	Fuse

REGULATED POWER SUPPLY WITH VARIABLE OUTPUT

Every hobbyist soon learns the value of a regulated power supply with an adjustable output. Without a regulated supply, any appreciable change in load current will usually change the output voltage. When this occurs, the supply voltage must be readjusted. Obviously this is disturbing.

With a regulated supply, once the voltage is set you can rest assured that a changing load will not alter the supply voltage setting. Power supplies with an excellent regulation adjustment output are not new in any way to the field of electronics, but because of their cost they are seldom found in the average electronics shop. This unit contains parts which are usually available in most electronics junk boxes (Table 10-2).

Table 10-2. Parts List for the Regulated Power Supply with Variable Output.

Resistors	Potentiometer
2 51 ½ w	1 10K ½ w linear
2 100 "	2 25K 2 w linear
2 1000 "	1 50K 2 w linear
1 10K "	**Capacitors**
1 82K "	9 .01 MFD 600 VDC
1 100K "	1 .1 MFD 400 VDC
1 360K "	1 .22 MFD 600 VDC
6 470K "	1 .47 MFD 400 VDC
1 500K "	1 16 MFD 450 VDC
1 300K 1 w	1 20 MFD 250 VDC
2 330K "	2 40 MFD 450 VDC
1 110K 4 w	1 80 MFD 450 VDC
1 6000 5 w	**Tubes**
1 23K 10 w	2 6L6 1 6AU6A
1 20K 10 w	3 0A2 1 12AX7A
R_s 300 times shunt for 1 ma meter	1 5651

Transformers
T_1 800 v, c.t. 200 ma.; 6.3 v, 5 amps.
T_2 6.3-v 2-amp filament transformer
T_3 6.3-v 1.2-amp filament transformer

Miscellaneous
1 chassis, 7" × 11" × 2"
1 panel, 9" × 11"
1 fuse holder
6 800 P.I.V., 750 ma. (Sarkes Tarzian F8)
1 pilot light
6 banana jacks
1 o-300 VDC meter
1 o-1 millimeter

The regulation of this supply is .03 %, which means that when the load is varied from 0-150 milliamperes, the output voltage changes less than one tenth of a volt.

Theory of Operation

The circuit is primarily a pair of 6L6s which operate as a variable resistance in series with the load. The 12AX7 senses any change in output voltage with reference to the fixed voltage drop across the 5651 regulator tube.

The 6AU6 tube amplifies the change and applies it as a change in bias to the 6L6s. The phase angle of changed bias alters the plate-cathode resistance of the 6L6s in the direction needed to restore the output voltage to its original point (Fig. 10-2).

The regulated DC supply has two variable outputs: "B supply" which ranges from 0 to 300 VDC at 150 milliamperes; and the "C supply" which ranges from 0 to – 150 volts at 5 milliamperes. The supply provides an excellent power source for most vacuum tube and transistor demonstrations, experiments and exercises.

Construction of the Supply

The supply is built on a 7″ × 11″ × 2″ aluminum chassis, with the control and meters mounted on a 9″ × 11″ × ⅛″ aluminum plate. The location of the components is not too critical; however, remember that the 6L6s put out a tremendous amount of heat, so do not mount the meter or critical parts directly over these tubes. When demonstrating the amplification properties of a triode or plotting the characteristic curve of a transistor, you'll soon learn to appreciate this handy unit.

LOW-VOLTAGE DC POWER SUPPLY

Comprising a make-it-yourself project, here is a simple, inexpensive prototype of an efficient, low-voltage DC power supply, designed to serve six projects at two completely different DC voltages.

The heart of the panel is a 110v/12v AC transformer that develops 12 volts at a maximum of 2.5 amps. A silicone diode rated at 2.5 amps gives half-wave, simulated direct current. Protecting the secondary circuit against damage due to shortage or overloading is a 3A slow-blow fuse. Large 500 μF capacitors act as filters and give direct current clean enough to operate any type of low-voltage transistorized equipment. See Table 10-3.

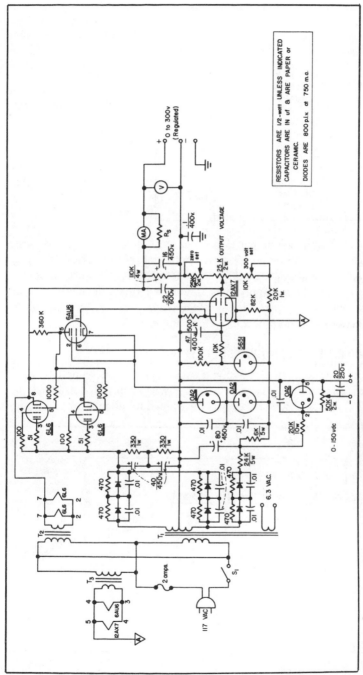

Fig. 10-2. Complete schematic of the regulated power supply with variable output.

379

no.	description
1	110-v/12-v a-c transformer rated at 2.5 amps
1	2.5 amp/12-v silicone diode
2	500-mfd at 25-w capacitors
1	3-amp fuse and fuse holder
2	1000-ohm 15-w wire-wound variable resistors
2	control knobs
6	red pin plugs
6	black pin plugs
1	SPST toggle switch
2	12-v d-c panel meters
	assorted wire, zip cord, line plug, and hardware

Table 10-3. Parts List for the Low-Voltage DC Power Supply.

A feature of the panel is its ease of modification. By limiting construction to the components to the left of the dotted line (Fig. 10-3), the unit can be built for three students for very little money.

This power supply was developed to fulfill a need for an inexpensive, rugged, low-voltage supply of DC power. Many pieces of test equipment draw high amperage.

EXPERIMENTER'S POWER SUPPLY

How many times have you wished for a power supply to test a new circuit? The experimenter's power supply, illustrated (Fig. 10-4) and described here, was developed following careful consid-

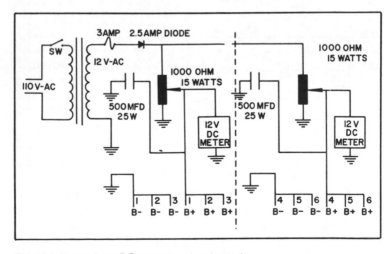

Fig. 10-3. Low-voltage DC power supply schematic.

380

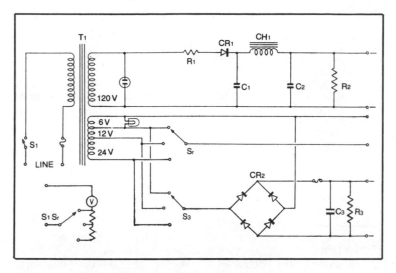

Fig. 10-4. Schematic of the experimenter's power supply.

eration of the voltage and current ratings that would supply most experimenter's needs. The transformer can be designed to meet your special needs, however.

To make this power supply, cut the laminations from sheet stock to wind the coils. For servicing or experimenting with AC-DC equipment, 110 VAC at 200W is available from an outlet on the panel. The low-voltage secondary is wound with No. 16 wire and it is tapped to give 6V, 12V and 24V of AC for lighting tubes, operating a pencil soldering iron, or small motors. This winding is also used in conjunction with a 3A rectifier to operate relays, small motors, HO gage electric trains, or slow-charge storage batteries.

The circuit has been fused to protect the rectifier against failure of the electrolytic capacitor as well as overload. The bleeder switch helps to stabilize the output when the switch is in the 24V position.

Voltage regulation was intentionally omitted. The addition of power transistors to regulate the low-voltage DC makes it an ideal power source for transistor experimentation.

In this model, the meters are separate, and they can be used to measure external voltages. The two-pole, three-position switches were wired as a single-pole, three-position switch to increase their current capacity to 4 amps. The dimensions of the cabinet are 4½" high, 13" wide, 8" deep, although these can vary. See Table 10-4.

R_1—100 ohms, 1 w
R_2—20,000 ohms, 10 w
R_3—100 ohms, 10 w
C_1—30 mfd, 150-v electrolytic
C_2—1,000 mfd, 150-v electrolytic
C_3—1,000 mfd, 50-v electrolytic
S_1—SPST toggle switch
S_2, S_3, S_4, S_5—two-pole, three-
    position lever switch
CR_1—100-ma selenium
CR_2—3-amp, 26-v, a-c, bridge type
CH_1—7Hy, 150 ma
T_1—Power transformer

Table 10-4. Parts List for the Experimenter's Power Supply.

CONVERTER BOX

A handy little box that will deliver DC current when you plug it into any standard AC house receptacle is simple to build and will usually make any small universal motor run smoother and at a higher speed. For instance, all electric hair clippers, shavers, small electric fans, etc., which are marked AC-DC will generally operate better on the current.

The best rectifier design for a converter box is the full bridge arrangement which requires four rectifiers. Check the nbameplate of the largest universal motor you wish to run on DC and note the maximum load. For instance, a small electric shaver is probably marked 10W. Divide the watts by the house-current voltage, which is, say 110V, and this gives 0.091 amperes, or multiplying by 1000 is 91 milliamperes. However, since no more than two rectifiers are working at any instance, thus allowing the other two to be cooling, it is only necessary to use the average current for obtaining the rectifier rating which, in this case, is only 46 milliamperes. The nearest standard 130V rectifier is rated at 65 milliamperes and four of the miniature type will handle the 10W load very easily.

The four rectifiers must be connected as shown in the schematic (Fig. 10-5) and will very conveniently assemble in a 4″ × 5″ × 3″ miniature utility cabinet with attached chassis, such as the Bud type C-1794 which can be obtained from most any radio supply house. Use No. 16 enameled copper wire for the connections because it is stiff enough to keep the units in alignment and the rectifiers can be mounted on brackets and bolted to the chassis as shown in the illustration. All joints should be soldered, and it is most important to be sure that the positive, or +, connections are all facing in the correct direction. A standard plug attached to about 2' of regular lamp cord is convenient for plugging in to the house receptacle. A standard female retainer, ring-type, receptacle is mounted at the DC outlet end to receive the appliance plug when operating on DC current.

A ZENER-REGULATED POWER SUPPLY

Of the several methods of operating portable radios and cassette recorders from a 12V automotive battery—the use of resistive networks, transistorized electronic voltage regulators, etc.,—the zener diode circuit gives excellent results and requires only a few straightforward calculations and a minimum of components. The circuit (Fig. 10-6) is a simple one. Resistor Rs will be calculated to a precise value, but in actual practice it can be replaced with the nearest commercial value with a tolerance of ± 20 %. Zener diodes are supplied in a range of values (both voltage and wattage ratings) readily available through local jobbers and mail-order sources at reasonable prices (see chart). Fortunately, zeners may be paralled to increase the power handling capability of the circuit. This circuit functions extremely well, but the designer/fabricator *must* insure that the wattage ratings of R_s and the zener, or parallel combination of zeners, are met or exceeded by the components used.

Fixed Voltage

The zener diode serves as an active solid-state device which holds a fixed voltage while the current through it changes. It functions as an automatically variable resistor. Since the voltage on the zener holds constant as the current through the zener changes, the power dissipated by the zener changes. In a properly designed circuit the current through R_s is a constant, and is always equal to

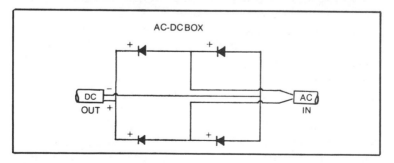

Fig. 10-5. Schematic of the converter box.

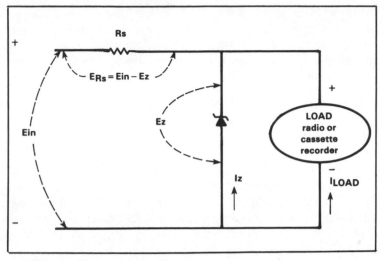

Fig. 10-6. A zener regulated power supply.

the sum of the zener current plus the load current at any instant. When the load is drawing no current, the zener current is forced to increase sharply. This condition must be considered in the design. If this condition, i.e., no load current flowing while E_{in} is connected, is ignored, the zener may burn out when the load is disconnected. The problem is not a difficult one to solve. The wattage rating of the zener used is simply selected to exceed the worst case condition.

Power supply specifications are needed to make the calculations for circuit design. In most instances the voltage of the battery pack is printed on a plate attached to the radio or cassette device. If this information is not available, determine the voltage by multiplying the number of cells by 1.5V. A tape recorder using four cells, for example, requires a 6V supply.

The current drain is seldom listed, since most units use class B audio amplifiers. It is the characteristic of such amplifiers to require heavier current as the volume level is increased. An advantage to this system is that the battery drain is reduced as the operating volume is lowered. An ammeter in series with the power supply will indicate the current drawn by the unit. One current reading is taken at normal volume and one current reading is taken at the highest volume level. These currents are designated as I_{Lmin} and I_{Lmax}, respectively. This data will be used in design calculations as outlined below.

384

I. Selecting the proper zener diode:
   A. *Voltage rating*
      1. A diode with a voltage rating most nearly matching the device to be powered is chosen.
      2. Refer to the zener diode chart.
   B. *Wattage rating*
      1. The equation:

      $$W_z = 2 I_{Lmax\, Ez}$$

      W_z = minimum wattage rating of diode
      I_{Lmax} = maximum load current in amperes
      E_z = voltage rating of zener, volts
      2. If a zener is not available with a wattage rating as calculated above, parallel the required number of zeners to meet or exceed the Wz number.
      3. When paralleling zeners the wattage rating of the combination is the sum of the individual wattage ratings.
      4. The result of 3. above is the numerical value to use for WZ_{reg} in watts.

II. The average load current is found:

$$I_{Lavg} = \frac{I_{Lmin} + I_{Lmax}}{2}$$

I_{Lavg} = average load current, amperes
I_{Lmin} = load current at normal volume, amperes
I_{Lmax} = load current at highest volume, amperes

III. The average allowable zener current is calculated:

$$I_{Zavg} = \frac{WZ_{reg}}{2\, Ez}$$

I_{Zavg} = average allowable zener current, amperes
W_{Zreq} = see I. B. 4. of outline, watts
Ez = commercial voltage rating of zener, volts (This was determined in I. A. of outline.)

IV. Rs is now determined:

$$Rs = \frac{Ein - Ez}{I_{Zavg} + I_{Lavg}}$$

Rs = resistance of resistor RS, ohms
Ein = input voltage to regulator, dc volts

V. Find the wattage dissipated by Rs:

$$W_{Rs} = \frac{(Ein - Ez)^2}{Rs}$$

W_{Rs} = power dissipated by Rs, watts

To mathematically confirm the proper operation of the circuit, good engineering practice requires that the zener wattage be calculated in the worst case situation, i.e., when E_{in} is connected and the load is disconnected completely.

VI. Calculate the maximum wattage to be dissipated by the zener:

$$W_{Zhi} = \frac{(Ez)\,(Ein - Ez)}{Rs}$$

W_{Zhi} = maximum wattage to be dissipated by the zener, watts

Using the Method

A practical example is now worked using the method outlined. A particular cassette radio/recorder has a 9V power supply. It draws 40 mA at normal volume and 120 mA at its highest volume. A voltage regulator is to be designed to allow the unit to be operated from a 12V battery.

The outline is followed, yielding:

I. A. Zener voltage rating = 9.1 V

    B. 1. $Wz = 2\,I_{Lmax\,Ez}$

        $Wz = 2(.120)\,(9.1)$

        $Wz = 2.184$

      2. Parallel three 1 W zeners

      3. $W_{Zreq} = 3\,W$

II.

$$I_{Lavg} = \frac{I_{Lmin} + I_{Lmax}}{2}$$

$$I_{Lavg} = \frac{.040 + .120}{2}$$

$$I_{Lavg} = 0.080\,A$$

III.

$$I_{Zavg} = \frac{W_{Zreq}}{2\,Ez}$$

$$I_{Zavg} = \frac{3}{2\,(9.1)}$$

$$I_{Zavg} = 0.164835\,A$$

IV.

$$Rs = \frac{Ein - Ez}{I_{Zavg} + I_{Lavg}}$$

$$Rs = \frac{12 - 9.1}{0.164835 + 0.080}$$

$$Rs = 11.844 \, \Omega$$

V.

$$W_{Rs} = \frac{(Ein - Ez)^2}{Rs}$$

$$W_{Rs} = \frac{(12 - 9.1)^2}{11.844}$$

$$W_{Rs} = 0.710 \, W$$

VI.

$$W_{Zhi} = \frac{(Ez)(Ein - Ez)}{Rs}$$

$$W_{Zhi} = \frac{(9.1)(12 - 9.1)}{11.844}$$

$$W_{Zhi} = 2.228 \, W$$

The proper circuit including specific values for the example problem just worked is shown in Fig. 10-7. Note that a parallel zener pack capable of dissipating 3W is used. The calculations indicate that the maximum power demanded from the zener pack in the worst case situation is only 2.228W. The design is sound, meeting requirements of the problem.

SOLID-STATE POWER FROM OLD TUBE POWER TRANSFORMERS

Don't throw away that expensive tube power transformer just because it's obsolete. Recycle it by making your own custom, high-quality solid-state power transformer.

Given a handful of givens, the old transformer can be converted to a lower output voltage with increased current capability, making an ideal transformer for your solid state requirements.

First, the given:

For maximum reliability recycle *only* transformers in apparently good condition. Check for any signs of defective or abusive operation. Look for burnt leads, heat blisters on the case, etc.

Determine the transformer's capability by weighing it. Figure 10-8 graphs the relationship between weight and power capability. Example: if the core and windings weigh 2 lbs., the transformer will have a recycled power capability of 40W.

Is the old transformer adequate for the intended application? For example, can you use, say, a 40W transformer if you want a new secondary of 10V at 2 amps? To find out, apply Ohm's Law, which tells us that 10V and 2 amps provides a product of 20W, within the capability of a 40W transformer.

$$P = IE = 2A\,(10V) = 20W$$

Make certain the transformer has a standard 115V, 60 Hz primary. You'll have to disassemble the outer case to examine the primary leads and the location of the winding in respect to the inner core. Transformers with the primary wound tightly around the core are the most versatile for conversion because they provide the most area for the new secondary. Transformers with the primary near the outer core are generally unsatisfactory for recycling.

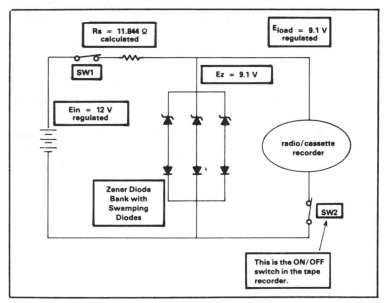

Fig. 10-7. The proper zener regulated circuit with specific values for the example problem.

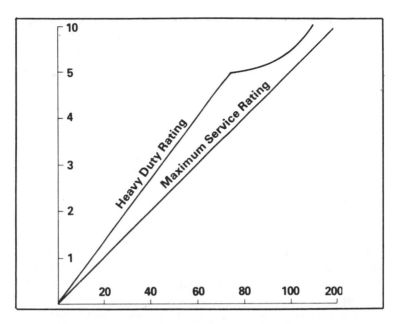

Fig. 10-8. The relationship between transformer weight and power capability.

How to Do It

Having assured yourself on all of the above you can begin the conversion. Use a hacksaw to cut and remove all secondary windings that are wound *around* the primary. Take extra care not to cut or damage the primary.

With the secondary winding removed, inspect the primary for damage and replace the old primary leads if necessary. Wrap the primary with one layer of plastic tape, making sure the leads are securely attached, insulated, and not subject to movement.

To determine the number of turns per volt for the new secondary, start by winding approximately five turns of, for example, No. 18 insulated wire around the primary as a temporary secondary, using the 10V, 2A secondary mentioned earlier. Then apply the normal primary voltage (115V, 60 Hz) to the primary leads and measure the output voltage of the secondary with a voltmeter. Say, for example, you measure 2V. Because it took five turns to produce the 2V, you now know that two and one-half turns will produce 1V. Thus, if you want 10V under no load, it will take 25 turns of the secondary (2.5 × 10) for this particular transformer.

Remove the temporary secondary and record the length of wire required for one turn. Later, you'll use this as a rough gauge for the total length of wire needed for the permanent secondary. Next determine the diameter of the copper wire needed for your application. Table 10-5 shows that a 2A secondary, for example, calls for a No. 18 wire. The center tap should be of the same diameter wire. If the correct wire size is not available, use the next largest diameter.

4X Safety Factor

Select a wire of the appropriate diameter and length *and with an insulation providing at least 4X the output voltage*. Example: if the secondary output is 10V, the insulation should provide protection up to at least 40V. In determining length, allow 50% extra wire for leads and possible additional turns.

If the secondary is to be operated in a circuit requiring a center tap, splice and solder—using resin core solder—a tap into the midpoint of the lead before winding. Be sure to insulate the splice with plastic tape. Skip this step if no center tap is required. Begin winding the secondary, keeping the windings neat and tight against each other.

If you haven't done so already, decide at this point if you want the output measured under full load or no load. If no-load, simply wind the transformer according to the turns-per-volt data referred to earlier. When the required number of turns is reached, apply normal voltage to the primary and measure the secondary. Voltage should be close to what you want, but you may have to add or take off a half turn or so to get an extra output.

If you want the output measured under full load, apply Ohm's Law to determine the load to be placed on the secondary before

Table 10-5. Wire Size.

Transformer Maximum DC output current in amps.	Required wire size A.W.G. copper	Copper wire Dia. in mils
.5	24	20.1
1	21	28.5
2	18	40.3
3	16	50.8
4	15	57.1
5	14	64.1
7	12	80.8

Note: If the correct wire size is not available, select the next larger diameter wire.

taking a measurement. Example: You require 10V at 2 amps. The load must be 5 ohms.

$$F = \frac{20V}{2\,amps} = 5\,ohms$$

Therefore, load the secondary with a 5Ω resistor and measure the output. Ohm's Law also requires the power dissipation of the resistor to be at least 20W.

$$P = 2\,amps\,(10\,v) = 20W$$

A smaller wattage resistor may be used if you work rapidly and do not allow the resistor to heat up.

If you require a substantial load you'll probably have to add a few extra turns to compensate for the resistance of the transformer. Finally, reassemble the transformer, apply a light coat of fast-drying paint to the case and label it.

ADJUSTABLE REGULATED POWER SUPPLY

This regulated power supply is intended to be a long-term (one semester or more) project for high school advanced electronics shop students. The power supply is designed for use with devices that require from 6 to 15 VDC at currents up to 1A. It will provide full load current at slightly better than 7% regulation. It can be used to recharge nicad batteries, operate and test transistor radios and cassette players, power experimental circuits, etc. Since the output can be adjusted down to 5V, the supply can also be used for TTL and other IC applications.

Most of the hardware—knobs, switches, cabinets, etc.—can come from surplus and scrap sources. In addition, rewound old TV power transformers can provide the required voltage and current ratings. Scrounging, plus careful shopping, keeps the cost to less than $10 in parts and consummable materials and supplies, such as chemicals for making the PC board.

Rewinding the Transformer

Old TV power transformers, if rewound, are suitable for the project. They are large and easy to handle, cost little, if anything, and the rewinding process teaches practical lessons along the way.

The first step is to determine the turns ratio of the old transformer. Remove the iron laminations and then remove the turns from a secondary of known voltage, counting the turns as they are removed. A low voltage winding (e.g. 5V filament) works best.

The turns ratio is:

$$\text{Turns ratio} = \frac{\text{turns of winding}}{\text{voltage of winding}}$$

Once this is done, you must then determine how many turns of wire you must wind on to achieve the correct secondary voltage by using:

$$\text{Turns} = (\text{turns ratio}) \times (\text{required secondary voltage})$$

The next step is to strip the transformer of all secondaries, leaving the primary intact. In some instances the primary may be split; i.e., part of the primary turns are wound on, then the secondaries and then the rest of the primary. In these instances you can leave all the windings intact if space will permit you to add on windings, or you can remove the outer layer of primary turns before removing the secondaries, and then wind back the *same number* of primary turns removed. Note that you may not necessarily wind back the same *length* of wire.

Finally, the correct number of turns of 18 or 20 ga. wire is wound over the primary with strips of "fish paper" between layers of wire. Excellent fish paper can be made by cutting old file folders into strips of the proper width. The final layer of wire is taped and the leads are insulated with spaghetti or heat shrinkable tubing so they can be brought out through the holes in the case. The whole thing can then be reassembled and tested.

PC Board Fabrications

The layout shown in Fig. 10-9 can be used to fabricate a board that will hold all the components. When wiring the PC board, it is advisable to solder on the resistors and capacitors first to reduce the risk of heat damage to the semiconductors. Both $Q1$ and $Q2$ must be adequately heat sinked. The heat sink for $Q1$ is a small clip-on sink, while that for $Q2$ can be made from a 1-⅜″ × 5-¾″ piece of 1/16″ soft aluminum and bolted to the circuit board with $Q2$ as shown in the photograph. See Table 10-6.

How it Works

Referring to Fig. 10-10, the line voltage is stepped down to 25V rms, bridge rectified and applied to filter $C1$. This provides about 35 VDC with no load. $R1$ and $Z1$ provide a stable 2.7 VDC which is used as a reference voltage. $R2$ serves as both collector load for the error amplifier $Q1$ and bias resistor for the pass element, $Q2$. $R3$, $R4$ and $R5$ make up a voltage divider that senses

392

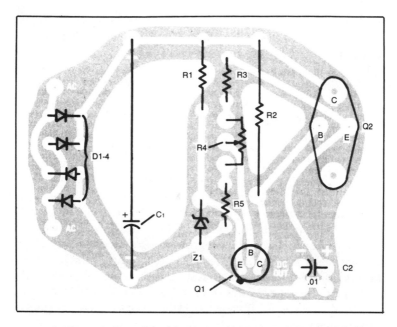

Fig. 10-9. PC board of the adjustable regulated power supply from the foil side.

any shift in the output voltage and applies a fraction of that shift to the base of *Q1*. *C2* filters out any transient spikes that may appear in the output.

To visualize how the regulator operates, it is helpful to remember that in a properly operating class A voltage amplifier, such as is used in the error amplifier, the base-emitter voltage will

Table 10-6. Parts List.

C1	50000µF 50 V DC
	(Sprague TVA-1313.7 or equiv.)
C2	.01µF 50 V dc ceramic
D1	
D2	2 A 100 piv silicon diodes
D3	
D4	
R1	560 Ω, 5 W wirewound
R2	820 Ω, 2 W, 10% carbon
R3	100 Ω, ½ W, 10% carbon
R4	5 K pot., linear taper
R5	820 Ω, ½ W, 10% carbon
Q1	2N1711 or equiv.
Q2	2N3055 or equiv.
Z1	2.7V, 1 W Zener
	(Sylvania ECG 5063)
T1	Power trans., 25 V 2 A sec.

be about 0.7 volts for silicon transistors. Since the emitter of *Q1* is fixed at 2.7 volts, the base will be about 0.7 volts higher, or 3.4 volts with respect to the negative buss. Any attempt to raise this voltage will increase the base current causing a corresponding increase in collector current. An attempt to lower the base voltage will have the opposite effect.

Assume the supply is operating with no load. If a load is placed across the output, the supply voltage will drop. This will cause a decrease in the *Q1* base current with a corresponding decrease in collector current. As a result, the collector voltage of *Q1* will rise, increasing the drive to *Q2*. The operating point of *Q2* will then move closer to saturation, and its collector-emitter voltage drop will decrease, raising the output voltage. The output voltage will continue to increase until the bias on *Q1* reaches its no-load value.

If the load is removed, the reverse process occurs. The output voltage tends to rise, increasing the bias on *Q1* which in turn lowers its collector voltage. This decreases the drive to *Q2*, moving its operating point back toward cutoff. The collector-emitter drop across *Q2* increases and the output voltage goes down to its proper value.

The minimum output voltage for the supply is determined by the voltage rating of the zener diode. Even when fully saturated, *Q1* cannot "pull" the *Q2* base voltage below the voltage of the zener, so *Q2* can never be fully cut off. This means that the zener voltage becomes the theoretical minimum output voltage. In reality, the minimum voltage is somewhat higher than this, being the sum of the zener voltage, the saturated collector-emitter voltage drop of *Q1*, and the base-emitter voltage drop of *Q2*. For this supply the minimum output voltage is about 4 volts.

A VERSATILE VARIAC

Let's consider what a Variac is and how it operates. Essentially, it is a variable autotransformer. An autotransformer is

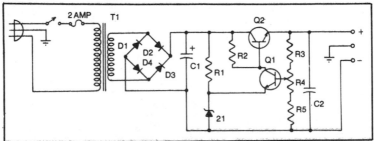

Fig. 10-10. Schematic of the adjustable regulated power supply.

defined as a transformer in which part of the winding is common to both the primary and secondary circuits. As with conventional transformers, the autotransformer transfers electrical energy from one circuit to another at a remarkably high efficiency; and, in the same way, the autotransformer is used to receive energy at one voltage and to deliver energy at another voltage. Like the conventional transformer, the autotransformer can be used to transform voltages from high to low and from low to high, economically.

A comparison of voltages and currents (at a given instantaneous polarity) is shown in Figs. 10-11A and B for the conventional transformer and autotransformer. Note that the current carried by the autotransformer winding is never as great as the current drawn by (and delivered to) the load, as is the case in the conventional transformer. Because of this there is a considerable saving in copper (and in iron) in the autotransformer over the conventional transformer.

The variable autotransformer, or Variac, is shown in Fig. 10-11C. Although it may appear to be a simple voltage divider, such is not the case. The voltage divider, or potentiometer, reduces voltage because of voltage drop across its elements. Consequently, a great deal of power is wasted in heat loss, whereas the autotransformer transfers practically all of the power from one circuit to another. It is because of this that, for the same KVA rating, the three-phase Variac, is shown in Fig. 10-11C.

The three-phase Variac requires fewer connections for operation, and, unlike the saturable reactor, requires no auxiliary d-c power source for its operation. In addition, an examination of the wiring diagram reveals that its three-phase wye connection (Fig. 10-12) is easily convertible to a parallel single-phase connection (Fig. 10-13). The dotted lines of Fig. 10-12 represent additional wiring required.

One difficulty remains to be resolved. This difficulty is inherent in the design and construction of a three-phase Variac. The type of Variac selected for conversion purposes was the ganged Variac type manufactured by the General Radio Company. This unit consists of three single-phase Variacs, mounted vertically—one on top of the other—having a common shaft controlled by a single knob at the top. As may be seen from the wiring diagram (Fig. 10-13) when connected in the single-phase parallel combination, it is essential that the brushes (which bear on the Variac commutator surface) of each Variac are mechanically

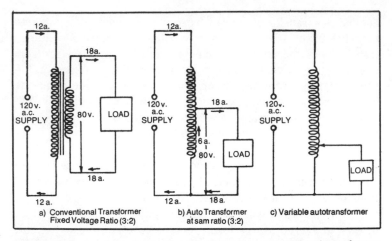

Fig. 10-11. Comparison of currents and voltages in a conventional transformer and autotransformer. A conventional transformer with a fixed voltage ratio of 3:2 is shown at A, an autotransformer at the same ratio of 3:2 is shown at B, and a variable autotransformer is shown at C.

adjusted so that the instantaneous potential at each brush is identical. Any mechanical difference in position between these brushes, however slight, would result in a potential difference between points 1, 2 and 3 of Fig. 10-13. This potential difference, despite its smallness, would result in large circulating currents between the Variac sections because the impedance of each Variac is relatively low. These circulating currents would seriously limit the capacity of the Variac when operated single phase, particularly in the smaller sizes.

Chokes Provided

To overcome this objection, the manufacturer has provided chokes which are installed at points X and Y (Fig. 10-13). The

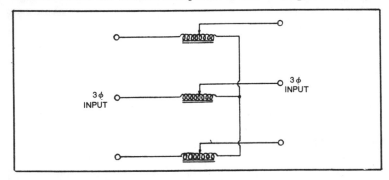

Fig. 10-12. A three-phase wye-connected variac.

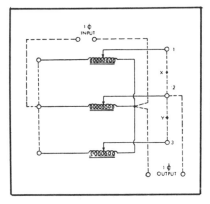

Fig. 10-13. Conversion of a three-phase wye connection to a single-phase partial combination.

purpose of the chokes is to limit the circulating currents caused by mechanical misalignment and consequent potential difference between the Variac sections. While it may be possible to mechanically align the Variac sections so that circulating currents may be kept at a minimum, it is advisable to install the chokes which can be obtained at a nominal cost. The complete single-phase parallel-combination wiring diagram, including the chokes for three-gang operation, is shown (Fig. 10-14).

Some mention should be made at this point about size and capacity as they might apply to particular needs. Ganged Variacs are supplied in two-gang or three-gang combinations. Each gang, or section, is rated, nominally, at 115 volts, although the maximum voltage rating is somewhat higher as will be seen later. Current ratings vary from 2 amps (the smallest size obtainable), 5 amps, 10 amps, 20 amps, and up to 50 amps (the largest size obtainable). Thus, the volt-ampere ratings of available single Variac sections, or gangs, are 230 vA, 575 vA, 1150 vA, 2300 vA, and 5750 vA.

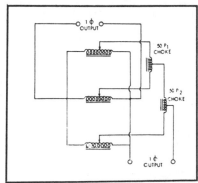

Fig. 10-14. A single-phase parallel combination showing choke connections.

Chapter 11
Solar

"Solar" is a word that's heard often in today's world. These simple projects should give you some understanding of what everyone is speaking about these days.

SOLAR ENERGY MODEL

Solar power has attracted considerable attention as a possible energy source. Although technologically developed, practical applications of solar energy have yet to be fully realized. In an attempt to increase the use of solar energy, the U.S. Department of Energy funded a project which brought electricity to Schuchuli, AZ, Papagos village whose remote location made the costs of a conventional electrical system prohibitive. Through the use of a photovoltaic array and a lead-acid battery storage unit, Schuchuli became the world's first solar-electric village. The photocell is capable of delivering 3500 watts—enough energy to power 15 four-cubic-foot refrigerators, a washing machine, a sewing machine, a 2 hp water pump and fluorescent lighting for the 96 villagers.

A series of components that simulate some of the operations in the village was developed. The simulation graphically teaches the principles of a complete solar energy system. The model involves a variety of processes.

The Set-Up

A series of silicon solar cells deliver approximately 2V at 400-500 mA—enough electricity to power a small flywheel motor, simulating the water pump and auxiliary equipment in the village. The solar cell array also generates enough energy to power a 1.5V-15 mA lamp that represents the lighting in Schuchuli. The storage cell system is simulated with a Nicad battery pack (4.8V-500 mAh) that can be recharged from a second photovoltaic array (Fig. 11-1) while powering a small motor-generator. The motor-generator system supplies power during long "sunless" periods.

Construction of the Model

To form the solar cell array, use cells 2″ in diameter, each with an output of 0.45V at 400-500 mA. Connect them in series as shown in Fig. 11-2, to produce 2V at 400-500 mA. The silicon cells (#PO4-2RD) can be obtained from Pitsco Inc., Box 1328, Pittsburg, KS 66762. See Table 11-1.

The negative terminal of the solar cell is located on the silicon side, and the positive terminal is located on the reverse glass side. Be careful when soldering the leads to the cells since too much heat can damage the silicon layers.

The cells are fragile and should be bonded with epoxy cement to a piece of ⅛″ thick plexiglass.

Solder a miniature plug to the leads of the solar cell array.

The control unit components are indicated inside the dashed lines in the schematic drawing (Fig. 11-3). You will need two single-pole, single-throw toggle switches, four miniature jacks and a miniature volmeter.

Placement of the components is not critical.

The components were housed in a plastic case with an aluminum cover. Use a chassis punch to make a neat opening for the volmeter.

Place two miniature jacks on each side of the case, as shown in Fig. 11-3.

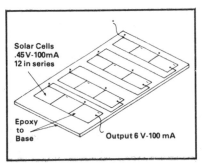

Solar Cells
.45V-100mA
12 in series

Epoxy to Base

Output 6 V-100 mA

Fig. 11-1. Bank of photovoltaic cells can be used to charge the nicad pack of the solar energy model.

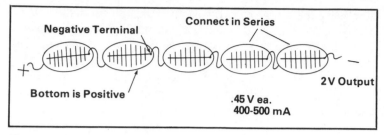

Fig. 11-2. Wiring diagram for the solar cell array of the solar energy model.

Be sure to observe the polarity of the components as indicated in the schematic of Fig. 11-3.

Connect one toggle switch to control the battery and motor-generator. The second switch provides for a voltmeter reading of the solar cell array.

The motor mount, flywheel, casing mount, and description tags are illustrated in Fig. 11-4. Construct the mounts and tags from brass sheet, and turn the brass flywheel on a lathe. Carefully follow the dimensioned drawings.

For the lighting system, use two 1.5V-15 mA lamps housed in a plastic case obtained from an electronic parts supplier. Secure the two lamps with rubber grommets and cover them with a piece of rubber tubing so they can be seen when the lamps are lit in daylight. Use a miniature plug to connect one of the lamps to the solar cell array. Connect the second lamp to the motor-generator.

Press fit two miniature motors together (such as the Aristo-craft model #RE-260), using the small gear provided with the motors. Attach a miniature plug to one of the motor's leads and connect the second motor's leads to the second lamp.

A third motor (such as the Aristo-craft model #RE-140) simulates the water pump used in the village. Attach the motor mounts and connect a miniature plug to the leads.

Mounting the Components

All of the model's components were mounted on a piece of 8″ × 15″ × ½″ cherry wood. Location of the items is not critical; however, Fig. 11-5 illustrates a compact format.

Secure the components with 2⅜″ brass wood screws.

Leave the solar array unattached to allow flexibility in obtaining the maximum radiation level from the sun.

Make item description tags for the components as illustrated in Fig. 11-4, containing the following information:

400

□ Title: The Papagos experiment

□ Flywheel motor: Water pump—2 hp for community well

□ Lighting case: Lighting for 15 houses, church, feast house, and domestic services building

□ Motor-generator: Motor-generator system provides for emergency use of utilities.

□ Battery: Storage cell system provides for use of utilities during the evening hours.

Bond the tags to the wood base with epoxy. Each description should be cut out and inserted into the tag and then covered with a piece of sheet acetate.

Operating the Model

The first phase of operation is charging the battery. A model airplane charger produces 6V at 40 mA and charges the battery in

Table 11-1. Parts list for the Solar Energy Model.

Part	Description*	Quantity
Lamps	Radio Shack #272-1139	2
Miniature plugs	Radio Shack #274-289	3 pkgs. (6 plugs)
Miniature sockets	Radio Shack #274-292	2 pkgs. (4 sockets)
Mini-meter	0-50 V Calectro #D1-952	1
Motor	Aristo-craft #RE-140	1
Motors	Aristo-craft #RE-260	2
Ni-Cad battery pack	Cox #800020 500 mAh	1
Control box	Radio Shack #270-230, 3-¼ × 2-⅛ × 1-⅛ Experimenter Box	1
Subminiature SPST toggle switch	Radio Shack #275-612	2
Brass sheet	K & S Eng. Co. #251 .010 × 4 × 10	1
Photovoltaic cells	see text	5
Plastic case	see text	1
Diode 1N914	Radio Shack #276-1122	1
Brass wood screws	#2,⅜"	1 box
Wood base	8" × 15" × ½"	1
Plexiglass	⅛" thick for mounting solar cells	1
Rubber grommets	Should come with lamps	2
Acetate sheet	8-½" × 11"	1

*Sources and trademarks listed are those the author used. Parts may, of course, be obtained elsewhere.

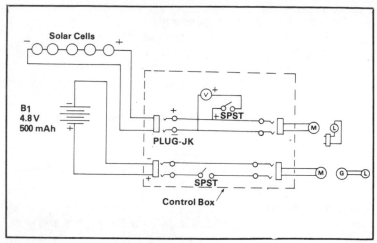

Fig. 11-3. Schematic of the solar energy model.

approximately 10-15 hours. A second method uses a 6V-250 mA tape recorder adaptor, such as the Panasonic #RP-663. This adaptor will quick charge the pack in two hours. The nicad battery can also be charged from the sun, using a bank of photovoltaic cells, which should produce 6V at 100 mA (see Fig. 11-1). Place a diode in the solar charging circuit to prevent damage to the battery (see Fig. 11-6). The battery charges in 8-10 hours using this method.

To activate the model, plug the solar cell array into the voltmeter jack of the control box. Turn on the toggle switch and measure the output of the photovoltaic array. The output will vary with respect to the sun's radiation level. Next plug the solar motor into the second jack on the opposite side of the control box to energize the simulated water pump.

To simulate the village's lighting, unplug the solar motor and plug the lighting unit into the control box. You will notice the 1.5V-15 mA lamp being powered by the sun. Demonstrate the storage cell lighting by plugging the battery and motor-generator into the control box. If you use the second set of jacks, you do not have to remove the solar array and solar motor when you demonstrate the lighting. Turn on the second toggle switch and the motor-generator energizes the second lamp, simulating the village's back-up system.

You may decide to recharge the storage cell supply from the sun, again measuring the output of the array through the voltmeter in the control box.

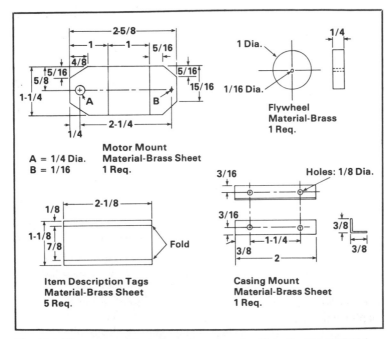

Fig. 11-4. Dimensions of the indicated components of the solar energy model.

A MINIATURE SOLAR COLLECTOR

An interesting topic in solar energy is the study of tracking devices—mechanisms that follow the sun's path to maintain

Fig. 11-5. The completed solar energy model.

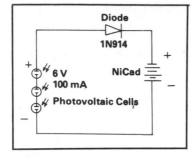

Fig. 11-6. Solar battery charger, showing the diode in the circuit which prevents damage to the battery.

maximum radiation of the face of solar collector units. Tracking devices have been designed to operate with several types of collector systems. One variety combines a tracking device with a trough collector which has been oriented either on an east-west axis and adjusted during the year, or on a north-south axis and rotated during the day to maintain maximum concentration. The second type connects the tracking unit to a disk collector that is adjusted every 15 minutes to keep the focal point of the unit at the desired location. With these tracking systems, the solar collector follows the sun!

Construction of a miniature solar collector tracking device can be a small group project that is a unit of study on solar energy. It could also correlate with a unit on solar collectors in general. Here's how the tracking device operates:

Two silicon solar cells are wired in series to provide an output of 1V at 100 mA. The two cells are placed in a frame and mounted on a miniature DC motor. A model collector (non-operating) is then mounted on the motor at a 90-degree angle to the silicon cells. When solar radiation strikes the silicon cells, they produce just enough electricity to turn the motor and model collectors. The motor and collectors will continue to rotate until the silicon cells no longer receive direct sunlight.

Since the output of the cells is matched to the torque of the motor, the system will only rotate 90 degrees or one quadrant; however, since the model collectors are mounted from the cells, this *partial* rotation brings the collectors into alignment with the sun. The collectors will remain in alignment with the sun until the silicon cells once again receive direct sunlight later in the day.

The model is designed to demonstrate the use of a tracking device to improve collector efficiency by receiving direct radiation instead of diffused sunlight. The heart of the tracking unit is the miniature DC motor, designed by Sagami and distributed by

404

Northwest Shortline Co. (P.O. Box 423, Seattle, WA 98111). These motors are available in a variety of sizes and require very little voltage and current to operate. The unit used in this project measures 20 × 32 mm (#20320-9) and will begin to rotate at 1V while drawing only 40 mA of current. See Table 11-2.

Constructing the Solar Frame

Following the diagram in Fig. 11-7, carefully wire two silicon solar cells (Radio Shack #276-120) in *series* to provide 1V at 100mA. Wire the cells as close together as possible since they will be placed into a frame, as shown in the assembly drawing.

Attach 2″ of wire as the positive and negative leads to the cells. Note: too much heat may damage the cells.

Lay out the frame for the silicon cells. The prototype model used .010 sheet brass. Easily bent aluminum flashing can be substituted.

Cut out frame using tin snips. File any rough edges.

Using a hand punch or drill, make two ⅛″ dia holes in the base of the frame.

Using a hand bender or box and pan press form the solar frame into the shape shown in the profile view.

Mount solar cells into the frame with spray adhesive. Use polyethylene foam to insulate the cells from the metal frame. The cells are packaged in this material. Spray both the polyethylene

Table 11-2. Parts List for the Miniature Solar Collector.

Item	Qty.	Description
Silicon solar cells	2	100 ma/.5V, Radio Shack #276-120
Sagami motor	1	12 V dc motor, 1 V starting /no load., 040 A, #20320-9
Solar frame	1 sheet	Sheet breas.,010″ × 4″ × 10″ #250, K&S Engineering Co.
Plastic insulation	1 sheet	¼″ thick flexible polyethylene or polystyrene foam
Base	1	Wood, 1-½″ × 3-½″
Acrylic plastic	1	1/16″-⅛″ thick
Collector covers	1	3″ × 5″ sheet
Basswood strip for collector frame	1	1/16″ × ⅛″ × 12″
Flat black paint	1 bottle	Testors #1149
Corrugated aluminum	1	Campbell scale model sheet
Washers, wire, solder, and epoxy		

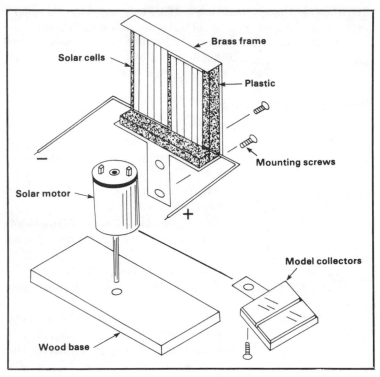

Fig. 11-7. The miniature solar collector assembly.

foam and the cells, but do *not* allow any of the adhesive to come into contact with the blue silicon surface. The adhesive works much like contact cement.

Constructing the Model Collector

Make collector base from sheet brass or aluminum flashing. After cutting out the base, file any rough edges. Locate and punch a ⅛" dia hole.

Bend the "tab" portion of the base to a 45-degree angle.

As shown in Fig. 11-7, the collector consists of the mounting base, absorber (foil), frame, and plastic cover. Campbell Scale Model Co. produces a "corrugated foil" (available at many hobby or craft stores) with a realistic appearance suitable for model collectors. Epoxy-bond a piece of this foil to the mounting base.

The frame of the collector is constructed from 1/16" and ⅛" basswood. Bond the pieces to the foil with epoxy cement.

When the collector is thoroughly dry, paint it with flat black enamel. The cover for the model collector can be made from the plastic case provided with the silicon solar cells. Simply cut the plastic to 7/16″ × 1″, sand the edges and epoxy-bond in place.

Mounting the Components

The entire tracking unit has only four components: the miniature motor, the solar cell frame, the model collector and the base.

The Sagami motor includes several mounting screws that fit into the base of the motor and the back of the unit. Mount the solar frame onto the back of the unit with two of these screws. Be sure to use two or three small washers with the screws, since turning the screws too far into the motor could interfere with the unit's operation.

Mount the model collector onto the base of the motor; again use several small washers to provide adequate spacing. Attach the positive lead from the solar cells to the positive lead of the motor and attach the negative lead from the cells to the negative lead of the motor. Do *not* solder the leads at this time, since you might want the motor to turn in the reverse direction.

The base for the solar tracking device is made from a 1-½″ × 3-½″ × ¼″ piece of wood. The prototype was mounted on ebony.

Fit the tracking unit's shaft into a 3/32″ hole drilled into the center of the base. It should fit snugly, and will not rotate or come loose when finished.

Operation

The solar collector tracking unit has been designed to work either in natural sunlight or indoors with a sunlamp. Hold the unit in your hand while orienting the solar frame toward the sun. The tracker will swing *away* from the sun; however, the model collector will now be in alignment with the direct radiation of the sun.

Try different orientations and watch the unit rotate, always bringing the model collectors into alignment. Remember that the unit will rotate every time the silicon cells are exposed to direct sunlight. Once you have determined the proper direction for your unit's rotation, solder the positive and negative leads in place.

Since it requires a considerable length of time for the sun to complete its path, you may prefer to use the unit indoors with a sunlamp. Place the tracker on a table and adjust the distance between sunlamp and tracker at a point where the unit just begins to rotate. With a little experimentation you can make the solar tracker follow the sunlamp, simulating the path of the sun.

Chapter 12
Test Equipment

Although it is usually beyond the average electronics hobbyist to make sophisticated electronic test equipment, it is often possible to build simple, everyday pieces of test gear. The following projects are for both the newcomer and the advanced hobbyist.

NEON TESTER AND ITS USE

This tester consists of two lengths of test lead wire (the insulation breakdown voltage must exceed 300V), two phono tips, a 220,000 Ω ½W resistor, an NE-2 neon bulb, a tube of silicone rubber bathtub caulk and a felt tip pen top. Push the end out of the top with a center punch, and cut ⅜″ from the narrow end so the neon bulb will fit.

Construction Hints

One lead of the NE-2 should be cut to compensate for the length of the 220K Ω resistor. Insulate the leads with tape just below the glass to avoid a short circuit. When attaching the phono tips to the test leads, a crimping tool works much better than soldering.

Insert the completed assembly into the pen top, allowing half of the NE-2 to protrude through the top, and secure by filling the top with silicone rubber bathtub caulk. The sealer adheres very well to plastic and will shockproof the components as well. Since the caulking provides very good electrical insulation, do no substitute. Allow the caulking to harden overnight, then insulate the phono tips with electrical tape. See Fig. 12-1.

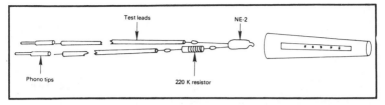

Fig. 12-1. Assembly sequence for neon tester.

The NE-2 will fire at approximately 90V, and the tester can be used for voltage measurements of up to 250V. Excess current flow in the tester circuit is limited by the 220K Ω resistor.

A SHOT FOR ELECTRONICS

Combine the operating principle of a squirt gun with a medical syringe to make test leads for your workbench. Used syringes can be obtained from hospitals. See Fig. 12-2.

☐ Attach a connector to one end of a length of test lead wire.

☐ Solder the other end to a small loop made in one end of a piece of 20 ga. phosphor bronze wire.

☐ Remove the gasket from the syringe inner barrel.

☐ Drill a hole through the plunger end and pass the wire through the hole.

☐ Put a spring (from a ball point pen) in the outer barrel and insert the wire and inner barrel so the wire passes through the spring and out the tip.

☐ Squeeze to compress the spring. Bend a hook in the end of the wire and cut off the excess.

VARIABLE RESISTANCE SUBSTITUTION BOX AND ITS USE

Where there is a constant need for resistances of varying values which may not be readily available, a variable resistance-substitution box can be an invaluable aid. Construction of the instrument also gives a newcomer an opportunity to learn series resistance principles, metal layout and soldering techniques.

The project's basic requirement is the linking together of three or more potentiometers so that resistance values from zero to infinity can be obtained. The potentiometers can be salvaged from any discarded piece of electronics gear that requires variable resistance in its circuitry, such as a television or radio. The wattage value of these components should be at least 2.5 watts and be wire-wound in construction. The components may be either linear or nonlinear, depending on availability, function, and accuracy. See Table 12-1.

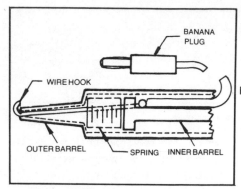

BANANA PLUG

WIRE HOOK

OUTER BARREL   SPRING   INNER BARREL

Fig. 12-2. Test lead assembly.

Constructing the Box

After obtaining and testing the ohmic resistance value and linearity of each potentiometer with an ohmmeter, lay out and bend up a box made of 18-ga. galvanized sheetmetal measuring 2″ × 3″ × 8″. Before drilling holes in the metal box, buff the galvanized surface with a power buffer and a compound similar to Tripoli or jeweler's rouge. Clean the buffed metal surface with a rag, and spray on a coat of lacquer or clear plastic film to provide a lustrous long-lasting finish.

Drill three ⅜″ evenly spaced holes in the front of the box to receive the shafts of the potentiometers. Drill four ¼″ holes on the top of the metal chassis to receive four banana jacks, making sure that the jacks are not grounded to the metal chassis. Use a continuity tester or ohmmeter to test for a possible ground by plugging one lead of the tester into each banana jack while touching the other to the chassis. Use this test for each jack as it is mounted.

Table 12-1. Parts List for the Variable Resistance Substitution Box.

Quant.	Description
1	10K ohm, 2½-w potentiometer, wire-wound
1	50K ohm, 2½-w potentiometer, wire-wound
3	Pointer-type plastic or hard rubber knobs
4	Banana jacks
1	Metal chassis—2″ × 3″ × 8″,, 18-g galvanized
Misc.:	Hook-up wire #18 gauge
	60-40 rosin core solder
	Tag board

Push the potentiometers up through the ⅜″ holes in the front of the chassis; their shafts should protrude no farther than ½″ above the surface. Secure the potentiometers in the chassis with the locking nuts found on the necks of each component.

It is desirable to have the resistance increase as the potentiometer shaft is rotated clockwise (Fig. 12-3). One easy method to determine which tabs protruding from the potentiometer are to be used is to place one lead of an ohmmeter on the center shaft of the potentiometer and the other lead on either of the two tabs on the potentiometer. While the ohmmeter is being connected, rotate the potentiometer shaft clockwise. If the resistance decreases, shift the ohmmeter lead on the one outside tab to the opposite outside tab.

Use this same procedure to locate the proper tabs of the remaining potentiometers. With tape, mark the proper tabs selected on each potentiometer so that clockwise rotation provides increased resistance. Refer to Fig. 12-4 for proper connections of the 18 ga. hookup wire. Solder with 60-40 rosin core solder. Secure the knobs on the shafts of the potentiometers.

How to Calibrate the Instrument

After construction is completed, the instrument is calibrated to provide a more accurate resistance reading. Glue three 2″ diameter circular pieces of tag board to the top surface of the instrument with the centers of these pieces cut out to receive each potentiometer shaft. For a more accurate resistance calibration, a vacuum-tube voltmeter should be used. Adjust the ohmmeter and rotate all three potentiometer shafts counterclockwise so that, when the ohmmeter leads are connected to banana jacks A and B,

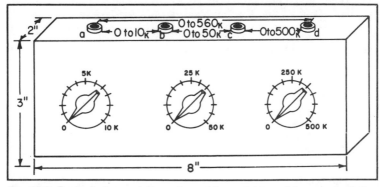

Fig. 12-3. Basic design and dimensions of the variable resistance-substitution box.

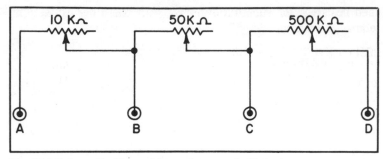

Fig. 12-4. Schematic of the variable resistance substitution box.

zero ohms of resistance is indicated. If the ohmmeter does not read zero, recheck your wiring against the schematic and reheat all solder joints.

Rotate the shaft of the 10K Ω potentiometer clockwise, observing the increasing resistance of the ohmmeter. With a soft lead pencil, mark evenly spaced lines on the tag board directly under potentiometer A. These marks may indicate multiples of 100, 150, 500 or 1000 ohms each, depending on the sequence desired. Return shaft A to zero via a counterclockwise rotation. Move the ohmmeter leads from banana jacks A and B to their new locations on jacks B and C. The ohmmeter should read zero. If not, rotate the 50K Ω potentiometer shaft counterclockwise.

Remember to relocate the ohmmeter leads to the corresponding jacks each time a new potentiometer is calibrated. It may be necessary to select a higher range on your ohmmeter when calibrating high resistance potentiometers.

Labelling

When you have completed your calibrations in pencil, retrace the lines with an ink pen and label the marks with ohmic values. Labels may be placed on the top of the chassis between each set of banana jacks to indicate the total amount of resistance obtainable between those two points. Labels may also be placed between combinations of these sets of jacks; for example, between jacks A and C, 0 to 60K Ω, and between jacks B and D, 0 to 550K Ω.

RESISTANCE SUBSTITUTION BOX—APPROACH I

This project is a low-cost, simple circuit of classic design that actually could be constructed in one or two hours. Use 1W resistors with a tolerance of 10%.

The box is an indispensible item that allows the hobbyist, the experimenter or the engineer to select any of a dozen commonly used resistors for substitution in a circuit when determining a required resistance. See Fig. 12-5.

RESISTANCE SUBSTITUTION BOX—APPROACH II

Learning to read and interpret schematic diagrams is a formidable task for the beginning electronics hobbyist. This open resistance substitution box project gives meaning and purpose to this portion of the training and prepares the student for advanced concepts in switching and control. The project can also be used as a manipulative skills test.

The unit is a piece of flanged aluminum on which are mounted two four-position rotary switches, a slide switch and two pin jacks. Wire leads of about 12″ are soldered to the connections of each component. A two-wafer rotary switch can be used instead of the two single switches.

The resistor substitution unit is shown in the schematic (Fig. 12-5). When the wiring is completed, the instructor plugs an ohmmeter into the pin jacks and checks the resistance of the various positions. The slide switch acts as a 10X multiplier switch.

The project offers an understanding and experience in: component identification, wire tracing and continuity testing, the resistor color code, reading schematics, operation of switches, breadboarding techniques, the time-saving value of the resistor substitution box in troubleshooting and how a variable attenuator functions with the use of a rotary switch; i.e., the vertical input to the oscilloscope. The open box can also be used to verify Ohm's Law.

CAPACITANCE SUBSTITUTION BOX AND ITS USE

For selecting the optimum value of cathode by-pass capacitance necessary to prevent degeneration, or choosing the correct

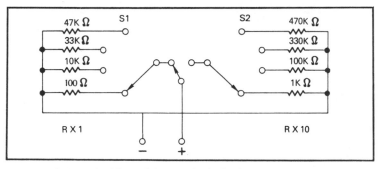

Fig. 12-5. Schematic of the resistance substitution box.

capacitance for coupling between stages of an RC coupled amplifier, or determining the capacitance required for a given inductance to filter certain frequencies, or to tune to certain frequencies, this capacitance substitution box will save you a lot of laborious calculations. For fast connect-disconnect and for hands-free operation, use alligator clips rather than test probes. A printed circuit board could have been used instead of the bifurcated terminal board. The box can be built as a companion unit to the resistance substitution box described in this chapter (Table 12-2).

With another type switch fewer capacitors might have been used, since capacitors connected in parallel, like resistors connected in series, are additive, and capacitors connected in series, like resistors connected in parallel, are equivalent to less than the smallest one (Fig. 2-6).

In doubling the *plate area* by paralleling capacitors, the equivalent capacitance is doubled. In Fig. 12-7A, a 2 μF capacitor in parallel with another 2 μF capacitor $(2 + 2) = 4 \mu$F.

In doubling the *thickness* of the *dielectric* or the distance between the plates, by series connecting capacitors, the equivalent

Table 12-2. Parts List for the Capacitance Substitution Box.

No.	Item	Qty.	Description
1.	C1	1	.0033μF capacitor
2.	C2	1	.0047μF capacitor
3.	C3	1	.0068μF capacitor
4.	C4	1	.01μF capacitor
5.	C5	1	.015μF capacitor
6.	C6	1	.022μF capacitor
7.	C7	1	.033μF capacitor
8.	C8	1	.047μF capacitor
9.	C9	1	.068μF capacitor
10.	C10	1	.1μF capacitor
11.	C11	1	.15μF capacitor
12.	C12	1	.22μF capacitor
13.	J1	1	Pin type jack (black).
14.	J2	1	Pin type jack (red).
15.	S1	1	Single Pole, 12 position, rotary, wafer switch.
16.	H	1	Instrument housing case 4" × 2-7/8" × 1-5/8"
17.	P	1	Matching aluminum panel for case.
18.	F	4	Pan head screws, 4/40 × 3/8"
19.	Tbt	1	Terminal board with biforcated (split) terminals.
20.	Misc.	-	Length of wire, solder, etc.

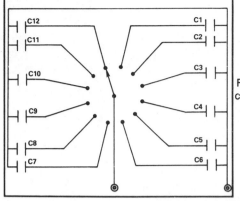

Fig. 12-6. Schematic of the capacitance substitution box.

capacitance is halved. In Fig. 12-7B, a 2 μF capacitor in series with another 2 μF capacitor $(2 + 2) = 1 \mu$F.

Continuity Tester and Its Use

This continuity tester can be easily constructed by even a novice. In addition to ease of construction, this project utilizes materials of an electrical nature which are often already available or are easily obtainable. Thus, the amount of construction and manipulation type of work is kept at a minimum. Another attribute of this project is the minimal amount of time required for consruction which allows you to progress rapidly to the next project or experiment without spending a large portion of time shaping materials.

It was found that, because this continuity tester is necessarily plugged into an electrical outlet, the rubber type, two-pronged plug is suitable housing for the few components (Fig. 12-8). An added advantage is that these plugs are easily obtained at a small cost if they are not already on hand. Being rubber, they will take a lot of punishment and are resilient enough to protect the components inside. The hole in the end of the plug where the electrical cord usually passes is utilized for the neon bulb. Two small holes are drilled opposite each other near the base of the plug for the probe wires to extend and the two 100K Ω, ½W resistors are within the plug cavity and eliminate any hazards from shock. A piece of ⅝" inner diameter aluminum tubing, ½" long is placed over the bulb end of the plug and then a ⅝" outer diameter, clear plastic lens is pressed into the tube. This enhances the appearance of the project and also protects the bulb.

415

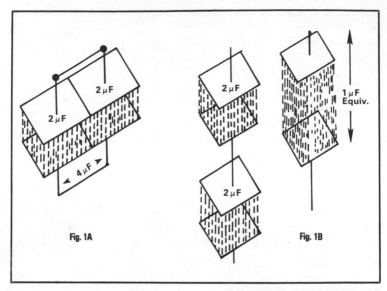

When assembling the project, be careful to keep parts and wires separate from each other and make the wires as short as possible. It is advisable to cover open wire with insulation from wire scraps. The probe wires can be made to any desired length, and the tester can be kept in the outlet near the workbench for

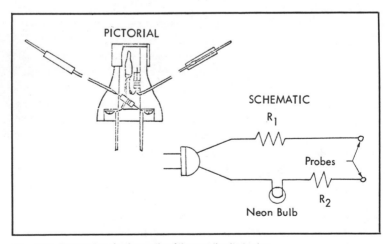

Fig. 12-8. Pictorial and schematic of the continuity tester.

ready use if provisions are made so that the probes are not left lying together. One solution is to drill two small holes into wood near the outlet so the probes may be placed into them when not in use.

A COMPACT CONTINUITY TESTER

This continuity tester is both small and inexpensive. As good as almost any commercial continuity tester, this one is a breeze to construct (Fig. 12-9).

HARMONIC GENERATOR AND ITS USE

This harmonic generator puts out a frequency of about 1400 hertz, making it useful in signal tracing a transistor radio audio section. This signal is also rich in harmonics so it can be used to signal trace the rf section as well.

Output is not sufficiently high to cause damage to component parts as is often the case with tube model signal generators. This harmonic generator has an output of approximately 1.5 volts peak-to-peak. It is a *standard* circuit with many variations.

The Circuit

This harmonic generator is a multivibrator circuit. It is an astable multivibrator circuit since the output of one stage is connected to the input of the other by capacitors (Fi. 12-10).

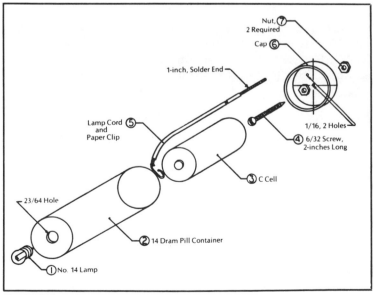

Fig. 12-9. Diagram of the compact continuity tester.

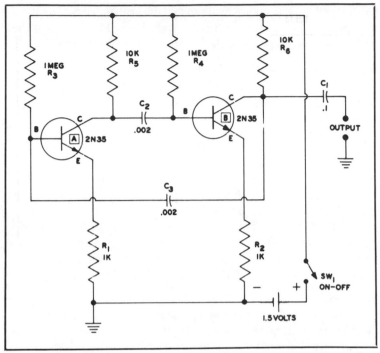

Fig. 12-10. Schematic of the harmonic generator.

Each transistor's emitter is grounded through a 1K Ω resistor. The 1 Meg resistors account for the biasing of the transistors. The 0.002 μF capacitors and the 10K Ω resistors form a coupling network. Note the transistors are NPN types; the resistors R1 and R2 are equal as are R3 and R4, and R5 and R6 (Table 12-3).

Construction Hints

Transistor sockets are used because of the ease with which the transistors can be switched. In a circuit such as this, it might be necessary to change transistors a number of times before obtaining the two that will work together to produce a usable output.

Placement of parts is not critical. Arrange the terminal strips so the components will fit the distance between connections without splicing.

A cell holder can be used to hold the penlite cell. The cell holder will make it easier to change cells when necessary. However, soldering the wire to the ends of the penlite cell is just as effective if you don't overheat the cell.

Table 12-3. Parts List for the Harmonic Generator.

1	Box ⅛″ × 2-⅞″ × 4″ Bakelite (Daka-Ware, Chicago)
1	Cover, 1/16″ × 2-⅝″ × 3-¾″ Bakelite (Daka-Ware)
6	Machine screws, ⅜″ #4-40
2	Nuts, #4-40
2	Transistor sockets
2	Transistors, 2N35 or equivalent (2N 169)
2	R1, R2 resistors—1-K, ½-w, +/− 10 percent
2	R3, R4, resistors—1 meg, ½-w, +/− 10 percent
2	R5, R6, resistors—10-K, ½-w, +/− 10 percent
1	C1 capacitor,.1mfd, any working voltage or 2 to .05 in parallel
2	C2, C3, capacitors, .002 mfd, any working voltage
2	TS1, TS2, terminal strips, 6 soldering logs
2	Leads, #18 twin lead lamp cord, 36″ long
1	Penlite cell, 1.5-v
1	SW1, switch, SPST, toggle
2	Switch-mounting nuts (part of switch)
1	Alligator clip, 2″ long
1	Battery clip, 5-amp rating

Two types of clips were used on the output to make them easily identifiable. The alligator clip is positive and the battery clip is negative.

All parts can be mounted on the cover, thus eliminating the necessity of drilling into the box. This makes it easier to remove when a cell needs replacing (Fig. 12-11).

Operation

Once the switch is turned on, there is no delay for warm-up—the unit is ready for use. The clips can be connected to the radio being tested and moved along as required by troubleshooting procedure. There is no danger of shock from the unit since it only has 1.5V power supply. The output is 1.5 volts peak-to-peak so there is little danger there. Since the 0.01 μF capacitor is in the line with the lead, there is no damage involved if the alligator and battery clips become shorted. Make sure to turn off the unit when not in operation.

BLOWN FUSE INDICATOR AND ITS USE

A frequent source of irritation when troubleshooting an inoperative device is the overlooking of a blown fuse. This happens often because the fuse may look all right. It is helpful to have a positive indication as to whether or not the fuse is blown so that one could either replace the fuse or go on to other areas to find the trouble.

The circuit shown here in Figs. 12-2 and 12-3 represents such a positive indicator for a blown fuse. Although this device is in the form of an extension cord for operating small appliances such as drills, etc., it could be very easily installed permanently in such devices as radio and television sets. In this case, the appliance must offer a continuous circuit such as a transformer or winding of a drill.

The basic circuit is simply a fused AC supply which has a neon pilot light inserted before the fuse. In normal operation, the pilot light will glow steadily so long as the fuse is intact. However, when the fuse is blown, the power to the output is interrupted and the appliance no longer will operate. The capacitor-resistor combination (the capacitor is now hooked in parallel with the neon lamp through the transformer or appliance still attached to the output) forms an RC relaxation oscillator which will cause the neon light to blink on and off in step with the charge-discharge rate of the RC oscillator. A blinking pilot lamp therefore indicates a blown fuse, and you can easily identify whether a blown fuse is the cause of the trouble.

In the case of a television set, the circuit may be installed directly, and permanently, in the housing—using the same pilot lamp that was originally installed in the set (if the pilot light is neon).

Although the extension cord type device shown in the photograph is built for light duty, the use of heavy-duty wire and

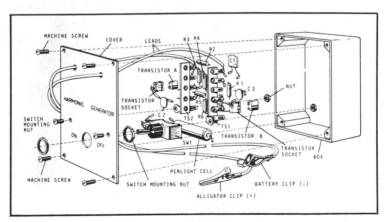

Fig. 12-11. Assembly of the harmonic generator.

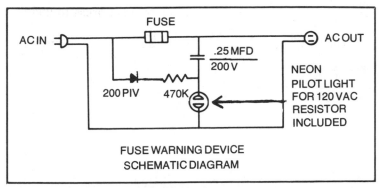

Fig. 12-12. Schematic of the blown fuse indicator. The 470K Ω resistor is ½W.

parts would allow you to construct a similar device for handling
larger current loads.

SIGNAL TRANSISTOR TESTER AND ITS USE

When building a project, it is often necessary to test a signal
transistor. This signal transistor tester will not only give a
forward-reverse resistance check but will also determine the
current amplification ability of the transistor, and can be used by
any beginning hobbyist in a matter of seconds.

The transistor is connected to the appropriate clips marked C,
B and E. If it is of the PNP type, the test leads are plugged into the

Fig. 12-13. The blown fuse indicator.

421

binding posts of the same color; for an NPN-type transistor, they must be plugged into the posts of the opposite color (Fig. 12-14).

The switch should be in the "low" position to start the test. When it is moved to "high" and forward bias is placed on the transistor, more current will pass through the meter. The amount of change in meter readings from "low" to "high" shows the transistor's ability for current amplification.

If the meter is to test a PNP-type transistor, but an NPN-type is connected, the meter will read the same in both positions. Simply reverse the test lead connections for the NPN test.

The in-out switch limits the amount of current which can flow from the battery through the meter. If a transistor passes so much current when the switch is in "high" that the meter reads past full scale, move the limit switch to "in." This places a 2200 Ω resistor in the circuit and the current can then be read on the meter scale. The battery may be shut off by removing the transistor and returning the switch to "low."

INDUCTOR CHECKER AND ITS USE

Shorted turns are fairly common in many types of inductors—such as ignition coils, transformers, chokes, etc.—causing circuits, of which they are a part, to malfunction or to cease functioning entirely. It is difficult, if not impossible, to detect a shorted turn in a coil by means of a conventional ohmmeter or any of the other equipment usually found on a workbench. However, an ordinary oscilloscope can be easily modified to enable one to determine whether a shorted turn (or turns) exists in an inductor or coil.

The modification is done by simply adding a jack to accommodate a test lead; a 20 pF, 600V mica or ceramic capacitor; and a

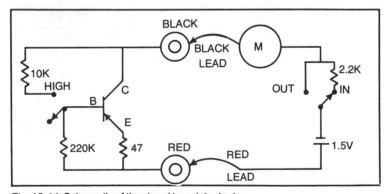

Fig. 12-14. Schematic of the signal transistor tester.

length of shielded cable. One side of the capacitor is connected to the jack, while the other side is connected to the cathode of the horizontal-sweep multivibrator by means of the shielded cable (Fig. 12-15).

To use the modified oscilloscope to test coils for shorted turns, a jumper is run from the added test jack to the vertical input, and the coil to be tested is connected between the vertical input and ground (Fig. 12-16).

Pulses from the horizontal-sweep multivibrator are thus fed into the vertical input of the oscilloscope and through the coil at a rate determined by the frequency setting of the internal horizontal sweep on the oscilloscope.

Damped Sine Wave

If the horizontal-sweep frequency is then varied, and if the coil being tested has no shorted turns, at a certain setting of the horizontal-sweep frequency—which depends on the resistance, inductance, and capacitance of the coil under test—a damped sine wave will appear on the oscilloscope. This wave form decays exponentially and is a result of the oscillatory currents in the coil.

Since the frequency of oscillation is inversely proportional to the inductance of the coil, coils with high inductances will generally require a low horizontal-sweep frequency to obtain the proper wave form, while coils with low inductances will generally require a high horizontal-sweep frequency. However, if the coil being tested happens to have one or more shorted turns, the shorted turn (or turns) will form a complete circuit or loop and thereby absorb the energy produced by the collapse of the magnetic field, thus preventing the coil from oscillating.

This will cause the wave form appearing on the oscilloscope to be damped almost immediately, appearing as in the photo. It is

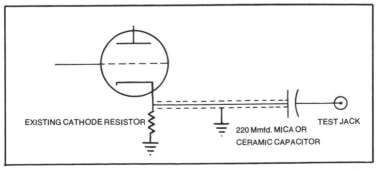

Fig. 12-15. Typical horizontal multivibrator showing added capacitor and jack.

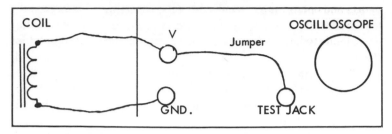

Fig. 12-16. Typical setup for testing coils with shorted turns.

impossible to get an exponentially decaying, damped sinewave to appear on the oscilloscope if the coil has one or more shorted turns.

Applications

Applications are many and varied. They include checking automotive ignition coils, checking transformers and rf coils in electronics, and checking high-voltage transformers in industry.

TRANSISTOR TESTER AND ITS USE

A simple tube filament/continuity tester has long been a very popular project for the beginning electronics hobbyist. While its construction and application continue to provide an opportunity for a variety of valuable experiences, a simple transistor tester today seems more appropriate for several reasons.

For one thing, you can now purchase good transistors at reasonable prices on circuit boards. Inexpensive bulk-package assortments of transistors are also readily available to young experimenters. The usual collection of components of the average electronics student contains many transistors. Often these transistors are unmarked or exhibit only special designations, the meanings of which are generally unknown and practically unobtainable.

The transistor/diode/continuity tester depicted by Fig. 12-17 is a completed project that will enable the user to test transistors for type and quality, check diodes for polarity and quality, and, of course, test for continuity.

Transistor Test

After attaching the "minigator" clips to the transistor leads, switch S1 on the schematic diagram (Fig. 2) is moved to "Type" position. With S2 in either NPN or PNP position, S3 is momentarily closed. Lamps L1 and L2 should light in only one of the positions of S2. In determining the type of transistor under test, relative shorts and opens in the base-emitter and base-collector diodes are also indicated.

A shorted emitter-collector (large EC leakage or I_{CEO}) test is made by moving to "leak" position, momentarily closing S3 and observing lamp S1.

The resistor in the base circuit limits current during the "type" test only. The momentary-contact switch, S3, prevents unnecessary and accidental battery drain. This feature of the tester makes it practical to use the popular, inexpensive and readily available 9V transistor radio battery.

Diode Test

The anode and cathode of the diode to be tested are attached to the emitter and base-clip leads. The polarity of this test attachment is unimportant. S1 is moved to "type" position. S2, which reverses the polarity of the test voltage, is placed in either of its positions. S3 is now closed. S2 is next moved to the opposite position and S3 is again briefly closed. L1 should light in only one of the positions of S2 for a "good" diode.

The polarity of the diode may be determined in the above test at the same time the quality is being checked. If L1 lights in the PNP position of S2, then the cathode of the diode must be

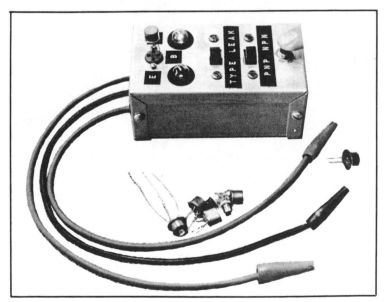

Fig. 12-17. The transistor tester.

connected to the base-clip lead. If L1 lights in the NPN position, the cathode is connected to the emitter-clip lead.

Continuity Test

The continuity-test procedure is the same as that for the diode, except for the position of S2. As polarity is not normally important in the test of continuity, S2 may be left in either position.

Construction Suggestions

The dimensions of the minibox used to house the tester are $2\frac{1}{4}'' \times 2\frac{1}{4}'' \times 4''$. The layout is flexible as component placement is certainly not critical.

Snug fitting $\frac{3}{8}''$ inner diameter grommets are used to hold the lamps. Conventional lamp holders and lenses may be substituted at a slight increase in cost.

As shown in Figs. 12-17 and 12-18, a transistor socket, S01, may be connected in parallel with the clip leads. The socket is convenient for testing those transistors with short leads, but it is an optional component.

Although No. 49 lamps are specified on the schematic, you might successfully employ other types, including the popular and inexpensive No. 47.

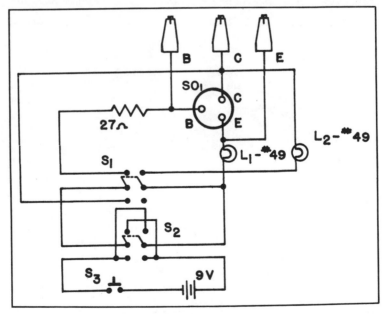

Fig. 12-18. Schematic of the transistor tester.

SIMPLIFIED COMBINATION TESTER AND ITS USE

While this tester does not represent an original idea, it does have distinct advantages over similar devices available in kit form. In commercial versions of this "multitester," the neon test lamp is combined with the voltage tester, usually resulting in several test leads. One has to select the correct two leads for the particular test and in some cases plug the spare lead into a jack in order to have the circuit correct. This can be confusing.

The feature suggested here is to incorporate a double-pole, double-throw switch to eliminate the need for extra test leads. By simply switching, the unit is converted from a neon test light to a voltmeter.

A quick check of the schematic (Fig. 12-19) will show that the switch in one position, the potentiometer is across the input and the neon light is connected to the center terminal of the potentiometer. Thus by rotating the dial one can select the point at which the proper voltage will light the neon bulb. To calibrate the voltmeter, turn the dial so the light is out when a known voltage is applied to the input leads. Then back the dial up to the point where the light just begins to glow. Mark this spot with the known voltage and proceed in a similar manner with other known voltages.

Conversion into a Neon Test Light

When the switch is in the other position the potentiometer is disconnected from the circuit and only the NE-2 with its series resistor is being used. It then can serve as a normal neon test light.

Since only one side of the lamp will glow on direct current the unit can be used to determine the polarity of a DC circuit. By using one red and one black lead, make a + mark on the side of the bulb which lights when the red lead is connected to a known positive source.

Naturally, miscellaneous hookup wire and hardware are needed in addition to the items on the parts list of Table 12-4. It might be suggested that the value of R1 is not too critical. This value may vary from the listed 68K to more than 200K. The potentiometer may be anywhere from 0.5 to 1 Meg. The main advantage of the smaller value in this case is that the scale divisions will have a wider spread.

VTVM-VOM OPERATION TESTER AND ITS USE

Electrical meters have to be completely operational if they are to be depended on. A VTVM, for example, has seven ranges for resistance check. The VTVM has also seven ranges for measuring

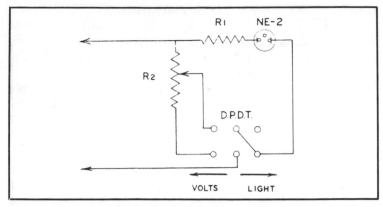

Fig. 12-19. By simple switching, the simplified combination tester can serve as either a neon test light or a voltmeter.

each of the following quantities: plus DC volts, minus DC volts and AC volts. This means there are a total of 28 different positions at which the function and range switches can be be set. If any of these 28 positions is not functioning properly the apparatus is not only limited but unreliable.

The operation (calibration) tester (Fig. 12-20) will enable you to make a complete check of your VTVM for proper range and functions in about two minutes. This calibrator is simply constructed and it is simple to operate. It should not take more than two hours to make. The parts are standard values, and they should be available at your electronic parts dealer (Table 12-5).

The parts values shown here will work well with most VTVM, VOM and multimeters. If you have any meters with special ranges or scales, however, you can substitute the part values to suit your particular needs. This can be done by calculating the voltage gradient of 1000-ohms-per-volt for the AC test. This totals 123,000 ohms across the AC line because the source voltage is 123 VAC.

The voltage appearing across the DC test resistors will depend on the size of the filter condenser. Within limits, the greater the capacity of the condenser, the higher the DC voltage

Table 12-4. Parts List for the Simplified Combination Tester.

		NE-2	Neon bulb
R1	68K resistor		D.P.D.T. slide switch
			Plastic case
R2	½ Meg potentiometer		Test prod wire—red
	Knob for R2		Test prod wire—black

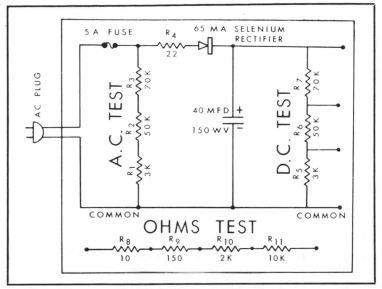

Fig. 12-20. Schematic of the VTVM-VOM operation tester.

will be. The limits are set by the peak value of the rectified AC sine wave and the time constant of the RC circuit formed by the filter condenser and the DC test resistors.

The parts chosen for the tester draw less than 5 milliamperes of current. This means the calibrator draws less than 1 watt of electrical power.

How This Meter Works

This tester is simplicity itself. The top side of the line goes to a fuse. This protects the calibrator tester in case of a short or overload. The AC voltage appears across R1, R2 and R3. The voltage appearing across each resistor is proportional to its resistance. The greater the resistance is, the higher the voltage is. There appears, therefore, 3V across R1, 50V across R2, and 70V across R3.

R4, the 22 Ω resistor, is a surge limiting resistor. It limits the current surges through the selenium rectifier to a safe value. This must be so because the discharged electrolytic filter condenser acts as a momentary short to the rectified current. If the current-surge limiting resistor was not in the circuit, current exceeding the current rating of the rectifier would flow through it and cause damage. The half-wave rectifier permits current to flow through it in only one direction. The 60-Hz AC, therefore, is supplied to the rectifier left side and pulsating DC appears on the right side.

These pulses have to be changed to steady direct current. The electrolytic filter condenser does this by charging up on the positive peaks and discharging during the intervals between the peaks.

This action produces DC having a low percentage of ripple. This slight ripple, however, is too small to have any adverse effect on the tester. The large electrolytic condenser charges up to peak value and, therefore, we have a higher DC voltage than the 60-Hz AC voltage.

The greater the capacity of the condenser is, the higher, within limits, the DC voltage will be; the limits being the rectified peak value of the AC sine wave and the time constant of the RC circuit being formed by the filter condenser and the DC test resistors. The DC voltage appears across R5, R6 and R7. The voltage appearing across each of these resistors is directly proportional to its resistance. There appears, therefore, plus 4V across R5, plus 63V across R6, and plus 98V across R7.

The ohms test is made by fixed resistors mounted on one side of the board. The resistance reading on the ohms scale of the meter is directly proportional to the resistance value of the resistors connected between the ohmmeter leads.

Construction Details

Any insulated base, 7½" square × ¾", can be used to mount the parts. Sand the base clean and smooth. Lay out the schematic diagram on the board. You might want to improve the appearance of the board and lessen the chance of error by using red ink for the DC

Table 12-5. Parts List for the VTVM-VOM Operation Tester.

1	Base: 7½" square × ¾"
	A-c line cord and plug
1	Fuse: Little No. 312, 3AG, 5 amps, 250v
1	Fuse holder
7	Insulated double tie points
1	Insulated single tie point
2	Resistors: 3K ohms, ½w
2	Resistors: 50K ohms, ½w
2	Resistors: 70K ohms, ½w
1	Resistor: 22 ohms, ½w
1	Resistor: 10 ohms, ½w
1	Resistor: 150 ohms, ½w
1	Resistor: 2K ohms, ½w
1	Resistor: 10K ohms, ½w
1	40mfd 150 wv electrolytic capacitor
1	65 Ma. selenium rectifier

test, green ink for the ohms test and blue ink for the AC test when laying out the diagrams.

Use a double-lug, insulated tie strip to mount each resistor to the board. These strips are fastened to the board with ½" roundhead woodscrews. Please note that one tie strip is insulated so that you can solder the ends of the line cord to it.

The small, glass, 5A fuse provides adequate protection in case of a short. It snaps into a fuse holder that is fastened in place with a woodscrew.

The selenium rectifier has a center hole which is ideal for mounting it with a 1" woodscrew. On one side of the rectifier is the letter K or letters POS stamped on it. This is the cathode side. This side should face up and it should be connected to the positive side of the filter condenser. The other lug on the selenium rectifier is used for connecting the 22Ω surge resistor. Solder one end of this surge resistor to this lug before you mount the selenium rectifier. This procedure should prevent burning the baseboard during soldering because the bottom lug is very close to the base.

A check of the ohms-test resistors in the accompanying schematic (Fig. 12-20) shows there are four carbon resistors connected in series. These resistors are isolated at one side of the board. They are not connected to either the AC test or DC test circuits. Resistors with a 10 % tolerance rating work well because accuracy depends chiefly on comparison readings taken at different calibration test times.

Preliminary Check

There is a mechanical zero-adjust screw on most meters. It is located on the lower center portion of the meter movement. When the meter is turned off, the needle should return to the zero position. If it does not, then turn the mechanical zero-adjust screw with a small screwdriver until the pointer is directly on zero. Moving the meter from a standing position to lying down might cause the needle to move away from zero.

Calibration Procedure

A record of tests should be kept. This record can be taped to the side of the meter. Make certain, however, that vents are not covered. A record such as this is important because the AC line voltage will vary depending on the time of day the test was made as well as on what day of the week.

Now you should be ready to begin calibrating voltage. First, clip the common or ground lead to the common point on the calibrator. The ground lead remains on the common point while the AC test prod is moved to the junction of R1 and R2 for the voltage reading across R1. The prod is then moved to the junction of R2 and R3 for the voltage reading across R1 plus R2. Next, the prod is moved to the junction of R3 and R4 for the AC line voltage which is the voltage reading across R1 plus R2 plus/R3. It is necessary to increase the voltage range before you move the test lead to a junction having higher voltage. *This must be done to protect the meter*. To test for DC volts a similar procedure is followed, *except* that the function switch is set to *plus DC volts* and you use the DC test prod.

The reason for making several checks on each function is simply to test the different ranges. For greater accuracy, it is wise to make calibration tests during the same hour of the day and day of the week because of variations in line voltage. Should you use the meter extensively—every day, for example—it is suggested that calibration be checked once a month. With infrequent use of the meter, a calibration test once every six months should suffice. It is well, too, that the meter be allowed on at least two to three minutes before beginning the test to allow the cathodes to become heated to a uniform temperature for the duration of the test.

This testing device is a comparison calibration tester. That is, the last calibration readings are compared with the previous readings. If there is any significant change or difference, check to learn the cause for the difference. These changes develop gradually, and it is possible to have a meter that is 10 to 20 volts in error because of changes to tubes, resistors, meter movements, etc. When the batteries age in meters, ohmmeter readings become erroneous.

The important point to bear in mind is that these changes occur gradually and very often they go unnoticed until the error becomes gross. When your meter is in need of adjustment, try calibrating it with the internal potentiometers in the VTVM. If this does not improve the situation, check the tubes. This will generally clear up the DC voltage and AC voltage deviations. After you have replaced the tubes, the internal potentiometer can be used to calibrate the meter according to the instructions from the manufacturer.

ELECTRICAL APPLIANCE TESTER AND ITS USE

All electricians have what is known as a test lamp. This device consists of a line cord, a light bulb and socket, and a pair of test leads. This device has been improved here by adding a fuse and an additional outlet, and installing the whole group of parts in a plastic box (Fig. 12-21).

The one part that needs fabrication is the panel (see Table 12-6). It can be made from wood, hardboard, or metal. The panel can be spray painted flat black. The lettering can be done by hand with a lettering pen, or decals can be used.

There is very little wiring to be done since all the parts are mounted on the front panel (Fig. 12-22). If you look carefully you can see in Fig. 12-23 that one side of the line cord and one wire from each lamp socket are twisted together and soldered. This should be the "hot" side of the line. This joint should be carefully taped with electrical tape. The other end of the line cord connects to one side of either receptacle. A piece of wire goes to the other receptacle and on to one of the pin jacks. The wire left on the socket marked "fuse" goes to the receptacle marked "line." The other wire from the socket marked "bulb" is connected to the receptacle marked "series" and a wire goes from there to the unconnected pin jack.

After the wiring is completed, a 40W light bulb is installed and a 10A fuse is screwed in. The reason that a 10A fuse is used is because this fuse, being smaller in size than the line fuses, will blow first and save many trips to the fuse box.

With the line cord plugged into a wall outlet, an electrical appliance plugged into "series" will place it in series with the light bulb. The test leads are used to check for shorts or opens when it cannot be plugged in. When anything is plugged into "line" it has the fuse in series with it and the line.

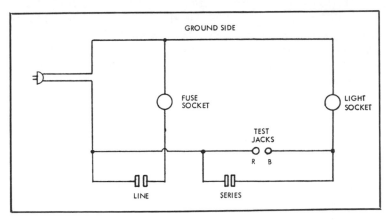

Fig. 12-21. Schematic of the electrical appliance tester.

Caution: Remember that there is approximately 115 VAC across the test prods and under no condition should you touch either end when standing on a concrete or metal floor; nor should you ever touch both ends of the test prods at the same time.

A MINI TROUBLESHOOTER FOR MINIATURIZED CIRCUITS

Miniaturization of electronic components has created a new generation of troubleshooting problems. Checking out printed circuit boards with an ohmmeter, for example, requires the removal of all but one lead of each components being tested. The desoldering and resoldering involved may locate the trouble, but it often creates new problems in the process, such as burned out devices, cold joints, loose pads and broken leads. Also, the board must be deenergized for testing. And, the power source of the ohmmeter is sometimes too great for the miniaturized—or microminiaturized—devices under test. In addition, after a thorough ohmmeter check, the technician often can't determine whether a suspicious inductor is shorted, partially shorted, or good.

To cope with this family of problems military electronic technicians developed an inexpensive in-circuit tester—the simple, versatile "octopus" used in conjunction with the ohmmeter. With it, components under test need not be removed from the board, eliminating most of the soldering problems. The octopus provides its own power source, but delivers less than 1.0 milliampere of current, energizing the microminiaturized components for a more reliable test without the risk of destroying them.

The device described here is a miniature version of the "octopus," performs the same functions. Construction and functions are shown in Figs. 12-25 through 12-26. By making comparison measurements on good and bad boards, and through continued use, you will become familiar with typical ohmmeter patterns indicating good and bad devices, including ICs.

If your budget won't allow a curve tracer, an impedance bridge, a transistor checker or some of the other niceties, the bug

Table 12-6. Parts List for the Electrical Appliance Tester.

1 Plastic box (Alled Radio cat. no. 86 P 287)	2 Pin jacks
	1 Pair of test leads
2 Porcelain sockets	1 40-w light bulb
2 Single receptacles	1 10-amp fuse
1 Line cord	1 Front Panel

Fig. 12-22. Top view of the electrical appliance tester.

Fig. 12-23. Bottom view of the electrical appliance tester.

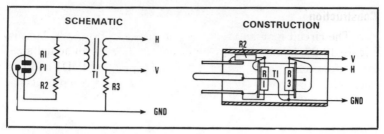

Fig. 12-24. Schematic and construction of the mini troubleshooter.

is a very useful piece of equipment. If your workshop has everything, the bug is a relevant, inexpensive project.

LOW-RANGE VOLTMETER AND ITS USE

Extensive use of semiconductors in today's electronic circuits demands voltage measurements in the 1 mV range. The typical VOM and most VTVMs are unable to obtain such low voltage. But with a minimum of components, you can construct a DC voltmeter with a range of 1 mV to 100 V and an input resistance of 100 kilohms/V.

Op amp IC1 (see Fig. 12-27) is connected as a current amplifier; that is, a change in current at the input gives a change in voltage at the output. Meter sensitivity and calibration is regulated by R1, while R3 and R4 form an offset bias network that will balance any differential voltage in the input. This insures a meter indication of zero output with no input signal. Capacitors C2 and C3 filter any stray alternating current that may be present in the supply voltage.

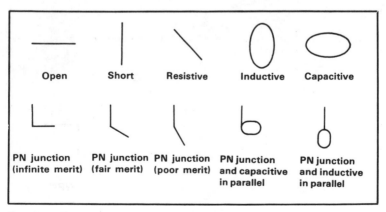

Fig. 12-25. Test functions with typical scope displays.

Construction

The circuit may be assembled on either a perf or PC board. Figure 12-28, which shows component location, is an actual size layout for etching a PC board. Be sure to install IC1, D1 and D2 correctly. If the circuit is assembled on perf board, keep leads as short as possible.

Calibration

Turn power on and adjust R4 until the meter is zeroed. With range switch S2 in the 10V position, connect a precision voltage source to the input. Any known voltage source between 1 and 10 VDC may be used. With the known input voltage connected to the input, adjust R1 until the known input voltage on the meter read M1.

To maintain the highest degree of accuracy, 1% resistors should be used on R6 through R11 (Table 12-7). If such resistors are unavailable, select and measure the resistance of 5 percent

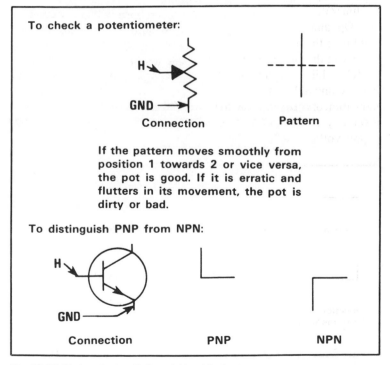

Fig. 12-26. Various tests with the mini troubleshooter.

resistors that will give the closest values to R6 through R11. Satisfactory results may thus be achieved; however, circuit accuracy will be somewhat impaired.

FILAMENT—CONTINUITY TESTER AND ITS USE

Of all electronic gear breakdowns, 85% are due to vacuum tube failure. Anyone can quickly isolate the offending stage and locate the defective tube if the filaments are parallel connected: an open filament just isn't lit. But what if they are series connected and none is lit? Use of this filament-continuity tester will enable testing for open filaments of vacuum tubes, flashlight bulbs, indicator lamps, etc., and fuses of all kinds.

Checking for Open

With the power disconnected, remove one tube or lamp at a time, insert it in the appropriate socket, and, if the filament is not open, the test lamp will light. The same method will test about any type of television or household fuse. Merely make certain that the function switch is in the proper position.

The three sockets (Table 12-8) used here will not accommodate all vacuum tubes; however, by consulting a tube manual and making use of test leads in lieu of a socket, all others may be checked for continuity. In those cases where an envelope houses more than one tube, the associated filaments should also be

Table 12-7. Parts List for the Low-Range Voltmeter.

IC1—741 op-amp or equivalent	*R9—10 K ¼ W (1%)
B1, B2—9 V transistor radio battery	*R10—1 K ¼ W (1%)
R1—10 K Trimpot #3006P-1-103	*R11—100 ohm, ¼ W (1%)
R2—100 K ¼ W (5%)	D1, D2—IN4002 or equivalent
R3—270 K ¼ W (5%)	C1—.001 mfd ceramic disc capacitor
R4—R K linear taper pot (10 turn)	C2, C3—.1 mfd (orange drop) capacitors
R5—1 K ¼ W (5%)	M1—(0 to 1 ma) meter
*R6—10 M ¼ W (1%)	
*R7—1 M ¼ W (1%)	S1—DPDT toggle switch
*R8—100 K ¼ W (1%)	S2—six position rotary switch
Misc.—suitable enclosure, battery holders, mounting hardware, etc.	

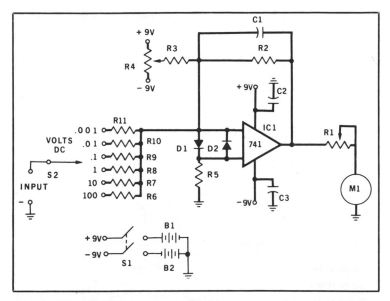

Fig. 12-27. Schematic of the low-range voltmeter.

checked. Since the voltage requirements of vacuum tubes and lamps vary, the intensity of the light will vary accordingly.

Versatile Tester

This self-powered unit will, in the event of a power failure, aid in getting service back. The tester also serves as a trouble lamp if a

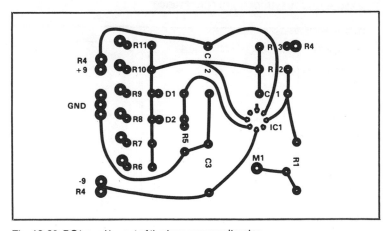

Fig. 12-28. PC board layout of the low-range voltmeter.

Table 12-8. Parts List for the Filament-Continuity Tester.

Part No.	Qty.	Description
A	1	Instrument case, 4 × 2-⅞ × 1-⅝ in., with fitted plastic panel
B	1	Slide switch S.P.D.T.
C	1	Single clip-in fuse holder (acdepts ½ × 1-¼ in. fuses)
D	1	Screw-in type fuse holder (accepts 10-30 A fuses)
E	1	Bulb, 2 V, 60 mA rating, either #48 or #49
F	1	Lamp base with bracket, either screw or bayonet type, to accept part E
G	1	Tube socket, octal
H	1	Tube socket, 7 pin miniature
I	1	Tube socket, 9 pin miniature
J	1	Test leads set (length of zip cord)
K	2	alligator clips with boots (one red, one black)
L	2	Pen-lite cells, "A" size
M	1	Clip-in type cell holder to accept part L
N	4	4-40 × ⅜ in. machine screws, binding head, cad. plated
O	11	SB4-2⅛ in. d button head rivets (blind rivets)

spare fuse is kept in the testing receptacle—just don't screw it in tight. The test leads may be utilized to check for a break in the copper run on a printed circuit board, or discontinuity of line cords.

In one switch position, the circuit consists of the cells and lamp only in series (Fig. 12-29). In the other position, the cells and lamp are in series as are a number of parallel branches, only one of which is completed through the device being checked; therefore, it also is a series circuit.

Construction is not critical. The components were positioned so as to allow for symmetry in appearance.

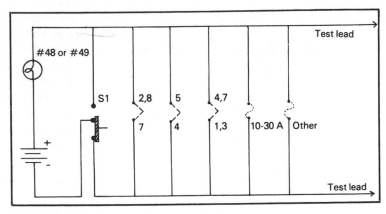

Fig. 12-29. Schematic of the filament-continuity tester.

HI-FI ANALYZER

In the case of this "audio analyzer," it was decided that the unit must contain the following: an oscilloscope, to give visual representation to the audio signals; a signal source, to provide a signal of known characteristics; an audio watt meter to load and determine the power output of the amplifier under test; and an intermodulation distortion meter, to supplement the oscilloscope.

Each section was then designed as an individual unit. Keep in mind the entire unit to avoid duplication. For example, one power supply is sufficient to operate all the individual units collectively or individually. And since the complete apparatus contains four individual units, there is a considerable saving in power transformers alone.

Care must be taken in layout because various components, or units, will interact and cause difficulties that are very hard to locate. Common practices in good design should always be followed. As an example, power transformers should be isolated from components which might be affected by the magnetic field of the transformer; and care should always be taken to assure minimum lead length in high-frequency circuits.

The purpose of the instrument is to provide a single instrument that will perform the necessary tests to determine the effectiveness of a high-fidelity amplifier or preamplifier. The circuit design is straight forward and no trick circuits are used. Improvements have been made in the original model of the analyzer which broaden its usefulness.

The block diagram (Fig. 12-30) and the following information are on the improved model. Each unit may be used as an independent test unit or the entire analyzer may be employed. It is important to regulate the power supplies so that voltages remain within +10 percent of original voltage as each unit is switched in or out of the testing process. It was determined best to switch B+ to place the unit in a standby condition, and switch a suitable load across B+ as the individual units are turned to standby. This helps to stabilize the B+ voltage by constantly loading the power supply.

The unique feature of this instrument is the electronic switch, which permits the operator to observe the input and output signals of the amplifier at the same time. A further refinement of the analyzer would be to provide more than two inputs to the electronic switch. However, as the number of traces on the oscilloscope screen increases, the difficulty of viewing them also increases. Therefore, the practical limit for a small screen scope seems to be two traces at one time. If, however, you increase the size of the CRT, the number of traces could be increased accordingly. The simple multivibrator switching system, however, would then be replaced by the more complex gated amplifier circuit.

SLAVE UNIT

This slave unit is designed to help you out of a jam that all of us find ourselves in at one time or another: power failure. The slave unit will turn on an emergency light over your fuse box the instant a fuse blows or a circuit breaker opens. The slave unit will also signal you if there is a general power failure.

Construction

The heart of the slave is a single-pole, single-throw relay with 110V coil. This coil must be energized at all times to keep its contacts open. This prevents the emergency light and buzzer from functioning. Figure 12-31 shows the main switch, S2, to be open so the relay contacts allow battery power to flow to the emergency light The buzzer, however, does not have power entering it because its switch, S1, is open.

If you do not want an audio-type warning, the buzzer part of the circuit can be eliminated, but this would cut down the efficiency

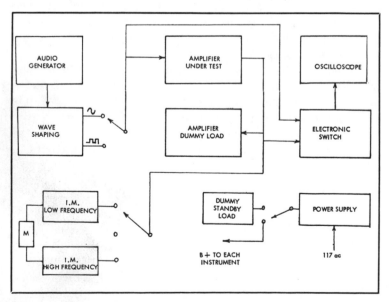

Fig. 12-30. Block diagram of the hi-fi analyzer.

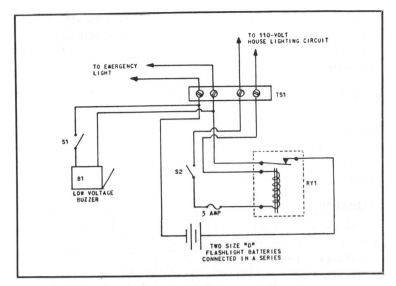

Fig. 12-31. Wiring diagram of the slave unit.

of the unit. It would signal power failure only by means of the light. Opening switch S1 would cut out the sound signal when it isn't needed.

Parts selection (Table 12-9) is not critical because the relay is the only unit that must meet required specifications. Most of the other parts can be salvaged items. The wiring, with one exception, is simple (Fig. 12-31). The exception is the wiring to the emergency-light fixture. It should not exceed 5′ in length because the voltage will drop if it is longer. This would result in a dimmer warning light.

Control Cabinet

The cabinet is very simple to construct. All of the joints are simple butts. The stock is 2½″ softwood, which is easy to work. The door is held shut with one wood screw (Fig. 12-32).

Testing

The unit can be tested by pulling or opening the main house switch or opening the main switch on the slave control cabinet. The relay should close the contacts and the emergency light should be turned on. The buzzer should sound. Turn the buzzer off by opening its switch. The emergency light should continue to burn. Now turn the main switch back on. you should hear the relay click

TS1	Tie strip, four terminal
B1	Buzzer, low voltage
S1,S2	Switch, SPST (cutler Hammer type 8280-16 or equivalent)
RY1	Relay, 110-v coil (advance No. GHA/IC/115 va)
1	Receptacle, miniature base porcelain cleat
1	Lamp, flashlight—screw base for size "D" cells
2	Batteries, flashlight, size "D"
1	Fuseholder (Buss type HKL or similar)
1	Fuse, 3 amp for above
	Line cord and attachment male plug to suit.
	Cabinet (see accompanying drawing or purchjase Budd No. CU 2110— Aluminum Minibox)

Table 12-9. Parts List for the Slave Unit.

and the emergency light will be extinguished. Make sure, however, to turn the buzzer back on.

Maintenance

The unit should be tested at least once a year by opening the main house switch or opening the main switch on the slave-unit control cabinet. The batteries should be replaced once a year even

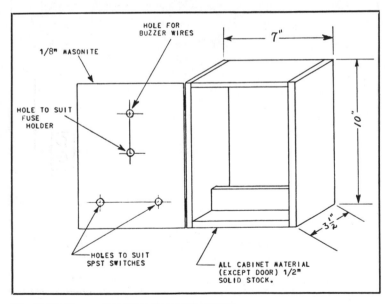

Fig. 12-32. Slave unit control cabinet dimensions.

444

if they show no sign of deterioration. Since batteries cost only a few cents it is cheap insurance so that you will not have to grope in the dark to change fuses and also you will be alerted when the power goes off.

PORTABLE POWER SUPPLY TEST PANEL

This panel will supply the following: 110V AC (fused or protected with a circuit breaker); low-voltage AC (6V, 12V, 18V); low-voltage DC (6V); and a built-in series tester (brought into use by flicking a three-way switch used in house wiring that will also serve as a single-power single-throw switch). See Fig. 12-33.

It should be noted that the high voltage and low voltage are on separate sides of the panel. This is a safety feature.

These panels may be built in a large quantity to match the number of benches available for them. They have two working surfaces—one on each side of the panel.

The panels are made of ½" stock. These panels do not require expensive meters or other costly parts. All parts should be obtainable locally. The transformer and the battery serve both sides of the panel.

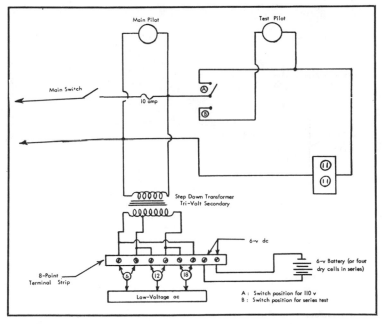

Fig. 12-33. Wiring diagram of the portable power supply test panel.

A DC power supply could be incorporated but this panel was designed to meet the following needs: high efficiency; ruggedness for portability; easily obtainable parts; and low cost.

IS THE UNIMETER FOR YOU?

How often have you been working on a circuit, only to insert your meter probes and discover that you are trying to read AC on the DC ranges? Have you ever wrapped your meter's needle around its stops by applying the wrong polarity? Then the universal meter (unimeter) is for you. This meter is practically infallible; it measures AC, DC, and DC reverse voltages with only one set of voltages to switch.

Operation

The unimeter is based on the conduction characteristics of a full-wave bridge rectifier (Fig. 12-34). If a positive voltage is applied to point A and a negative voltage to point B, diodes D2 and D3 conduct in the forward direction producing a positive voltage at point C and a negative voltage at D. If the polarity of the voltage between A and B is reversed, diodes D1 and D4 conduct to again produce a positive voltage at point C and a negative voltage at D. An AC voltage applied between A and B will also produce a positive voltage at C as the polarity of the ac swings back and forth

Table 12-10. Parts List for the Portable Power Supply Test Panel.

5	Receptacles, concealed cleat
2	Strips, terminal—8 point
2	Switches, three-way toggle—(SPST)
1	Switch, single-pole toggle
2	Receptacles, duplex
3	Plates, single-gang toggle switch (metal suggested)
2	Plates, single-gang duplex receptacle
1	Battery—6v
1	Transformer—110v primary, tri-volt secondary (6v, 12v, 18v)
1	Fuse, 10-amp plug type (circuit breaker may be used if cost is no problem)
1	Handle, chrome
2	Panels, Masonite—⅛″ × 15″ × 19″ Solid Stock—½″ thick (bottom, top, and ends)
8′	Cord, supply—SJ Cord—18/2
1	Attachment Cap, heavy-duty rubber Necessary interior wire—110s will use No. 14-type T; low voltage—No. 18 Bell
4	Lamp Bulbs (2 red for main pilots—2 white for test pilots)

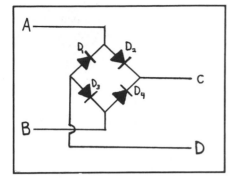

Fig. 12-34. Full-wave bridge rectifier.

between positive and negative. Thus, regardless of the polarity or type of voltage being applied between A and B, only a positive voltage will appear at C.

Due to the nature of a diode junction a certain forward bias must be reached before the diode will conduct (Fig. 12-35). This is due to what is known as a space charge (Fig. 12-36). The space charge is a region of reverse polarity that is opposite to the flow of current across the junction. This phenomenon is caused by the diffusion of P material into the N region and vice versa.

In general, for germanium diodes, it takes about 0.3 v to overcome this space charge and allow the diode to conduct. For this reason, the meter scale is calibrated to about four-fifths of the full-scale voltage. After conduction begins, there is a small area of non-linear conduction. Once this point is reached, the conduction of the diodes—and therefore the scale as well—will be linear.

Fig. 12-35. A certain forward bias must be reached before the diode will conduct.

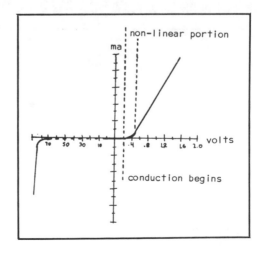

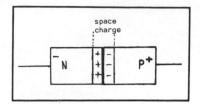

Fig. 12-36. Space charge of a diode.

Construction

By attaching a meter and suitable resistors to the bridge rectifier, a unimeter is created (Fig. 12-37). Component arrangement is not critical, but polarities of the meter and diodes must be strictly observed. Heat sinks should be used while soldering the diodes and resistors since excess heat can destroy the diode junctions and alter the values of the resistors. A lock nut should be used on R5 to prevent its setting from being altered (Table 12-11).

The meter face will need a new scale (Fig. 12-38). Then, to calibrate the unimeter, apply a known, fixed voltage to the meter and adjust R5 until the meter needle rests over the scale marking which corresponds to the voltage being measured. Tighten the lock nut on R_5.

Do not substitute other diodes since their conduction characteristics differ from those specified and will cause the scale to be incorrectly calibrated.

Using the Unimeter

The unimeter is a very easy instrument to work with. Simply set S1 to the scale which will cover the level of the voltage to be

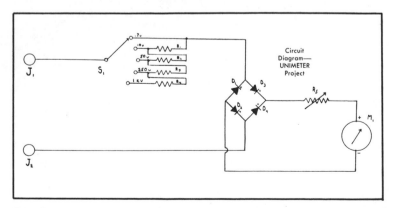

Fig. 12-37. Schematic of the unimeter.

D1, D2, D3, D4—1N34A
J1, J2—pin or banana jacks, depending on
   the type of probe used
M1—0.50 a, 4½" meter (Allied 52 A 7205
   or equiv.)
R1—180 K, 1 percent resistor
R2—800 K, 1 percent resistor
R3—4 meg, 1 percent resistor
R4—15 meg, 1 percent resistor
R5—25 K linear taper potentiometer
S1—5 position rotary switch
1—7" × 5" ×3" cabinet

Table 12-11. Parts List
for the Unimeter.

measured and insert the probes into the circuit being tested.
Polarity and/or type of voltage does not matter. The unimeter will
read anything.

TTL LOGIC PROBE AND ITS USE

Transistor-transistor logic (TTL) probes are extremely handy
tools to have in the digital lab. They are used to check the states of
logic circuits. Malfunctions are found and their cause easily
determined. This can be done in much less time than is possible
when using a meter or scope. The indicators are located at the
probe tip, which allows you to determine the logic state, pulse
polarity, and duty cycle without lifting your eyes from the contact
point. This eliminates the risk of shorts due to the probe slipping
when the technician takes his eyes from the circuit to check the
meter.

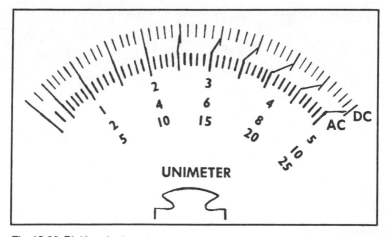

Fig. 12-38. Dial face for the unimeter.

Economize on Time and Cost

Although logic probes are commercially available, their cost can dent any hobbyist's budget. To construct a probe, normally you must use printed circuit techniques, such as a double-sided board. This probe has a different means of construction. The recent availability of a wiring pencil [the Vector P173] suggested a possible solution. With the wiring pencil, many new miniaturized IC circuits can be built without the time-consuming problems involved in making layouts and printed circuits.

The probe can be put together in a few hours. It will detect a logical "0" (0 to .7V) by lighting the green LED. A logical "1" (1.8 to 5 V) lights the red LED. The pulse extender circuit (IC-1, Fig. 12-39) will catch a pulse as short as 50 ns and light the yellow LED briefly. The pulse catcher switch (S1) will lock the yellow LED on when a pulse passes by. By observing the pattern of LEDs lit, you can estimate the duty cycle and polarity of the pulses present at the probe tip (Fig. 12-40).

Probe Construction

Feed specially insulated #36 ga. wire from a lightweight hand-held tool with a hardened feeding tube at the end. Hold the

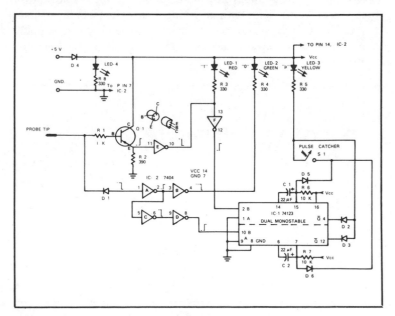

Fig. 12-39. Scematic of the TTL logic probe.

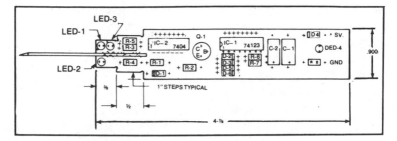

Fig. 12-40. TTL logic probe board dimensions.

loose end of the wire by the wooden wire hold-down tool (Fig. 12-41). Using the tip of this tube, wrap the wire around the desired component and route to another component. If desired, ICs may be directly wired without using a socket. Cut the wire by applying pressure to the tip, and solder using a 50W iron with about a ⅛ in. dia. tip. The higher heat of this tip melts the special insulation. After it burns away, hold the iron another 5 to 10 seconds while adding more solder to the wire wrap. The flux core will clear away the char, and the solder will form a good bond.

Mount components in a board which has holes punched on a .1 in. grid. Unpunched phenolic board 1/16″ thick may be used but #58(.042) holes must be drilled at the part locations shown in Fig. 12-40. Secure the components by first bending the leads to the correct width to fit the hole pattern. Insert the part and hold it against the board. Cut the leads about 3/16″ from the board. Use the lead bending tool (Fig. 12-41) to bend the lead 45 degrees to the side, and then press the lead down against the board to make the tool and lead perpendicular to the board. This "Z" bend results in a

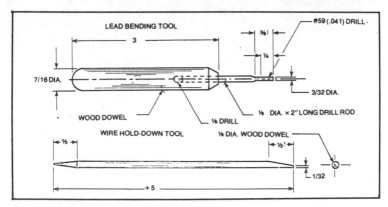

Fig. 12-41. Tools used in constructing the TTL logic probe.

short length of component lead wire that is perpendicular to the board. Use the wiring pencil to wrap the #36 wire. Wire routing may be direct. The use of the small hold-down tool makes routing easy. The insulation on the wire allows any number of wires to cross without shorting.

Electronics Construction

Cut out board as shown in Fig. 12-40. Mount the probe tip wire by "sewing" it to the board using the holes provided and a single strand of tinned wire. Solder in place.

Mount the LEDs at the tip end. Connect all anodes together and solder. Be sure there is at least ⅛″ wire from the soldered anode junction to any LED. The heat of a soldering iron too close to the LED may destroy it. Use a low wattage iron for this operation.

Mount the LED resistors, R3, 4, and 5, and connect and solder each LED cathode lead to the corresponding LED resistor. Check for operation of LEDs by grounding each LED resistor.

Table 12-12. Parts List for the TTL Logic Probe.

Part No.	Description
R1	1 kilohm, ¼ W
R2	390 ohms, ¼ W
R3,4,5,8	330 ohms, ¼ W
R6, 7	10 kilohms, ¼ W
C1,2,22	mfd. (approx.), 16 V TS #272-1003. Axial type
D1,2,3,4	1N914 or 1N4148 diode
LED 1,4	Miniature light emitting diode,.125 in. dia., red, XC209R or equiv.
LED 2	Miniature light emitting diode,.125 in. dia., green, XC209G or equiv.
LED 3	Miniature light emitting diode.,124 in. dia., yellow, XC209Y or equiv.
IC-1	74123N hex inverter
S1	SPST miniature push button switch (N.O.) RS #275-1547*
Q1	NPN switching transistor, 2N3904 or equiv.
probe tip board	1/16 in. dia. holes by 100 spacing RS #276-1394
acrylic rod	1.100 in. dia. by 3 in. long (for tip and end cap)
acrylic tubing	1 in. od nominal by 1/16 in. wall

*Note: RS = Radio Shack part number.

Install and secure with the bending tool all other resistors, diodes, capacitors, and ICs. The ICs may be secured by bending the four outer leads outwards.

Begin Wiring

Use the hold-down tool to hold the end of wire while making the first wrap. Feed the wire by sliding the index finger forward on the wire toward the cone tip of the wiring pencil. Route loosely to the next position. Never route tightly because some wires will have to be moved out of the way during soldering of adjacent wires. Cut the wire with the wiring pencil tip. Solder using a 50W iron and rosin core solder of about .020 in. dia. Cut off excess wire on the first wrap using a knife against the phenolic board. To keep track of the wiring, be sure to mark the schematic diagram with a colored pencil as each wire is soldered in place and cut. Do this consistently and few errors will result. After wiring in IC-2, check proper "1" and "0" operation of the probe. Then wire in IC-1.

To save time when wiring with the wiring pencil, plan out the route the wire will take. When routing for power, a good practice is to start at one component (i.e., pin 14, IC-2), wrap around Vcc, and then go to the other component (pin 16, IC-1). This method saves time in soldering and routing and allows each component to have the current it needs.

Use #22 ga. stranded wire, 30" long, and terminated with miniclips for the power leads. Use a red miniclip for the positive side of the power and black for the ground side.

Probe Check-Out

Attach clip leads to the positive and ground of a 5V power supply. Touch probe tip to ground. The green ("0") LED should light. At the same time, the yellow ("P") LED should pulse. Upon removal of the tip, the green LED will extinguish and the yellow LED will pulse again. Repeat procedure for the positive side of the power. The red LED should operate, and the yellow LED will again pulse. A pulse will be indicated for every transition of state. The duty cycle of the pulse train can be estimated by observing the combination of LEDs lit. If the yellow LED remains on and both the red and green LEDs are also on, then you can assume that a pulse train with 50 percent duty cycle is being experienced. If the yellow LED is lit along with the green LED then you can assume that a positive pulse train of low duty cycle is present.

Logic Probe Operation

The power for the operation of the probe is taken from the circuit being tested. The diode D4 protects the probe from reverse polarity. When the probe tip is floating, Q1 is off and allows a "0" to enter pin 11 of inverter E. The "0" is inverted to a "1" which does not allow LED-1 to light. The output of inverter A is low, forcing the output of B to be high. Therefore LED-2 remains off also.

When the probe tip is high, Q1 is turned on and acts as a high input impedance buffer. Inverter E goes low and turns on LED-1 indicating a "1." When the probe tip is low, Q1 remains off but inverter A sees the low through diode D1 and forces the inverter B low. This turns on LED-2 indicating a "0."

Each time LED-1 or 2 turns on, a positive transition is generated which fires IC-1, the dual monostable multivibrator. The outputs of the multivibrator are AND'd with diodes D2 and D3 to LED-3. The length of time LED-3 remains on is dependent upon the timing resistor and capacitor. Switch #1 prevents the timing circuit from timing out after the monostable is triggered. D5 or D6 allows the low output of the monostable to be coupled back to its associated timing capacitor, yet isolates it from the other half of the monostable.

The Probe Case

The biggest problem in constructing a logic probe is finding a suitable case in which to house the electronics. The use of the wiring pencil eliminates the problem of trying to design a probe to fit into an available case. Any suitable case will do, and it is not absolutely necessary to follow this design, especially if the plastic specified is not available in your area. The design shown was selected, first, because of the availability of the tubing and acrylic rod, and second because it allows high visibility of the components and wiring technique. It can double as a useful teaching aid.

The Body

The acrylic rod and tubing can be purchased at any plastics supply house. If the sizes specified in the drawings are not available, then dimension changes and perhaps part layout changes will have to be made. Be sure the ID of the tubing is no less than .900 in. if the same parts layout is to be used.

Assembly

Cut a small slot at one end of the tubing to allow power leads to pass through. Insert pushbutton switch SI in cap and mount, using nut supplied. Slide board through tubing. Slide probe tip over probe wire. Push all plastic parts together. Probe is now ready to use.

AN INEXPENSIVE TV TUNER SUBSTITUTER AND ITS USE

This tuner substituter is a useful and economical piece of test equipment for the television repair technician and hobbyist. It can be built with a junked tuner, a case and a minimum of parts, and may be used as a portable or bench model. The finished product can be used for troubleshooting tuner or IF/AGC problems and will allow a full range of fine tuning.

A discarded, but functioning, tuner can usually be found in any spare parts supply, or in the junk pile of a TV service technician. A solid-state, color VHF tuner is preferred. The power supply is easier to design with a solid state model, because a tube model requires higher voltages and more current. A color tuner is preferable because of its wider frequency bandpass and usually higher gain. A VHF tuner reaches greater distances than UHF in most places, and a VHF tuner is necessary for a UHF tuner to work. If you want to make a more complex piece of equipment, both UHF and VHF tuners may be used.

Power Supply

With a transistorized VHF tuner, the sub can be used as a portable or bench model. Only the AC power supply is included here, but a battery pack can easily be included for a portable unit; both AC and DC supplies will be needed for a portable AC operated model.

Three voltages are needed for most tuners: RF/AGC voltage, automatic fine tuning (AFT) voltage, and B+ voltage. Voltage values are obtained from the schematic of the receiver from which the tuner was taken. The main power supply is designed with a bridge rectifier, a capacitor filter and a voltage divider network. A 12V transformer supplies 17V of B+ voltage when rectified and filtered. This was sufficient to power our tuner, though it was specified as requiring a B+ of 20 volts.

The voltage divider is connected across the main supply to derive the necessary tuner AGC voltage. The two resistor values were calculated to be 1850 and 150 Ω. The AGC current drain would be minimal (μA), so a larger voltage divider current for voltage regulation was arbitrarily set at 10 mA.

With a 10 mA current in the divider, Ohm's Law was used to calculate the total divider resistance and the resistance required to provide the approximate 1.2V AGC voltage. The 150 Ω resistor provides the AGC voltage source. For AGC voltage substitution, another voltage divider was placed across the power supply outlet.

This divider contains a potentiometer which allows a 0-12V variable AGC bias to be used for AGC voltage substitution. Switch S2 allows the unit to function as either a tuner substituter or as an AGC bias supply. With slight modification, both functions could be used at once.

To use the tuner sub in troubleshooting, first disconnect the IF/tuner coaxial link from the tuner. Then connect the IF/tuner link to the inner sub unit and turn on both the tuner sub and the receiver. It's really very simple!

The tuner sub is used to troubleshoot possible tuner or IF/AGC problems. It is difficult to determine between these two defects without an external signal that can be injected between the two sections. To distinguish between tuner and IF/AGC problems, inject a signal at point A(Fig. 1). If the signal passes through the IF and gives video and audio, then the IF/AGC is functioning correctly. The assumption would then be that the tuner or the antenna system is defective. If the signal does not pass through the IF section, it must be determined whether the problem is IF or AGC. IF amplifier conduction is controlled by a voltage from the AGC circuit, so lack of IF conduction can be caused by the IF itself or by a defective AGC circuit.

To troubleshoot for a defective AGC, included in the same enclosure is a variable AGC bias power supply. By supplying the proper AGC correction voltage at point B (Fig. 12-42), the AGC can be eliminated as a source of trouble if sound and video are observed. If no sound or picture is observed, the IF is defective.

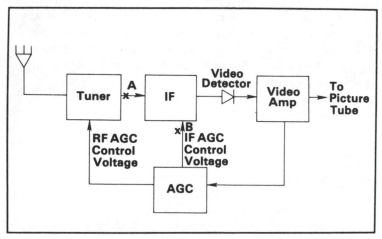

Fig. 12-42. IF/AGC system.

Electronic Fine Tuning

In the original tuner construction, a number of gears were used to provide mechanical control of fine tuning. This mechanical fine tune control could not be used because of the size of the junk-box cabinet. Instead, an electronic fine tune was designed.

With the tuner operating a variable power supply was connected to the tuner AFT input and its voltage was varied until the tuner was driven from perfect tune to totally out of tune. This required 3 volts. If the tuner was fine tuned, a variable voltage connected to its AFT could be varied $\pm$ 3 volts to provide a full range of fine tuning. The current drain was measured to be 30μA so the electronic fine tuning could be provided by two 1-½V penlite batteries and a 5K Ω potentiometer (Fig. 12-43). A DPDT switch was used to reverse the voltage polarity. Batteries were used because their output could be reversed easily without regard to ground and interference with other power sources.

The only adjustment needed before using this project is to disable the electronic fine tuning and mechanically fine tune each channel to be used. This gives the electronic fine tuning an equal range of adjustment on either side of channel center frequency. Of course, if the tuner had an intact fine tuning system this would not be necessary. In such an instance, the AFT circuit might be left open, or it might need to be "tied" to a particular voltage as specified in the schematic.

To prevent an unwanted "shocking experience," no voltage should exist between any two pieces of equipment. This can be

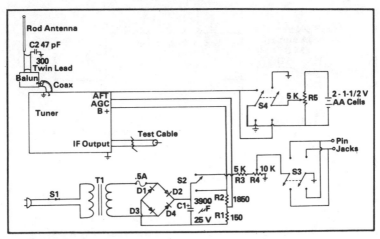

Fig. 12-43. Tuner substituter schematic.

prevented by using an isolation transformer in the tuner sub, by using an external isolation transformer for the television being serviced, or by connecting an AC voltmeter between the two and making sure no voltage difference exists. A transformer power supply was used here.

When using the voltmeter test method, the AC plug can be reversed if voltage differences are measured. Usually the difference will disappear when the plug is reversed. However, in some receivers with the chassis operating at an above-ground voltage, this doesn't work and an isolation transformer *must* be used, if the substituter is not transformer isolated.

AN INEXPENSIVE SIGNAL TRACER AND ITS USE

When inexpensive cassette or eight-track tape players break down—frequently because of a defective head or a mechanical malfunction—owners tend to discard them rather than bear the cost of repairs, thus making them easily obtainable at no cost.

However, the amplifier, in all likelihood, is still in excellent condition. The high gain and sensitivity of these amplifiers make them ideally suited for use as the heart of a practical audio signal tracer. Even if the amplifier is stereo, it can still be used in our project as long as one channel remains operative.

As seen in Fig. 12-44, the circuit for the signal tracer is simple and straight-forward. The input signal is fed through a single-pole-double-throw switch which either passes the signal directly to

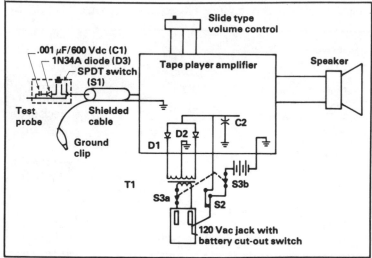

Fig. 12-44. Schematic of the signal tracer.

the amplifier or routes it through a detector circuit for demodulating AF-RF signals. The amplifier then produces enough amplification to operate a small speaker. The output level is controlled by a slide potentiometer. The power supply was left just as it was taken from the recorder case.

Construction is relatively simple, but calls for some thought and prior planning. Obtain a box in which the amplifier and associated circuitry will easily fit. Plan accurately, on paper, the front panel layout, component layout and parts placement (Table 12-13). If this phase is neglected, sloppy appearance and wasted time may result from misplaced holes, components, or sub-assemblies. For best results, always plan ahead.

The probe you select should allow comfortable use of the tracer, thus the probe cable should be as light-weight and flexible as possible. In the prototype, about 4′ or RG-59/U coaxial cable was used with a BNC connector. The connector was used so the probe can be disconnected for convenient storage, or with other pieces of equipment. A discarded VOM probe was used.

Using the Signal Tracer

Refer to Fig. 12-45 for the explanation. Set the AF-RF switch (S1) to the RF position, connect the ground lead to the receiver ground and the probe to point A. *Use an isolation transformer or be absolutely certain that there is no potential between the chassis of the two pieces of equipment. Otherwise, a dangerous shock hazard exists.*

Tune the receiver until a clear local station signal is heard. For simplicty, point A is shown as the input from the antenna. In reality it is the input to the converter stage and the tuning section precedes it. To complete receiver servicing, set the AF-RF switch to the af position and touch the probe to point D. If the signal is good at point D, the problem is in the af amplifier stages, or the speaker. If the signal does not check good at point D, the fault lies in either the detector, if amplifier, or converter sections. By checking the signal at points A, B, C and E, the faulty stage can be quickly isolated.

When observing a signal before the detector stage, be certain the AF-RF switch is in the rf position. The switch must be in the af position for any stage following the detector.

Defective Components

After the faulty stage is found, a similar method can be used to isolate the defective components. Transistors should be tested for quality of signal flow and amplification from base to collector.

Table 12-13. Parts List for the Inexpensive Signal Tracer.

D3 general purpose crystal diode
    (1N34A or equivalent)
C1 .001 μF, 600 Vdc capacitor
S1 SPDT switch
S3 DPST switch
D1, D2, C2, and S2 (were from
    salvaged cassette player, as
    were the volume control, and the
    s-ealer)
Enclosure
Probe
Amplifier (from salvaged cassette
    or 8-track player)
Miscellaneous: Shielded cable, hookup wire, hardware, transfer
    lettering kit, contact paper,
    ground clip, etc.

Coupling capacitors should be checked from input to output for distortion and degree of signal loss.

Whenever a component is found which has a signal flowing into it and that signal is distorted or missing at the component output, that component should be removed for futher testing to determine whether or not it is actually faulty.

Troubleshooting audio amplifiers is accomplished in much the same manner, except that only the AF position of the function switch is used.

Check for a proper signal at the center of the amplifier (point D in Fig. 12-46). Whether or not the signal is correct at this point will show which half of the unit is at fault. The next test would be to eliminate half of the faulty stages as causing the trouble. To do this, check the signal at either point B or E, depending on which side of point D the fault was first determined to exist. In this manner you are dividing and conquering, so to speak, which will lead to quick detection of the defective stage.

The tracer may be crafted as elaborate or as simple as you desire. A dual power supply may be used for AC or battery portability. A signal lever meter to provide a useful visual indication of signal strength may be incorporated. The AF-RF switch may be built into the probe, as in some commercial models. Or, any other refinement which would make the signal tracer easier to use or better suited to your use may be employed. In any event, this is a sure-fire economical way to add a valuable piece of test equipment to your service bench.

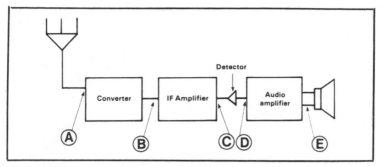

Fig. 12-45. Block diagram of the AM superhet radio.

Three-Function Generator And Its Use

Make the Intersil 8038 function generator your next electronics project. It's an advanced integrated circuit that houses practically all of the circuitry required for a three-function (sine, square, triangle) signal generator. The simplicity lends itself to easy construction and the small number of required components is easy on the budget. The IC itself can be purchased from a discount house for about $4.50

A signal generator is valuable not only in basic electronics as a signal source for experimental amplifiers, but also in audio troubleshooting, where the generator would be used to locate a defective amplifier stage.

Furthermore, the Intersil generator can provide the source of the digital electronic circuit's signal—the square pulse. This square pulse can also be used in audio work to determine the frequency response of amplifiers. Though limited in its practical use, the triangular wave can help demonstrate sweep voltages in electrostatic deflection circuits or in applications that call for a linearly increasing voltage.

Construction Notes

Fig. 12-47 shows a schematic diagram of the generator. The majority of components (Table 12-14) are in the power supply

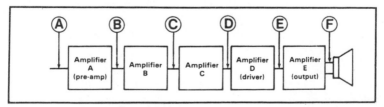

Fig. 12-46. Audio amplifier functional block diagram.

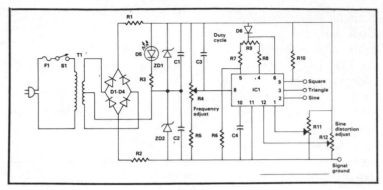

Fig. 12-47. Schematic of the Intersil 8038 generator.

Desig.	Description
R1	250 Ω ½ W
R2	250 Ω ½ W
R3	100 Ω ¼ W
R4	10 kΩ pot linear taper
R5	20 kΩ ¼ W
R6	8.2 MΩ ¼ W
R7	3.6 kΩ ¼ W
R8	3.6 kΩ ¼ W
R9	1 kΩ trimmer
R10	3.3 kΩ ¼ W
R11	100 kΩ trimmer
R12	100 kΩ trimmer
C1	470 µF Vdc
C2	470 µF 16 Vdc
C3	.1 µF 25 Vdc disk
C4	.0039 µF Mylar
D1-D4	1N4001 diodes
D5	LED pilot light
D6	1N914A silicon diode
ZD1	12 V zener diode ½ W
ZD2	12 V zener diode ½ W
IC1	Intersil 8038 16 pin DIP
TI	Transformer 110 V primary 24 V secondary center tapped
S1	SPST switch
F1	⅛ A fuse
Miscellaneous	
	Printed circuit board
	Line cord
	Fuse holder
	Output jacks
	Function switch
	Chassis
	Hook-up wire

Table 12-14. Parts List for the Three-Function Generator.

section and only a handful of parts need be clustered around the IC to operate it. Note that *C4* assists in keeping the output frequency stable and thus should be of good quality Mylar.

Use an IC socket or Molex strip to mount the IC, rather than soldering it directly to the printed circuit board. This will make removal of the IC a bit easier if it becomes necessary. The generator shown here is equipped with a three-position switch that allows the user to select the sine, square, or triangle output. Another feature that could be added to the generator is an output amplitude adjust control. This would simply be a potentiometer connected between the output and ground with the adjustable output taken from the potentiometer wiper.

Adjustments

Before putting it to use, adjust the completed project as follows: Apply power. With an oscilloscope connected to the triangle output, adjust *R9* for minimum distortion. Move the oscilloscope to the sine wave output and alternately adjust *R11* and *R12* for minimum distortion. If desired, use a distortion meter here as a final check. Finally, move the scope to the square wave output and readjust *R9*, if necessary, for a 50 percent duty cycle.

SIGNAL PROBE AND ITS USE

Here's a signal probe which eliminates much of the inconvenience and potential danger to delicate electronic circuits normally prevalent when using the conventional vacuum tube audio-R.F. signal generator for troubleshooting. A straightforward transistorized multivibrator circuit is used in the probe (Fig. 12-48). Some deviation can be made from values indicated; however, it is suggested that values given be adhered to as closely as possible. For example, resistors R1 and R4 could vary from 1500 ohms to 2200 ohms, transistors Q1 and Q2 can be almost any small signal germanium type. Transistor types 2N35, 2N209 and 2N170 seem to work equally well.

In the prototype 2N584 medium-speed switching transistors were used. This particular transistor was available from surplus (military) stock and was the most economical to use under the circumstances. It does not matter whether the transistors used are of the PNP or NPN types, except that when using NPNs the battery polarity will be the reverse of PNPs.

The fundamental multivibrator frequency is approximately 1000 Hz. The output voltage waveshape is essentially square. The square wave contains harmonics of the fundamental to 2 MHz or

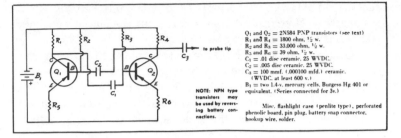

Fig. 12-48. Schematic of the signal probe.

beyond. The output from the probe, when applied to an operating receiver antenna, IF amplifiers, detector or audio stages, results in an audible tone from the receiver speaker.

An inoperative receiver can be tested by starting at the last audio amplifier grid and working back to the plate and grids (bases and collectors) of the preceding stages until the tone from the speaker is lost. The probe *cannot* be used for alignment of a receiver and is not intended for that purpose. The conventional, accurately calibrated signal generator is the only practical test instrument for alignment.

Although construction of the probe is not difficult, it does require some patience, ingenuity and skill in working with small components in a limited amount of space. The entire unit is built into a discarded two cell "penlite" type flashlight. The penlite case provides just enough space to accommodate the multivibrator circuit and two mercury cells. The case itself needs little modification other than adapting a pin plug tip into the hole in the plastic housing that originally accommodated the pre-focused bulb.

The multivibrator is wired on a piece of perforated phenolic board (½″ hole spacing) cut to proper width and length. The size of the board in the prototype was approximately ⅝″ × 2⅛″. This length was the maximum compatible with case dimensions and still has good positive contact with the cells in the top end of the case. The board was cemented into the plastic housing for rigidity. However, these dimensions will vary depending upon the type case involved.

Another challenge is the design of a contact between the cells and the multivibrator that will make it unnecessary to solder the cells to the unit. In the prototype a snap type battery connector was cemented to the extreme end of the phenolic board and aligned so it came in contact with the cells when the unit was inserted in the case. The original switch on the case then served as an "on-off" control for the probe. Two 1.4V mercury cells were selected in order to provide at least three volts and still have maximum space for the multivibrator section. The mercury cells should last indefinitely due to low current requirements. The multivibrator draws about 1.5 to 2mA.

The placement of components is not critical as far as their relationship to one another is concerned, however, because there are several involved, care must be taken so that the completed unit fits in the case. In the prototype, standard ½-W resistors were used. Using ¼-W, values occupies less space and eliminates some of the layout problems. C1 and C2 of necessity, must be the low-voltage disc ceramic types. The conventional 5..0- to 1000-V ceramics are physically too large to use conveniently. Any disc ceramic with a voltage rating of 6 to 25 volts works nicely. It is suggested that all components be mounted in place first by pigtailing the connections *without soldering.* Once the unit is assembled and there is assurance that it will slide into the case, the connections can be soldered. Be sure all joints are heat sinked while soldering to prevent damage to components.

The fundamental frequency of the unit can be changed by varying C1 above or below the value indicated. Increasing C1 decreases frequency, and vice versa.

The unit can be tested by touching the tip of the probe to the input of an audio amplifier. If the probe is operating properly a tone will be heard in the speaker. Placing the tip near the antenna of a receiver will result in an audible tone from the speaker. The same results should be found when the tip is touched to the grids of a areceiver IF amplifier, or to the bases of transistors in a receiver. If the probe fails to operate, check battery polarity or look for circuit wiring errors.

TEST LEAD RACK

An 8′ length of 1½″ perforated angle iron can provide an excellent test lead rack. Figure 12-49 shows how the angle iron is utilized as a lead rack. The accesss slots are hacksawed out and filed lightly to remove sharp edges. The protruding ends of the rack are rounded and filed smooth.

An 8′ rack provides space for 50 pairs of leads with good separation between pairs. Most common size coaxial test leads will also fit in the elongated holes. The rack is mounted on the wall with suitable screws. The height above the floor is determined mainly by the average lead length. The rack will accommodate

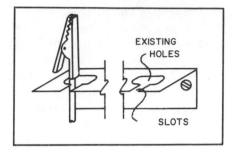

Fig. 12-49. Test lead rack slots.

leads with almost any type of clip (alligator clips, banana plugs, etc.).

A 0-999 COUNTER

Time was when we used only meters to indicate voltage and current. Digital electonics has changed this. With integrated circuits we can now easily and cheaply construct a display in discrete decimal quantities. The display described here is capable of counting from 0 to 999.

The display (Fig. 12-50) consists of three decade counting units (DCU), one for each digit. The most significant digit (MSD) is the hundredths place, and the least significant digit (LSD) is the units's place. A reset input permits the display to be reset to 000. An overload indicator (see Fig. 12-51) shows when we reach a count greater than 999.

Each DCU is the same, so we show only one. The DCU circuit (Fig. 12-52) consists of a seven-segment light-emitting diode (LED) display unit (type MAN-1), a binary-coded decimal (BCD) -to-seven-segment convert (type 7447) and a decade counter (type 7490).

The LED display consists of diodes arranged in a seven-segment array. When the cathode of any of any diode is grounded, current flows through the diode and it emits light. For example, if all the cathodes are grounded, all the diodes will be lit showing the figure "8." The type 7447 IC provides the seven-segment code

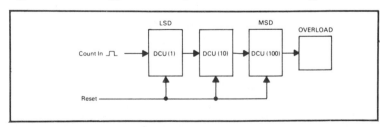

Fig. 12-50. Block diagram of the digital counter.

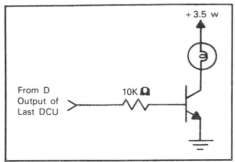

Fig. 12-51. Overload indicator circuit.

output to the display to ground the necessary display diodes. The input of the 7447 is a binary code consisting of 1s and 0s. *Binary* means counting in the base 2 with 1s and 0s. Hence, the 7447 IC converts the BCD code to seven-segment code.

Decade Counter

The type 7490 IC is a decade counter, counting 10 input pulses and indicating this with a BCD output code at the A, B, C and D

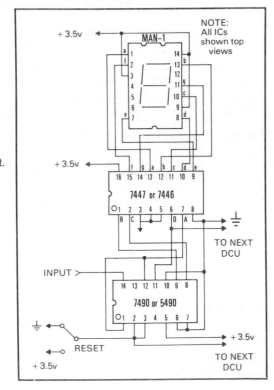

Fig. 12-52. DCU circuit.

467

outputs. The 7490-IC also has inputs which will reset it to a count of 0 or 9. We use only the reset-to-zero inputs.

The ICs are of the TTLV TYPE (transistor-to-transistor logic) and will operate from any DC power supply between 3-7V. The power supply should be well regulated. We use 3.5V to limit the maximum total display current to less than 1A. If a higher voltage is used, put limiting resistors (typically 100 ohms) in series with each input to the MAN-1 display.

The input should be a square-wave signal between 3-7V peak-to-peak. The overload indicator is a low-current lamp controlled by an NPN transistor. The input is taken from the D output of the DCU. Any high gain NPN transistor that can handle the lamp current will work.

Construction Tips

Construct the display on IC breadboards on which up to three digits can be wired quickly and easily. No. 22AWG solid wire interconnects the ICs directly with no soldering required. The displays and ICs are available from surplus electronics suppliers at low cost.

The power supply should be a regulated type capable of supplying up to 1A at 3.5 VDC. The signal source should be a square wave generator with fast rise and fall time. If the square wave is not good, it will result in false counting. A switch may be used as a pulse input, but requires some filtering to prevent the 7490 IC from counting the bounce of the switch contacts.

Suggested Sources For Components

Breadboards:

    ☐ E & L Instruments, Inc., 61 First Street, Derby, CT 06418.

    ☐ Continental Specialties, 325 East Street, Post Office Box 1942, New Heven, CT 06509.

ICs:

    ☐ Poly Paks, P.O. Box 942, South Lynnfield, MA 01940

    ☐ Solid State Systems, P.O. Box 773, Columbia, MO 65201.

    ☐ Solid State Sales, P.O. Box 74E, Somerville, MA 02143.

TRI-FUNCTION GENERATOR AND ITS USE

Could your shop use a signal generator? Here is how to build one which will also show you the capability and practical use of integrated circuits in linear applications. Using a minimum of discrete electronic components and one IC, this tri-function generator can produce a variable amplitude square wave, a triangle wave, and a sine wave output with a frequency range of 1-100 kHz.

The heart of the generator is an Intersil 8038 voltage—controlled oscillator (VCO) integrated circuit. The controls are mounted on the front panel, with the power supply and remaining components (Table 12-15) mounted on a perforated board for ease of assembly. Completely assembled, the generator is small enough to be held in the palm of the hand.

Theory of Operation

The frequency of the VCO is determined by an external RC combination circuit. When the voltage across the capacitor rises to $\frac{2}{3}$ of the supply voltage, a comparator triggers a flip-flop which changes states. The flip-flop allows the capacitor to discharge. When the voltage across the capacitor falls to $\frac{1}{3}$ of the supply voltage, another comparator triggers the flip-flop to the initial state, allowing the capacitor to charge again. A triangle waveform is developed across the capacitor, fed to a buffer and is available at pin 3. The triangle waveform is fed to a sine converter, and a satisfactory sine wave is provided at pin 2. The output of the

Table 12-15. Parts List for the Tri-Function Generator.

Qty.	Description
1	2500 mF, 10 VDC capacitor
3	10 mF, 10 VDC capacitors
1	1 mF capacitor
1	.1 mF capacitor
1	.01 mF capacitor
1	.001 mF capacitor
1	33 ohm, ½-w resistor
1	47k ohm, ½-w resistor
1	68k ohm, ½-w resistor
1	100k ohm, ½-w resistor
1	2.5k ohm, potentiometer
1	15k ohm, potentiometer
1	Intersil 8038 integrated circuit
1	2N3568 transistor
1	SPST switch
1	.5-amp fuse and holder
1	6.3-v filament transformer
1	4.5-v lamp
1	full-wave bridge rectifier module
1	line cord

Miscellaneous: knobs, perforated board, nuts and bolts, chassis, paint and decals.

flip-flop is fed to a buffer, and a square waveform is available at pin 9.

A more complete description of the theory of operation of the VCO appears in Intersil's *Aplication bulletin A012*. The connection and functional diagrams (Figs. 12-53 and 12-54) are reproduced with Intersil's permission.

Circuit Description

The 8038 IC simultaneously produces the three waveforms from pins 2, 3, and 9. The frequency is determined by externally selected capacitor (range select) and a 15K Ω potentiometer (frequency select) as shown in Fis. 12-55. The resistor connected between pins 11 and 12 can be made variable to adjust the sine wave output for minimum distortion. The 68k Ω resistor was selected to provide an approximate 5 volts maximum output for the square wave to be used in triggering digital circuits. The triangle wave output is approximately 3 volts, maximum, while the sine wave output is approximately 1.5 volts, maximum. The function (fun.) select switch determines which waveform will be fed to the output via the emitter follower. The loading effect is negligible for most applications.

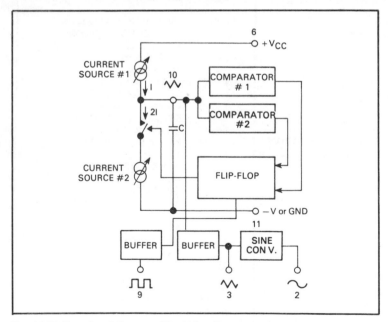

Fig. 12-53. Functional diagram of the tri-function generator.

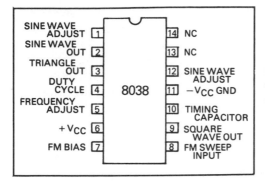

Fig. 12-54. Connection diagram of the Intersil 8038 IC used in the tri-function generator.

The power supply consists of T1 (a 6.3V filament transformers), D1 (a full-wave bridge rectifier module, a 4.5V lamp with limiting resistor, and a 2500 mF capacitor rated at 10VDC.

Calibration

The tri-function generator can be calibrated with a good standard audio generator and oscilloscope using the lissajous figure method.

The frequency output is divided into five ranges, with the upper limit shown on the front panel: 10=1-10 Hz, 100=10-100 Hz, 1K=100 Hz-1 kHz, 10K=1-10 kHz, and 10K=10-100 kHz. When the frequency select knob is turned fully counterclockwise, the lowest frequency for the range selected will appear at the output.

The frequency select control is not linear because of the fixed 1.5K Ω resistor and the resulting nonlinear voltage gradients of the 15K Ω potentiometer. This arrangement was selected to keep the circuit simple and in order to use commercially available components.

MEASURING BREAKDOWN VOLTAGE

Electrical properties of materials used in industry are generally over-looked in technology-level literature compared to the emphasis placed on mechanical, physical, and thermal properties. The purpose here is to discuss an aspect of conductivity, one of the subdivisions of the electrical properties category, and to describe a tester that you can build and some related procedures. Specifically, the article will examine breakdown voltage of an insulation material.

The subdivision of conductivity can be further divided into:

☐ Conductors. Generally understood to be those materials having a volume resistivity of less than 3×10^{-6} ohm-centimeters. They will not support an electric strain, i.e., charge movement readily occurs when an electric field is applied.

☐ Insulators. Those materials which will support an electric strain; i.e., a charge is not readily communicated through or across the material when an electric field is applied.

☐ Semiconductors. Those materials having volume resistivities falling between the extremes of conductors and insulators. Some of these materials are frequency altered in composition to make them useful, such as in diodes and transistors.

Some Theory

The breakdown voltage of an insulation material is the voltage at which the *insulation between two conductors* will break down. Another term, dielectric strength, means the maximum voltage that a dielectric can withstand without rupture. This may be also referred to as insulating strength, and is expressed in units of voltage per unit thickness, such as volts/meter.

When speaking of breakdown voltage or dielectric strength we are not necessarily concerned with the dielectric constant of a material. The dielectric constant indicates the material's relative worth toward making a desired capacitance when the material is placed between two conductive plates. Thus, for electric insulation we may be concerned only with breakdown voltage, while for establishing a desired capacitance we may be concerned with both the material's breakdown voltage and its dielectric constant.

The breakdown voltage of an insulation material is dependent on such factors as:

☐ Interatomic or intermolecular bonding. The "attraction" by thich the adjacent atoms or molecules of the material are bonded together. Covalent bonding; i.e., the shaving of adjacent elements' electrons, is involved as bonding in plastic materials, for example.

☐ Crystaline lattice structure. The arrangement of adjacent atoms or molecules which form one cell of the material.

☐ Moisture content or moisture absorption. The presence of H_2O molecules within the material. While most common insulation materials do not absorb appreciable moisture, some, like asbestos used in power cords or some heat-generating appliances, can absorb moisture quite readily.

☐ Nature of applied E-field. The E-field may be unidirectional (produced by a direct current source), periodically or intermittently varying in magnitude and direction (produced by an alternating current source), or a combination of the two.

Regardless of how the E-field is produced, the field is a stress which causes a strain on the material's atoms or molecules. This

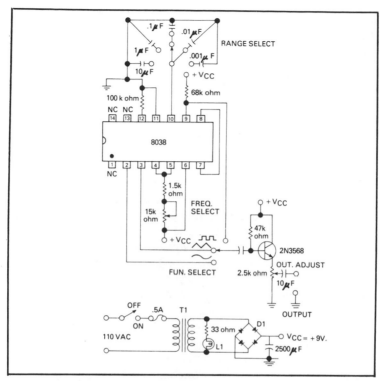

Fig. 12-55. Schematic of the tri-function generator.

stress distorts the orbital paths of the material's electrons, as illustrated in Fig. 12-56. If the magnitude of the E-field becomes sufficient to overcome the attraction of the negative electrons toward the positively charged nucleus, "leakage" will begin to occur and eventual rupture of the material will result. Thus, the breakdown voltage of the material has been reached or exceeded. If the applied E-field is alternating in direction, the resultant strain on the material's atoms or molecules changes direction also. The orbital path distortion is first in one direction and then, when polarity changes, in the other direction. This "flexing" of orbital paths results in hysteresis loss: energy which is dissipated as heat in the insulation or dielectric.

Numerous listings of the breakdown voltage of common insulation and capacitor dielectric materials exist. However, it is important to note that the breakdown voltage of a material may be different for a direct current E-field than for an alternating current E-field. The latter is somewhat lower in magnitude and is inversely proportional to the frequency of the source voltage. Thus, for example, capacitors, and other circuit elements, may have different maximum voltage ratings for radio frequency than for audio frequency or direct current applications.

Tester Design Considerations

The breakdown voltage tester is a specialized instrument. The tester shown schematically in Fig. 12-57 was designed and built because of this writer's belief that students should be given opportunity to explore as many as possible of the seven engineering properties of materials categories.

The design concept embraced these features:

☐ Safety. The tester must be safe for responsible student use. This must include instrument shielding connected to earth ground, a materials specimen insertion method which prevents fingers from reaching any high voltage point, a current-limiting method so that when breakdown voltage of a specimen is reached the discharge current cannot exceed some reasonably low level, automatic shutdown of the high voltage portion of the tester as soon as specimen rupture occurs, and provision for locking the tester to prevent its nonsupervised use by students or others.

☐ Voltage capabilities. The tester must be capable of reaching a voltage level sufficient to rupture common insulation materials of typically available thicknesses.

☐ Mechanical design. The tester must be sufficiently strong to withstand ordinary use, be easy to assemble, and all controls must be readily accessible and easy to operate.

☐ Cost. The tester should be made of components (Table 12-16) which are readily available with limited expenditure, preferably using parts often found in surplus equipment or in the "junk box" of many electronics programs.

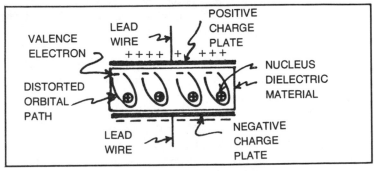

Fig. 12-56. Electrostatic stress-caused distortion of dielectric atoms.

474

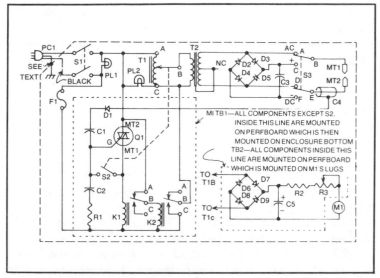

Fig. 12-57. Schematic of breakdown voltage tester and metering circuits.

Circuit Electrical Operation

The operation of the tester will be explained by going through one cycle of testing procedure using references to specific components in the schematic of Fig. 12-57. Assuming the power cord *PC1* is plugged into a 117 VAC source, the sequence is:

□ Turn switch *S1* ON. This powers the circuit, readying it for the testing sequence. *PL1* is lighted and will remain so unless *S1* is turned OFF.

□ With the variac *T1* output set to zero, switch *S2* (mounted on the slide arm of *T1*) is closed, thus shorting the gate and *MT1* terminals of the triac *Q1*. Therefore, the following conditions exist:

     —The triac *Q1* is OFF.
     —Therefore, *K1* is de-energized and *K1* bis at a.
     —Therefore, *Ks* is energized and *K2* is a b.
     —Therefore, *T1* is powered and *PL2* is lighted.

□ Advancing the output control of the variac *T1* opens switch *S2* (mounted on the slide arm of *T1*). However, the triac *Q1* remains OFF.

□ Further advancing the variac *T1* control increases its output voltage, and therefore the input voltage to and output voltage from the high-voltage transformer *T2*. Thus the voltage

475

across the arc terminals *AT1* and *AT2* and the specimen is increased.

☐ When rupture of the specimen occurs, the following happens:

—The arcing across *AT1* and *AT2* through the specimen causes a pulse which is picked up by *C4* and fed through *D1* and *C1* to the gate of the triac *Q1*.

—This positive pulse (polarity assured by rectifier *D1*) appears at the gate with respect to *MT1* or *Q1* and turns *Q1* ON. *Q1* will now stay ON until the second step occurs again.

—The triac current energizes *K1*, thus *K1* b goes to c.

—The action of *K1* allows *K2* to be deenergized, thus *K2* b goes to a.

This action of *K2* shuts OFF the power to *T1* and thus to the high-voltage section of the circuit. Also, *PL2* extinguishes.

☐ The control of the variac *T1* must now be returned to the zero output position so as to close switches *S2* to repeat the cycle for a second specimen test.

Component Descriptions and Functions

Some of the components shown in Fig. 12-57 have specific functions. Others are, or may be, of unique character, primarily to minimize the cost of constructing the tester. Therefore, comments regarding some of the components will be given:

Table 12-16. Parts List for the Breakdown Voltage Tester.

Desig.	Description
C1,2	1.0µF, 40 V non-polarized capacitor
C3	0.1µF, 25 kV capacitor
C4	Capacitor (see text)
C5	3µF, 3 V electrolytic capacitor
D1	IN4006 rectifier diode
D2-5	VARO No. VF 25-40 high-voltage rectifier diode
D6-9	IN4004 rectifier diode
F1	4 A slow-blow fuse with holder
K1	177 V SPST relay, contacts: 117 V 1 a minimum, inductive
K2	117 V SPST relay, contacts: 117 V 6 A minimum, inductive
M1	0-1 mA meter, scale: recommended 4-½" and 100° arc angle
MT1	Arc terminal, upper (see text)
MT2	Arc terminal, lower (see text)
PC1	18 ga. 3 conductor power cord
PL1,2	117 V pilot light assembly
Q1	RCA No. SK 3507 triac
R1	250Ω, 10 W resistor
R2	330 kΩ ½ W resistor
R3	100 kΩ, ½ W, screw driver adjust, potentiometer
S1	DPDT, 6 A toggle switch
S2	N.O., momentary contact, lever-actuated microswitch (see text)
S3	DPDT knife switch (see text)
T1	117: 0-140 V 5 A variac
T2	117: 10 kV C-T 23 mA transformer (see text)
TB1,2	0.1" grid terminal board, size to accommodate components

C3   Filters the *T2* output (for dc operation of the tester) and charges to v_{omax} of the *T2* secondary. If a 0.1 μN, 25 kV capacitor cannot be obtained, several series-connected capacitors producing equal or greater voltage rating and somewhat lower capacitance may be used.

C4   Picks up the pulse at specimen rupture. See Fig. 3 for consturction.

F1   Protects the circuitry, primarily *T1*, when rupture occurs. The *T2* primary current rises to approximately 4 A if the *T2* secondary is shorted.

PL1   Lights when *S1* is turned ON.

PL2   Lights when *T1* is powered thus indicating that the tester is readied to increase voltage across *AT1* and *AT2* and the specimen, and extinguishes when the specimen rupture occurs.

Q1   Acts to shut down the circuit when the pulse produced by specimen rupture is received at its gate. This traic must be a large pellet type (such as housed in a TO-66 or TO-3 case) so as to be sufficiently sensitive to the pulse produced by specimen rupture in this tester.

S2   This switch is a normally-open, momentary-contact, lever-actuated micro switch. It is mounted on a ⅛" thick fiberglass strip, which in turn is mounted on the slide arm of variac *T1*. The mounting is such that when *T1* is turned to zero output voltage the switch lever encounters a pin (made of insulative material) which is mounted on the frame of the variac *T1*, the switch *S2* is closed. The needed action of closing this switch is explained elsewhere in this article.

T2   A very suitable high-voltage transformer for the tester is a used oilfurnace burner-ignition transformer. These are usually of 10 kVac center-tapped output. The center tap is connected to the case or housing, and, thus, the case *must* be sufficiently isolated from and not touch any other part or the enclosure of the tester. A feature of such a transformer, needed to fulfill a design requirement of the tester, is its limited secondary current, usually in the range of 20 mA to 25 mA. This current level will not be exceeded even if the secondary is short-circuited. This current limiting is provided by having a shading coil (one-turn shorting band) around one leg of the transformer core. If the secondary is shorted, the transformer's primary current increases substantially, and it is for this reason that *F1* and *T1* must have the current ratings given in the parts list.

Other Construction Details

Construction of the tester is quite simple, though some precautions are needed in parts placement and materials selection.

Compact construction is not advisable. An enclosure 10" widy by 12" high by 16" deep inside dimensions will suffice. If a metal enclosure is used, it *must* be connected to earth ground, as illustrated by the dashed line surrounding the schematic of Fig. 12-57. The metal parts of the switch *S1* and the pilot *PL1* and *PL2* would then be grounded by connection to the metal enclosure.

It is recommended that the bottom be made of ¾' plywood covered with ¼" thick tempered hardboard (S2S) to provide a strong base for the heavy transformer *T2*. With this bottom, *T2*, of which the case is the secondary centertap, does not need additional insulation from the bottom, provided its mounting screws do not extend too deep into the bottom.

It is recommended that the front be made of ¼" plexiglass to permit viewing of the specimen during insertion and other components within the tester. Since the switch *S1*, the pilot lamps *PL1* and *PL2*, the fuse holder for *F1*, the meter *M1*, the variac *T1*, the *S3* switch actuating lever or rod, and the specimen holder assembly (see Fig. 12-59) are ideally located on the front, grounding of any metallic portions of these items which are external to the front may be accomplished by placing ⅛" or ¼" wire mesh over the front, cutting a slot slightly larger than the specimen insertion slot (see Fig. 12-58), and connecting the wire mesh to the metal sides of the enclosure.

The high-voltage diodes *D2-5* and the high-voltage filter capacitor *C3* can be mounted on a rack of insulation material, such as ¼" plexiglass or tempered hardboard (S2S) at the rear, but well above the transformer *T2*. This rack should be supported by legs, made of insulative material, setting on the enclosure bottom.

Switch *S3 must* be of high voltage rating. A ceramic-base knife switch having poles of at least 1-½" *length and spacing* will suffice. This switch should be mounted in a vertical plane, possibly above the high-voltage rectifiers, and be switchable using an insulative rod (⅜" dowel rod, for example) which protrudes through the front of the tester. Because most knife switches are made for heavy current applications, their blade- or pole-receiving terminals make tight spring contact. To ease the switching of *S3*, these pole-receiving terminals (AC and DC terminals) may be spread slightly. A loose contact will suffice for the low current application of this switch in this tester.

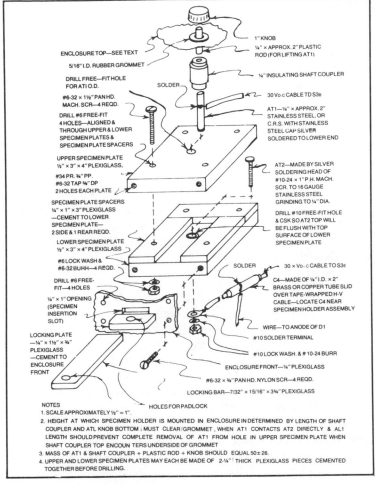

Fig. 12-58. Specimen holder, arc terminals and pulse pickup capacitor used in the voltage breakdown test.

It is recommended that the enclosure bottom be provided with large rubber feet to prevent sliding, and that the power cord *PC1* enter the enclosure through the bottom.

Construction of the specimen holder, shown in Fig. 12-58, is quite simple. The main objectives are to have the arc terminals *AT1* and *AT2* back approximately 4″ inside the holder, to assure that *AT1* (the upper terminal) cannot be lifted totally out of the hole in the upper specimen plate, to have the specimen contacting surfaces of both terminals smooth, made of good-quality material such as stainless steel to minimize pitting and small enough to prevent arcing *around* the specimen, and to provide some locking method, such as that shown, whereby direct contact of the terminals is impossible when the locking bar is in place.

The construction of the pulse pickup capacitor C4, shown in Fig. 12-58, is somewhat critical because the pulse needed to turn the triac Q1 ON must be of some minimum voltage value. The lowest voltage pulse will occur when testing materials of low breakdown voltage and/or small thickness. It is recommended that a brass tube of 2" minimum length be used to assure that Q1 turns ON.

If these suggestions and several of the comments under the "Component Description and Functions" section of this article are followed, components and parts will be mounted on the front and bottom of the tester. Only the AT1 terminal lifting rod will enter through the top. Therefore, assembly, and disassembly if needed, of the tester is simplified.

Output Metering and Calibration

The specimen breakdown-voltage indicating meter, shown in Fig. 12-57, is not connected to the high voltage part of the circuit. To do this would require more insulation precautions than are needed for metering. Instead, the meter is connected, through its own rectifier and filter network, to the output of variac T1 which is also the input to the high-voltage transformer T2. This arrangement works well because the linearity between the T2 primary voltage and secondary voltage, whether using the ac or the dc output voltage for testing, is very good. Thus, the meter provides an accurate indication of specimen breakdown voltage.

After tester construction is completed, calibration of the meter can proceed as follows:

Set swtich S3 of the tester to the dc position and insert the plastic locking bar into the specimen holder to assure that arcing will not occur between AT1 and AT2.

Set R3, shown in Fig. 12-57, for maximun in-circuit resistance.

Connect a high-voltage voltmeter (25 kVDC minimum) as a "standard" across AT1 and AT2 and position this voltmeter so it and the tester meter can be viewed at the same time.

Turn S1 ON and then slowly advance the variac T1 control to maximum. Note the voltage indicated by the "standard" voltmeter.

Using an insulated screwdriver or alignment tool, adjust R3 until its needle deflection agrees with that of the "standard." For example, if the "standard" voltmeter indicates 19.6K VDC, *R3* should be adjusted so the tester meter deflection is at 19.6/20 = 0.98 of full scale. This means that full-scale deflection equals 20K VDC, half-scale deflection equals 10K VDC, etc.

If a high-voltage ac voltmeter (rms reading) is available, steps 3 and 4 above can be repeated to determine the tester meter full-scale value of AC voltage. However, *R3* should not be adjusted for AC voltage indication. Rather, either a multiplying factor of the relationship between the AC and DC values should be determined and marked on the meter, or the meter scale can be removed and completely relabeled. If no high-voltage AC voltmeter is available for determining the tester's ac output voltage, the known relationship of Vrms = VDC/1.414 (where VDC is the just-determined scale values) will provide an accurate basis for marking of ac voltage for the tester meter scale.

Test Specimen Considerations

The preparation of specimens for testing breakdown voltage is not critical. However, some suggestions may help you to use this tester:

Cut the material specimens so their width is slightly less than that of the specimen insertion slot (see Fig. 12-58) and to a length which will permit cutting off the end that is "ruptured" by one test trial and reinserting the remainder for another test trial.

Some materials, such as plastic films, wax paper, etc., which are available primarily in very thin form, should be layered and taped to produce a thickness which will require higher voltage; i.e., nearer quarter-scale or greater deflection of the meter at breakdown. This will produce a more accurate measure of the breakdown voltage indication via the meter.

Test Procedure

Actual testing of specimens should follow the following sequence:

Record the type, trade-name, and any other descriptive information pertaining to the test material.

Accurately measure (using a micrometer) and record the individual layer thickness, number of layers and the total thickness of the specimen.

Ready the tester: Check to be certain that switch *S1* is OFF and that the variac *T1* control is set to minimum. Plug the power cord into a 117 Vac source. Set switch *S3* to the dc testing position. Unlock the tester and remove the locking bar.

Lift the arc terminal AT1 while inserting the specimen all the way to the back of the specimen holder assembly and then lower AT1 so it rests on the specimen.

Turn switch S1 ON. Slowly increase the applied voltage by uniform and slow CW rotation of the variac T1 control while watching the tester voltmeter. A spark will be heard the instant rupture of the specimen occurs. The meter will go to zero just after rupture. This indicates that shutdown of the high-voltage section has occurred. Record the voltage attained at the instant of specimen rupture.

Turn switch *S1* OFF (as a precautionary measure) and then rotate the variac *T1* control CCW to the zero position.

Remove the specimen and cut off the ruptured part. *AT1* may rest on *AT2* whle the specimen is being readied for another trial so long as *S1* is turned OFF.

Repeat the last five steps four more times to permit later determination of an average.

Using the trial values of breakdown voltage and the total thickness of the specimen, calculate and record the dielectric strength of the material for each trial.

Set switch *S3* to the ac testing position and determine the specimen's ac dielectric strength by repeating the last seven steps.

Proceed to the next "type" of the material by repeating the previous steps.

Chapter 13
Workshop Devices

Every electronics hobbyist should have his own workshop. These projects are especially designed for the hobbyist who wants to make sure his workshop is the best.

FLEXIBLE POWER SUPPLY

Modernized in every respect the shop was designed for high flexibility in serving its purposes without requiring detailed plans for machine arrangement. To cover the 2,500 sq. ft. of shop area efficiently, power distribution is provided in 65' of 100-amp bus duct suspended above the machinery. The bus duct is centered over areas where the electric-powered machine tools are placed, and has two plug-in outlets for every 2' of length. Only the area where work benches for manual tools are located is without bus-duct electrical outlets.

To serve the machine tools most efficiently in the 45' × 56' shop, the electrical contractor hung the bus-duct system in an L-shape, with a short run extending out of a T from one of the legs. Its position and length were adjusted between light fixtures by means of 2' lengths of duct.

The bus-duct system takes power from 2" conduit at one end of the L where the conduit emerges from the ceiling over the woodworking machines. The conduit runs from a breaker panel in an adjoining room. The bus duct delivers current at 120/208-v, 3-phase, 4-wire. Since the duct carries four bus bars and has a 277/480-v rating, it permits future changes whenever higher voltages might be required by future machinery or electrical equipment.

Costs were minimized by the simplicity of bus-duct installation and provision for enough power outlets to avoid the necessity of running new feeders. In addition, each drop cable has its own fuse box which prevents overloads or shorts from shutting down all machines in the shop.

As a safety measure, the entire run of bus duct can be de-energized and all machines stopped by pushing any of four "panic buttons" that are strategically located in the shop. This operates a contactor on the panelboard that feeds the bus duct.

It is a simple matter to extend the existing bus duct into areas where it is not now required. Why not try this project?

SELECTING TEST EQUIPMENT

Good test equipment in adequate quantity is vital to successful electronics.

An introductory course in electronics shop should be at least one semester in length, including between 80 and 90 clock hours of instruction. There will be a considerable emphasis on exploration.

Certain minimum specifications should be sought in this equipment to get the most out of each project. However, expense and budgetary provisions often play a large part in any decision as to specifications in equipment to be purchased. In this respect, AC and DC voltmeters represent a typical example. Ideally, a DC voltmeter should have a high sensitivity (ohms-per-volt rating) to minimize circuit loading. A typical industrial quality meter will have a sensitivity rating of 20,000 ohms per volt. Yet a multi-range d-c voltmeter of this sensitivity of 1,000 ohms per volt may cost only about $30, but would have far less desirable effect on circuit operation.

The same factor is true of AC voltmeters: 5,000 ohms per volt is a typical industrial meter sensitivity. But such an instrument may cost $80, whereas a 1,000 ohms-per-volt meter can be bought for about $30. Typical prices for the other single-purpose meters mentioned above include for a DC ammeter about $25; for an ohmmeter, about $35; and for a DC milliameter, about $25.

TEST INSTRUMENTS FOR ADVANCED ELECTRONICS

The equipment used by advanced hobbyists will be considerably more sophisticated than that used by beginners.

In a relatively advanced course, such as the second and third semester offerings, demonstrations are of tremendous value, furnishing as they do a much simplified "first look" at circuit operation. But it is not enough to demonstrate.

Some additional comment on the oscilloscope may be helpful to those who have only a slight familiarity with this instrument. An oscilloscope can be considered a form of voltmeter, since once calibrated it can be used to measure DC voltages (DC scope only). But it can do far more than this; it can be used to measure complex waveform peak values, to examine the shape of time-varying signals, and verify the presence or absence of various types of signals, to mention only some of its uses. In the study of basic circuits—amplifiers, oscillators, and special-purpose circuits—amplifiers is invaluable as a teaching and study tool.

The vacuum-tube voltmeter has not been mentioned previously. Its characteristics are different from the VOM chiefly in that no matter what voltage range is selected the input resistance remains 11 megohms, and consequently the circuit loading is very low. This type of meter is valuable for making resistance measurements in high-resistance circuits and voltage measurements in oscillator and amplifier circuits in which meter loading is very critical. In most current, voltage, and resistance measurements made in electronic circuitry, the 20,000 ohms-per-volt VOM is more than adequate, but for critical applications the VTVM should be used. Note, however, that the VTVM cannot be used to measure current, only voltage and resistance.

This is a limited list of tools and equipment used in the general electrical trades. It is from a list prepared by the Bureau of Trade and Technical Education, New York State Department of Education. It is considered a minimum list and the quantity of each item should be determined by local conditions.

EQUIPPING YOUR ELECTRONIC SHOP

ELECTRICAL TRADES TOOLS

Blowtorch—gasoline, one qt.
Calipers—6″ inside, spring steel
Calipers—6″ outside, spring steel
Chisels—assorted popular sizes, cape and cold
Chisels—wood
Bender—sizes of pipe shoes, ½″-2½″
Dividers—6″ spring steel
Die-machine-screw—sets of standard die sizes no. 6-32 to no. 10-32; also no. 10-24, ¼″-20
Die stock—sizes for above dies
Drill stand—three sizes, no. 1-60, 1/16″-½″, A-Z drills
Drills, electric—¼″ and ½″, chuck sizes, heavy-duty type
Drills, star—¼″, ⅜″, ½″, ¾″
Drills, tungsten carbide—¼″, ⅜″, ½″

File card—steel back, wire bristles
Files—assorted mill and bastard; flat, round, triangular
Wire gauge
Gas torch—propane, with replaceable cans
Gage—60°, V-type thread
Hammer, machinist—12 oz., ball peen, wood handle
Hammer, claw—carpenter's type, wood handle
Hammer, soft face—rawhide or plastic tipped
Hickey heads—½", ¾", and 1"
Hole saws and mandrels for ½" to 2" pipe
Micrometer—outside, calibrated to .001", range 0-1"
Oil stone—¾" × 2" × 5", medium grit
Pipe dies and stocks with reversible dies, ½" to 1¼"
Pipe dies—ratchet with dies ¾" to 2"
Pipe vise and standard
Pliers, assorted types—needle, duckbill, and slipjoint, 6"
Pliers—channel lock, adjustable, 10"
Pliers—diagonal cutter type, 5" and 7" sizes
Pliers—lineman's 8"
Punch—assorted sizes of long, aligning, starting, center, and pinpunches
Punch—gasket-cutting type, assorted sizes
Reamer, pipe
Rule—steel, 6" and 12" sizes, steel-hardened type
Rule—folding, 6'
Saws—carpenter's, crosscut and rip
Saw—compass with removable blade
Saw, hand back—8" to 12" adjustable frame, pistol grip
Screwdriver—assorted types and sizes; i.e., general purpose, electrician, insulated
Screwdrivers—Phillips, offset
Screw extractor—straight shank, fluted blade, assorted sizes
Scriber—5", replaceable blade type
Snips—tin, straight 3", hardened tool steel
Soldering iron—electric, 120-v, with assorted tips
Soldering gun—2 speed, 300-w, 120-v
Speed counters for timing rpm
Square, combination—tempered 12" blade with center head and protractor
Stones—commutator, assorted sizes ¼"-2"
Taps—machine screw, hand tap, plug type, assorted sizes, no. 4-40 to no. 12 N.C. and N.F.
Taps—fractional sizes, N.C. and N.F. ¼"-¾"
Tap wrench—"T" handle, adjustable chuck, hardened jaws, sizes to take machine screw and fractional size taps
Tap wrench—bar type, adjustable vise jaw ¼" to ¾" taps
Wrenches—adjustable, 6", 8", and 12" (crescent pattern)
Wrenches—pipe, 10" and 18"
Wrenches—spinner, assorted sizes ⅛" to 1"
Wrenches—Allen, hexagonal offset, assorted sizes 1/16" to ⅜"

ELECTRICAL TRADES EQUIPMENT

Air compressor—small, portable
Annunicators—assorted drops and types
Arbor press for motor bearings
Alternator—single-phase, run by DC compound motor
Alternator—three-phase, run by DC compound motor
Bells

Breakers—air, circuit
Breakers—oil
Buzzers
Can—gasoline
Controller—cam, DC
Controller—automatic, DC, dashpot or counter E.M.F.
Controller—drum
Compensator—three-phase, 220-v, manual
Compensator—three-phase, 220-v, automatic
Drill press and drills
Fixtures—fluorescent
Fixtures—incandescent
Fixtures—neon tube
Generators—compound, DC, 120-v
Grinder—bench
Hydrometer
Hoist—chain, 1 ton
Magnets—magnetic experiments and demonstration units
Motor channel box
Microphones—carbon, crystal, and dynamic
Motor—compound, DC
Motor—shunt, DC
Motor—generator unit, 120-v, 10-kw, master unit
Motor-squirrel-cage induction, three-phase DC with compound generator
Motor—wound rotor induction
Motor—Selsyn
Motor—squirrel-cage induction, three-phase with special terminal board for different internal connections
Motor—single-phase, resistance-reactance type
Motor—single phase, capacitance-start type
Motor—single-phase, repulsion-start induction run
Motor—single-phase universal
Motor—Saint Louis, for tests and demonstration
Motor—a-c and d-c demonstration unit for generators and motors
Oscillator—beat frequency, audio range
Oscillator—r.f. and 400-c modulated
Panels—switch board
Panel plugs and receptacles for dead-face panels
Photo-electric unit—commercial, with light source
Power supply—electronic circuit with meters for filament, B. and C. supplies
Push buttons—motor start-stop, start-stop and reverse, emergency
Receivers—telephone for tests
Rectifiers—dry disc and tungar
Relays
Resistors—start connected for three-phase wound rotor motor, speed control
Rhesotats—field
Rheostats—load 4 kw carbon stack panel board
Rheostats—load, portable, wire wound with convenient switches
Service box—4 circuit, fused, including master and stove switch
Service box—breaker-controlled, 4-circuit, master and stove switch
Speakers—permanent and electro-dynamic
Storage batteries
Starter—across the line
Starter—across the line, reversing
Switches—single-pole, double-throw knife
Switches—double-pole, double-throw knife

Switches—three-pole, single-throw, fused, panel
Switches—two-pole, single-throw, fused, panel
Starting box—three-point, DC
Starting box—four-point, DC
Starting box—four-point with speed-regulating DC
Telephones
Thermometers
Terminals—insulated terminal posts suitable for motor connections
Thermostats
Transformers—1 K.V.A. with taps on primary and 220, 110-v secondary
Transformers—auto, 120-v
Transformers—auto, with adjustable sliding contacts, 0-150-v
Transformer—neon, tube
Transformer—potential for meters
Transformer—current with positive safety secondary
Tube checker
Lathe
Benches
Portable drills (light and heavy)
Undercutting machine
Coil-winding machine

EQUIPPING A RADIO AND TV SHOP

The following list of tools and equipment for radio and television is adapted from a list prepared by the Engineering Extension Service of the Texas A. and M. College System.

RADIO AND TELEVISION
TEST INSTRUMENTS

Item
Volt-ohm-milliammeter 100 ohms per volt (kit form)..................
Volt-ohm-milliammeter, 20,000 ohms per volt........................
Vacuum-tube-voltmeter, (kit form) ....
Tube tester, dynamic mutual conductance type ....
Picture tube tester-rejuvenator...
Signal generator ....
Audio oscillator...
Condenser tester..
Condenser substitution box (kit form) ....
Resistor, substitution box (kit form) ....
Oscilloscope (kit form) ....
*Sweep and marker generator (combination, with 3″ oscilloscope).
Isolation transformer, variable...
Bench, radio, work, with instruments and 115-v outlets..as needed
Bench, television, work, with instruments and 115 V and outlets as
    needed

Panel, electronic-circuit demonstration ....as needed
Table, instructor's demonstration, with instruments 1
 *This generator is necessary only if alignment procedures are to be taught. May be purchases later in the program if desired.

STOCKROOM TOOLS

Drill motor, ¼" portable ....
Drill set, 1/32" to ¼" in 32nds..
Drill set, ¼" to ½" in 32nds...
Hand drill ....
Combination square...
Center punch...
Work stand, 28" square plywood top, steel frame 28" high, on 3"
 rubber casters (shop made)...
Socket punch set, ⅝", ¾", 1", 1⅛", 1 3/16" ....
Hacksaw and blades ....
Tin snips, 12" ....
Chisel (cold, ½") ....
Set of files...
Set of taps, small, 4-40 to ¼-20...
Hammer, 8 oz. ball-peen ....
Mallet, plastic..
Wrench set, open end to 1"...
Nut driver set, no. 6 to 18 ....
Socket wrench set, small, ¼" drive ....
Crescent wrench 6" ....
Allen wrench set...
Spline wrench set..
Tube pin straightener, 7 and 9 pin...
TV mirror, on stand with casters, shop made ....
Wire gage, American Standard...
Wire strippers, automatic hand type ....
Micrometer, 1" ....
Scribe ....

RADIO & TV
BENCH TOOLS

Drill press, ½" chuck, table model ....
Vise, 3" jaw...
Grinder-buffer, 6" combination ....
Box bender, 24" ....

SHOP MACHINES ARE POWER GLUTTONS!

The first step in becoming energy conservation-conscious is to learn about power consumption. All your shop needs is power equipment that operates on a 110 to 120 V single phase electrical current (such as a drill press, table saw, lathe, etc.), and an electrical measurement device which you can build.

Construction

If the meter box is to be fabricated in the shop, establish the size and build the box. Then wire meters using the diagrams supplied with them or use the following instructions: Mount meter or meters in box. Cut extension cord at approximate middle of cord. Put cable connectors in box. Strip back cable housing approximately 6 in. from plug side of cable. Install through cable connectors and crimp the terminals to wire.

Install white and black wires (hot and neutral) of the plug end of cable to the voltmeter. Remove 6 in. of cable housing from the receptacle end of the cord. Install through cable connector and secure the white lead to the white wire on the voltmeter. Cut 3 in. of the black wire from the cable and connect it to the other terminal of the voltmeter and to one on the ammeter.

Connect the remaining wire of the cord to the other terminal of the ammeter. Connect terminals to green groundwire and connect to metal box. Check all connectors. Assemble box.

How To Use The Device

Safety dictates that the measuring device be located away from the curious eyes of the operator and outside the safety parameter of the machine.

IS YOUR WORKSHOP BOOBY-TRAPPED?

Shop safety is of prime concern to an electronics hobbyist. Reports in recent months indicate that the number of cases of electrical shock and short-circuit damage are increasing. Many of these incidents have resulted in personal injury and extensive damage to equipment.

When projects are limited to low-voltage battery-operated equipment and selfpowered test instruments, the hazards are much less serious. In many workshops, however, you must work with equipment operated from power-line voltage sources. Such equipment includes motors, generators, industrial-control circuits, appliances, and radio and television equipment. Projects with these devices require foolproof procedures and periodic checks.

Hazards of Direct Power-Line Operation

Most serious of all hazards is the operation of equipment directly from the power line without the aid of a transformer. In this case, the available voltage and current are sufficiently high to be fatal, as well as to cause serious equipment damage. Transformerless equipment can include all the devices mentioned above.

This type of circuit arrangement is commonly used to supply DC power from an AC line for industrial equipment, radio and television, and appliances. Under normal operating conditions, this circuit does not present a hazard because it is installed so that the operator cannot contact any part of the circuit or common metalwork. If the protective covers or circuit assembly are removed for testing, however, a considerable hazard awaits the careless or unwary.

The hazard of a "hot" common circuit is considerable. For example, there is a severe shock hazard between this point and any other point in the circuit, and to any grounded object—such as the shop floor, metal work benches, shelving, and pipes—in the work area. The shock hazard can also exist between this circuit and the test leads and metal cases of any adjacent equipment. The same danger exists in this circuit when the power switch is set to "off" because the switch opens only the "cold" side of the power line. This arrangement is common in much commercial equipment.

It should be remembered, therefore, that when the ground or common test lead of any voltmeter, oscilloscope or signal generator is connected to a test point that has any voltage on it, the metal instrument case automatically presents a hazard if the case is connected directly to the common ground circuit of the test instrument.

CRITERIA FOR SELECTING METERS

The selection of meters for your workshop can be a confusing and never ending procedure of interpreting instrument manufacturers' specifications. The choice of the proper instrument for your shop depends on many factors other than those purely mechanical, such as size, style, color, and quantity, although these factors are also important.

Why a Meter Is Used

Before establishing criterions for the selection of meters, it is important to establish the purpose of meters. Since it is not practical to count electrons visibly as they pass through a conductor, meters measure current by the effects these electrons produce. The effects of current, can be categorized into the following types: chemical, physiological, photoelectric, piezoelectric, thermal, and electromagnetic. The first four effects of current—chemical, physiological, photoelectric, and piezoelectric— are of no observable importance in the consideration of meters, because in most cases the measurement of the amount of current would call for other allied equipment to interpret the findings. For this reason the latter two effects of current, thermal and electromagnetic, are used in meters.

Types of Meter Mechanisms

In the hot-wire ammeter the current being measured causes a wire to heat and expand. Through an arrangement of springs, pulleys, and pointer, plus scale, current could be measured. Hot-wire mechanisms are now obsolete as such. They have been replaced by the more sensitive, more accurate, and better compensated "heating-element and permanent-magnet moving-coil combination" known as the thermo instruments. Thermo instruments measure the effective value of current and can be used on dc, ac, or audio or radio frequencies. They are used extensively where wave forms or frequencies encountered are such as to cause errors in other types of instruments.

Electromagnetic-meter movements fall into four basic types of mechanism: permanent-magnet moving-coil, polarized iron-vane, moving iron-vane and electrodynamometer. Another type of mechanism that is encountered is the electrostatic mechanism which is the only meter movement that measures voltage directly, rather than the effect of current which operates on the principle of the electroscope.

Accuracy

Meter accuracy is normally expressed in a percent of the full-scale value. This accuracy can be classified into four degrees of accuracy: meters of low accuracy—2 percent to 5 percent; medium accuracy—1 percent to 2 percent; high accuracy—.5 percent; and precision accuracy—.1 percent. Various degrees of accuracy can be obtained in most meter mechanisms. As the accuracy increases, however, so does the cost of the instrument.

Sensitivity

The sensitivity of a meter movement is usually expressed as the amount of current necessary to give full-scale deflection. In addition, it may also be expressed as the number of millivolts across the meter when full-scale current flows through it. A meter movement, whose resistance is 50 ohms and requires 1 milliampere for full-scale reading, is often described as a 50 millivolt, 0-1 milliammeter.

Because meters have resistance, they affect the circuit in which they are placed. The sensitivity of a meter is an indication of how much the meter affects the circuit and, therefore, the accuracy of the reading. When selecting a meter, keep the following in mind: An ammeter must have considerably less resistance than the circuit with which it is in series. A voltmeter must have considerably more resistance than the circuit with which it is in parallel. The reason an ammeter must have very low resistance is obvious for accurate readings. Since the meter is placed in series with the load, it increases the total resistance of the circuit, resulting in less current than under normal operating conditions.

If a meter is used that has 20,000 ohms-per-volt sensitivity and is placed on the 0-100-v scale, the resistance offered would be 2,000,000 ohms or 2 megohms. The total resistance across would be 95,239 ohms; therefore, the circuit is not loaded as much, the reading is more accurate, and it gives a truer indication of the circuit under normal operating conditions.

ELECTRICAL LABORATORY

One difficulty any hobbyist will probably have is a lack of facilities. The answer to this problem is a portable DC lab which is selfcontained. It consists of a cabinet, project boards, instruments, tools and a built-in power supply.

With this arrangement any workshop with a wall plug can become an electricity-electronics laboratory in minutes.

The storage cabinet houses all the elements. It has locking swivel casters on one end and fixed casters on the other, large enough to jump the gap in the elevator door or easily ride over a threshold. The cabinet was constructed of usually available materials. The outside is birch plywood, ¾" thick, with a natural finish. The dimensions of the unit are 4' long, 2' deep, and less than 4' high. These dimensions afford maximum utilization of the 4' × 8' plywood sheet.

The partition and shelves inside of the cabinet are made from salvaged fir plywood. The top is made of a sheet of ¾" plywood to which was bonded a ⅛" sheet of tempered masonite. Because the cabinet will be constantly moved about, particular care is taken to glue and screw all joints. The corners are reinforced from the inside, using 1" by 1" hardwood corner posts. A handle is bent from electrical thinwall and affixed to the top.

Index

Edited by Raymond A. Collins